OUTLIVE

OUTLIVE

THE SCIENCE & ART OF LONGEVITY

PETER ATTIA, MD

WITH BILL GIFFORD

RETHINKING MEDICINE TO LIVE BETTER LONGER

1

Vermilion, an imprint of Ebury Publishing
20 Vauxhall Bridge Road
London SW1V 2SA

Vermilion is part of the Penguin Random House group of companies
whose addresses can be found at global.penguinrandomhouse.com

First published in Great Britain in 2023 by Vermilion

This edition published by arrangement with Harmony Books, an imprint of Random House, a division of Penguin Random House LLC.

www.penguin.co.uk

A CIP catalogue record for this book is available from the British Library

CENTENARIAN DECATHLON is a trademark of PA IA, LLC.
MARGINAL DECADE is a trademark of PA IP, LLC.

Book design by Andrea Lau
Jacket design by Rodrigo Coral Studio

Hardback ISBN 9781785044540
Trade Paperback ISBN 9781785044557

Printed and bound in India by Replika Press Pvt. Ltd.

The authorised representative in the EEA is Penguin Random House Ireland, Morrison Chambers, 32 Nassau Street, Dublin D02 YH68

Penguin Random House is committed to a sustainable future for our business, our readers and our planet. This book is made from Forest Stewardship Council® certified paper.

For my patients.

And for Jill, Olivia, Reese, and Ayrton … for your patience.

AUTHOR'S NOTE

Writing about science and medicine for the public requires striking a balance between brevity and nuance, rigor and readability. I've done my best to find the sweet spot on that continuum, getting the substance right while keeping this book accessible to the lay reader. You'll be the judge of whether or not I hit the target.

CONTENTS

OUTLIVE

INTRODUCTION

In the dream, I'm trying to catch the falling eggs.

I'm standing on a sidewalk in a big, dirty city that looks a lot like Baltimore, holding a padded basket and looking up. Every few seconds, I spot an egg whizzing down at me from above, and I run to try to catch it in the basket.

They're coming at me fast, and I'm doing my best to catch them, running all over the place with my basket outstretched like an outfielder's glove. But I can't catch them all. Some of them—many of them—smack on the ground, splattering yellow yolk all over my shoes and medical scrubs. I'm desperate for this to stop.

Where are the eggs coming from? There must be a guy up there on top of the building, or on a balcony, just casually tossing them over the rail. But I can't see him, and I'm so busy I barely even have time to think about him. I'm just running around trying to catch as many eggs as possible. And I'm failing miserably. Emotion wells up in my body as I realize that no matter how hard I try, I'll never be able to catch all the eggs. I feel overwhelmed, and helpless.

And then I wake up, another chance at precious sleep ruined.

We forget nearly all our dreams, but two decades later, I can't seem to get this one out of my head. It invaded my nights many times when I was a surgical resident at Johns Hopkins Hospital, in training to become a cancer surgeon. It was one of the best periods of my life, even if at times I felt like I was going crazy. It wasn't uncommon for my colleagues and me to work for twenty-four hours straight. I craved sleep. The dream kept ruining it.

The attending surgeons at Hopkins specialized in serious cases like pancreatic cancer, which meant that very often we were the only people standing between the patient and death. Pancreatic cancer grows silently, without symptoms, and by the time it is discovered, it is often quite advanced. Surgery was an option for only about 20 to 30 percent of patients. We were their last hope.

Our weapon of choice was something called the Whipple Procedure, which involved removing the head of the patient's pancreas and the upper part of the small intestine, called the duodenum. It's a difficult, dangerous operation, and in the early days it was almost always fatal. Yet still surgeons attempted it; that's how desperate pancreatic cancer is. By the time I was in training, more than 99 percent of patients survived for at least thirty days after this surgery. We had gotten pretty good at catching the eggs.

At that point in my life, I was determined to become the best cancer surgeon that I could possibly be. I had worked really hard to get where I was; most of my high school teachers, and even my parents, had not expected me to make it to college, much less graduate from Stanford Medical School. But more and more, I found myself torn. On the one hand, I loved the complexity of these surgeries, and I felt elated every time we finished a successful procedure. We had removed the tumor—we had caught the egg, or so we thought.

On the other hand, I was beginning to wonder how "success" was defined. The reality was that nearly all these patients would still die within a few years. The egg would inevitably hit the ground. What were we really accomplishing?

When I finally recognized the futility of this, I grew so frustrated that I quit medicine for an entirely different career. But then a confluence of events occurred that ended up radically changing the way I thought about health and disease. I made my way back into the medical profession with a fresh approach, and new hope.

The reason why goes back to my dream about the falling eggs. In short, it had finally dawned on me that the only way to solve the problem was not to get better at catching the eggs. Instead, we needed to try to stop the guy who was throwing them. We had to figure out how to get to the top of the building, find the guy, and take him out.

I'd have relished that job in real life; as a young boxer, I had a pretty mean left hook. But medicine is obviously a bit more complicated. Ultimately, I realized that we needed to approach the situation—the falling eggs—in an entirely different way, with a different mindset, and using a different set of tools.

That, very briefly, is what this book is about.

PART I

CHAPTER 1

The Long Game

From Fast Death to Slow Death

There comes a point where we need to stop just pulling people out of the river. We need to go upstream and find out why they're falling in.

—Bishop Desmond Tutu

I'll never forget the first patient whom I ever saw die. It was early in my second year of medical school, and I was spending a Saturday evening volunteering at the hospital, which is something the school encouraged us to do. But we were only supposed to observe, because by that point we knew just enough to be dangerous.

At some point, a woman in her midthirties came into the ER complaining of shortness of breath. She was from East Palo Alto, a pocket of poverty in that very wealthy town. While the nurses snapped a set of EKG leads on her and fitted an oxygen mask over her nose and mouth, I sat by her side, trying to distract her with small talk. *What's your name? Do you have kids? How long have you been feeling this way?*

All of a sudden, her face tightened with fear and she began gasping for breath. Then her eyes rolled back and she lost consciousness.

Within seconds, nurses and doctors flooded into the ER bay and began running a "code" on her, snaking a breathing tube down her airway and injecting her full of potent drugs in a last-ditch effort at resuscitation. Meanwhile, one of the residents began doing chest compressions on her prone body. Every couple of minutes, everyone would step back as the attending physician slapped defibrillation paddles on her chest, and her body would twitch with the immense jolt of electricity. Everything was precisely choreographed; they knew the drill.

I shrank into a corner, trying to stay out of the way, but the resident doing CPR caught my eye and said, "Hey, man, can you come over here and relieve me? Just pump with the same force and rhythm as I am now, okay?"

So I began doing compressions for the first time in my life on someone who was not a mannequin. But nothing worked. She died, right there on the table, as I was still pounding on her chest. Just a few minutes earlier, I'd been asking about her family. A nurse pulled the sheet up over her face and everyone scattered as quickly as they had arrived.

This was not a rare occurrence for anyone else in the room, but I was freaked out, horrified. *What the hell just happened?*

I would see many other patients die, but that woman's death haunted me for years. I now suspect that she probably died because of a massive pulmonary embolism, but I kept wondering, what was really wrong with her? What was going on before she made her way to the ER? And would things have turned out differently if she had had better access to medical care? Could her sad fate have been changed?

Later, as a surgical resident at Johns Hopkins, I would learn that death comes at two speeds: fast and slow. In inner-city Baltimore, fast death ruled the streets, meted out by guns, knives, and speeding automobiles. As perverse as it sounds, the violence of the city was a "feature" of the training program. While I chose Hopkins because of its excellence in liver and pancreatic cancer surgery, the fact that it averaged more than ten penetrating trauma cases per day, mostly gunshot or stabbing wounds, meant that my colleagues and I would have ample opportunity to develop our surgical skills repairing bodies that were too often young, poor, Black, and male.

If trauma dominated the nighttime, our days belonged to patients with vascular disease, GI disease, and especially cancer. The difference was that these patients' "wounds" were caused by slow-growing, long-undetected tumors, and not all of them survived either—not even the wealthy ones, the ones who were on top of the world. Cancer doesn't care how rich you are. Or who your surgeon is, really. If it wants to find a way to kill you, it will. Ultimately, these slow deaths ended up bothering me even more.

But this is not a book about death. Quite the opposite, in fact.

More than twenty-five years after that woman walked into the ER, I'm still practicing medicine, but in a very different way from how I had imagined. I no longer perform cancer surgeries, or any other kind of surgery. If you come to see me with a rash or a broken arm, I probably won't be of very much help.

So, what *do* I do?

Good question. If you were to ask me that at a party, I would do my best to duck out of the conversation. Or I would lie and say I'm a race car driver, which is what I really want to be when I grow up. (Plan B: shepherd.)

My focus as a physician is on longevity. The problem is that I kind of hate the word *longevity.* It has been hopelessly tainted by a centuries-long parade of quacks and charlatans who have claimed to possess the secret elixir to a longer life. I don't want to be associated with those people, and I'm not arrogant enough to think that I myself have some sort of easy answer to this problem, which has puzzled humankind for millennia. If longevity were simple, then there might not be a need for this book.

I'll start with what longevity isn't. Longevity does not mean living forever. Or even to age 120, or 150, which some self-proclaimed experts are now routinely promising to their followers. Barring some major breakthrough that, somehow, someway, reverses two billion years of evolutionary history and frees us from time's arrow, everyone and everything that is alive today will inevitably die. It's a one-way street.

Nor does longevity mean merely notching more and more birthdays as we slowly wither away. This is what happened to a hapless mythical Greek named Tithonus, who asked the gods for eternal life. To his joy, the gods granted his wish. But because he forgot to ask for eternal youth as well, his body continued to decay. Oops.

Most of my patients instinctively get this. When they first come to see me, they generally insist that they *don't* want to live longer, if doing so means lingering on in a state of ever-declining health. Many of them have watched their parents or grandparents endure such a fate, still alive but crippled by physical frailty or dementia. They have no desire to reenact their elders' suffering. Here's where I stop them. Just because your parents endured a painful old age, or died younger than they should have, I say, does not mean that you must do the same. The past need not dictate the future. Your longevity is more malleable than you think.

In 1900, life expectancy hovered somewhere south of age fifty, and most people were likely to die from "fast" causes: accidents, injuries, and infectious diseases of various kinds. Since then, slow death has supplanted fast death. The majority of people reading this book can expect to die somewhere in their seventies or eighties, give or take, and almost all from "slow" causes. Assuming that you're not someone who engages in ultrarisky behaviors like BASE jumping, motorcycle racing, or texting and driving, the odds are overwhelming that you will die as a result of one of the chronic diseases of aging that I call the Four Horsemen: heart disease, cancer, neurodegenerative disease, or type 2 diabetes and related metabolic dysfunction. To achieve longevity—to live longer and live better for longer—we must understand and confront these causes of slow death.

Longevity has two components. The first is how *long* you live, your chronological lifespan, but the second and equally important part is how *well* you live—the quality of your years. This is called *healthspan,* and it is what Tithonus forgot to ask for. Healthspan is typically defined as the period of life when we are free from disability or disease, but I find this too simplistic. I'm as free from "disability and disease" as when I was a twenty-five-year-old medical student, but my twenty-something self could run circles around fifty-year-old me, both physically and mentally. That's just a fact. Thus the second part of our plan for longevity is to maintain and improve our physical and mental function.

The key question is, Where am I headed from here? What's my future trajectory? Already, in midlife, the warning signs abound. I've been to funerals for friends from high school, reflecting the steep rise in mortality risk that begins in middle age. At the same time, many of us in our thirties, forties, and fifties are watching our parents disappear down the road to physical disability, dementia, or long-term disease. This is always sad to see, and it reinforces one of my core principles, which is that the only way to create a better future for yourself—to set yourself on a better trajectory—is to start thinking about it and taking action *now.*

One of the main obstacles in anyone's quest for longevity is the fact that the skills that my colleagues and I acquired during our medical training have proved to be far more effective against fast death than slow death. We learned to fix broken bones, wipe out infections with powerful antibiotics, support and even replace damaged organs, and decompress serious spine or brain injuries. We had an amazing ability to save lives and restore full function to broken bodies, even reviving patients who were nearly dead. But we were markedly less successful at helping our patients with chronic conditions, such as cancer, cardiovascular disease, or neurological disease, evade slow death. We could relieve their symptoms, and often delay the end slightly, but it didn't seem as if we could reset the clock the way we could with acute problems. We had become better at catching the eggs, but we had little ability to stop them from falling off the building in the first place.

The problem was that we approached both sets of patients—trauma victims and chronic disease sufferers—with the same basic script. Our job was to *stop the patient from dying,* no matter what. I remember one case in particular, a fourteen-year-old boy who was brought into our ER one night, barely alive. He had been a passenger in a Honda that was T-boned by a driver who ran a red light at murderous speed. His vital signs were weak and his pupils were fixed and dilated, suggesting severe head trauma. He was close to death. As trauma chief, I immediately ran a code to try to revive him, but just as with the woman in the Stanford ER, nothing worked. My colleagues wanted me to call it, yet I stubbornly refused to declare him dead. Instead, I kept coding him, pouring bag after bag of blood and epinephrine into his lifeless body, because I couldn't accept the fact that an innocent young boy's life could end like this. Afterwards, I sobbed in the stairwell, wishing I could have saved him. But by the time he got to me, his fate was sealed.

This ethos is ingrained in anyone who goes into medicine: nobody dies on my watch. We approached our cancer patients in the same way. But very often it was clear that we were coming in too late, when the disease had already progressed to the point where death was almost inevitable. Nevertheless, just as with the boy in the car crash, we did everything possible to prolong their lives, deploying toxic and often painful treatments right up until the very end, buying a few more weeks or months of life at best.

The problem is not that we aren't trying. Modern medicine has thrown an unbelievable amount of effort and resources at each of these diseases. But our progress has been less than stellar, with the possible exception of cardiovascular disease, where we have cut mortality rates by two-thirds in the industrialized world in about sixty years (although there's more yet to do, as we will see). Death rates from cancer, on the other hand, have hardly budged in the more than fifty years since the War on Cancer was declared, despite hundreds of billions of dollars' worth of public and private spending on research. Type 2 diabetes remains a raging public health crisis, showing no sign of abating, and Alzheimer's disease and related neurodegenerative diseases stalk our growing elderly population, with virtually no effective treatments on the horizon.

But in every case, we are intervening at the wrong point in time, well after the disease has taken hold, and often when it's already too late—when the eggs are already dropping. It gutted me every time I had to tell someone suffering from cancer that she had six months to live, knowing that the disease had likely taken up residence in her body several years before it was ever detectable. We had wasted a lot of time. While the prevalence of each of the Horsemen diseases increases sharply with age, they typically begin much earlier than we recognize, and they generally take a very long time to kill you. Even when someone dies "suddenly" of a heart attack, the disease had likely been progressing in their coronary arteries for two decades. Slow death moves even more slowly than we realize.

The logical conclusion is that we need to step in sooner to try to stop the Horsemen in their tracks—or better yet, prevent them altogether. None of our treatments for late-stage lung cancer has reduced mortality by nearly as much as the worldwide reduction in smoking that has occurred over the last two decades, thanks in part to widespread smoking bans. This simple preventive measure (not smoking) has saved more lives than any late-stage intervention that medicine has devised. Yet mainstream medicine still insists on waiting until the point of diagnosis before we intervene.

Type 2 diabetes offers a perfect example of this. The standard-of-care treatment guidelines of the American Diabetes Association specify that a patient can be diagnosed with diabetes mellitus when they return a hemoglobin A1c (HbA1c) test result[*] of 6.5 percent or higher, corresponding to an average blood glucose level of 140 mg/dL (normal is more like 100 mg/dL, or an HbA1c of 5.1 percent). These patients are given extensive treatment, including drugs that help the body produce more insulin, drugs that reduce the amount of glucose the body produces, and eventually the hormone insulin itself, to ram glucose into their highly insulin-resistant tissues.

But if their HbA1c test comes back at 6.4 percent, implying an average blood glucose of 137 mg/dL—just three points lower—they technically don't have type 2 diabetes at all. Instead, they have a condition called prediabetes, where the standard-of-care guidelines recommend mild amounts of exercise, vaguely defined dietary changes, possible use of a glucose control medication called metformin, and "annual monitoring"—basically, to wait and see if the patient actually develops diabetes before treating it as an urgent problem.

---

* HbA1c measures the amount of glycosylated hemoglobin in the blood, which allows us to estimate the patient's average level of blood glucose over the past ninety days or so.

I would argue that this is almost the exact wrong way to approach type 2 diabetes. As we will see in chapter 6, type 2 diabetes belongs to a spectrum of metabolic dysfunction that begins long before someone crosses that magical diagnostic threshold on a blood test. Type 2 diabetes is merely the last stop on the line. The time to intervene is well before the patient gets anywhere near that zone; even prediabetes is very late in the game. It is absurd and harmful to treat this disease like a cold or a broken bone, where you either have it or you don't; it's not binary. Yet too often, the point of clinical diagnosis is where our interventions begin. Why is this okay?

I believe that our goal should be to act as early as possible, to try to *prevent* people from developing type 2 diabetes and all the other Horsemen. We should be proactive instead of reactive in our approach. Changing that mindset must be our first step in attacking slow death. We want to delay or prevent these conditions so that we can live longer *without* disease, rather than lingering *with* disease. That means that the best time to intervene is before the eggs start falling—as I discovered in my own life.

On September 8, 2009, a day I will never forget, I was standing on a beach on Catalina Island when my wife, Jill, turned to me and said, "Peter, I think you should work on being a little less not thin."

I was so shocked that I nearly dropped my cheeseburger. "Less not thin?" My sweet wife said *that*?

I was pretty sure that I'd earned the burger, as well as the Coke in my other hand, having just swum to this island from Los Angeles, across twenty-one miles of open ocean—a journey that had taken me fourteen hours, with a current in my face for much of the way. A minute earlier, I'd been thrilled to have finished this bucket-list long-distance swim.* Now I was Not-Thin Peter.

Nevertheless, I instantly knew that Jill was right. Without even realizing it, I had ballooned up to 210 pounds, a solid 50 more than my fighting weight

* This was actually my second time making this crossing. I'd swum from Catalina to LA a few years earlier, but the reverse direction took four hours longer, because of the current.

as a teenage boxer. Like a lot of middle-aged guys, I still thought of myself as an "athlete," even as I squeezed my sausage-like body into size 36 pants. Photographs from around that time remind me that my stomach looked just like Jill's when she was six months pregnant. I had become the proud owner of a full-fledged dad bod, and I had not even hit forty.

Blood tests revealed worse problems than the ones I could see in the mirror. Despite the fact that I exercised fanatically and ate what I believed to be a healthy diet (notwithstanding the odd post-swim cheeseburger), I had somehow become insulin resistant, one of the first steps down the road to type 2 diabetes and many other bad things. My testosterone levels were below the 5th percentile for a man my age. It's not an exaggeration to say that my life was in danger—not imminently, but certainly over the long term. I knew exactly where this road could lead. I had amputated the feet of people who, twenty years earlier, had been a lot like me. Closer to home, my own family tree was full of men who had died in their forties from cardiovascular disease.

That moment on the beach marked the beginning of my interest in—that word again—longevity. I was thirty-six years old, and I was on the precipice. I had just become a father with the birth of our first child, Olivia. From the moment I first held her, wrapped in her white swaddling blanket, I fell in love—and knew my life had changed forever. But I would also soon learn that my various risk factors and my genetics likely pointed toward an early death from cardiovascular disease. What I didn't yet realize was that my situation was entirely fixable.

As I delved into the scientific literature, I quickly became as obsessed with understanding nutrition and metabolism as I had once been with learning cancer surgery. Because I am an insatiably curious person by nature, I reached out to the leading experts in these fields and persuaded them to mentor me on my quest for knowledge. I wanted to understand how I'd gotten myself into that state and what it meant for my future. And I needed to figure out how to get myself back on track.

My next task was to try to understand the true nature and causes of atherosclerosis, or heart disease, which stalks the men in my dad's family. Two of his brothers had died from heart attacks before age fifty, and a third had succumbed in his sixties. From there it was a short leap over to cancer, which has always fascinated me, and then to neurodegenerative diseases like Alzheimer's disease. Finally, I began to study the fast-moving field of gerontology—the effort to understand what drives the aging process itself and how it might be slowed.

Perhaps my biggest takeaway was that modern medicine does not really have a handle on when and how to treat the chronic diseases of aging that will likely kill most of us. This is in part because each of the Horsemen is intricately complex, more of a disease *process* than an acute illness like a common cold. The surprise is that this is actually good news for us, in a way. Each one of the Horsemen is cumulative, the product of multiple risk factors adding up and compounding over time. Many of these same individual risk factors, it turns out, are relatively easy to reduce or even eliminate. Even better, they share certain features or drivers in common that make them vulnerable to some of the same tactics and behavioral changes we will discuss in this book.

Medicine's biggest failing is in attempting to treat all these conditions at the wrong end of the timescale—after they are entrenched—rather than before they take root. As a result, we ignore important warning signs and miss opportunities to intervene at a point where we still have a chance to beat back these diseases, improve health, and potentially extend lifespan.

Just to pick a few examples:

- Despite throwing billions of dollars in research funding at the Horsemen, mainstream medicine has gotten crucial things dead wrong about their root causes. We will examine some promising new theories about the origin and causes of each, and possible strategies for prevention.
- The typical cholesterol panel that you receive and discuss at your annual physical, along with many of the underlying assumptions behind it (e.g., "good" and "bad" cholesterol), is misleading and oversimplified to the point of uselessness. It doesn't tell us nearly enough about your actual risk of dying from heart disease—and we don't do nearly enough to stop this killer.

- Millions of people are suffering from a little-known and underdiagnosed liver condition that is a potential precursor to type 2 diabetes. Yet people at the early stages of this metabolic derangement will often return blood test results in the "normal" range. Unfortunately, in today's unhealthy society, "normal" or "average" is not the same as "optimal."

- The metabolic derangement that leads to type 2 diabetes also helps foster and promote heart disease, cancer, *and* Alzheimer's disease. Addressing our metabolic health can lower the risk of each of the Horsemen.

- Almost all "diets" are similar: they may help some people but prove useless for most. Instead of arguing about diets, we will focus on *nutritional biochemistry*—how the combinations of nutrients that you eat affect your own metabolism and physiology, and how to use data and technology to come up with the best eating pattern for you.

- One macronutrient, in particular, demands more of our attention than most people realize: not carbs, not fat, but *protein* becomes critically important as we age.

- Exercise is by far the most potent longevity "drug." No other intervention does nearly as much to prolong our lifespan and preserve our cognitive and physical function. But most people don't do nearly enough—and exercising the wrong way can do as much harm as good.

- Finally, as I learned the hard way, striving for physical health and longevity is meaningless if we ignore our emotional health. Emotional suffering can decimate our health on all fronts, and it must be addressed.

Why does the world need another book about longevity? I've asked myself that question often over the last few years. Most writers in this space fall into certain categories. There are the true believers, who insist that if you follow their specific diet (the more restrictive the better), or practice meditation a certain way, or eat a particular type of superfood, or maintain your "energy" properly, then you will be able to avoid death and live forever. What they often lack in scientific rigor they make up for with passion.

On the other end of the spectrum are those who are convinced that science will soon figure out how to unplug the aging process itself, by tweaking some obscure cellular pathway, or lengthening our telomeres, or "reprogramming" our cells so that we no longer need to age at all. This seems highly unlikely in our lifetime, although it is certainly true that science is making huge leaps in our understanding of aging and of the Horsemen diseases. We are learning so much, but the tricky part is knowing how to apply this new knowledge to real people outside the lab—or at a minimum, how to hedge our bets in case this highfalutin science somehow fails to put longevity into a pill.

This is how I see my role: I am not a laboratory scientist or clinical researcher but more of a translator, helping you understand and apply these insights. This requires a thorough understanding of the science but also a bit of art, just as if we were translating a poem by Shakespeare into another language. We have to get the meaning of the words exactly right (the science), while also capturing the tone, the nuance, the feeling, and the rhythm (the art). Similarly, my approach to longevity is firmly rooted in science, but there is also a good deal of art in figuring out how and when to apply our knowledge to you, the patient, with your specific genes, your history and habits, and your goals.

I believe that we already know more than enough to bend the curve. That is why this book is called *Outlive.* I mean it in both senses of the word: live longer and live better. Unlike Tithonus, you can outlive your life expectancy *and* enjoy better health, getting more out of your life.

My goal is to create an actionable operating manual for the *practice* of longevity. A guide that will help you Outlive. I hope to convince you that with enough time and effort, you can potentially extend your lifespan by a decade and your healthspan possibly by two, meaning you might hope to function like someone twenty years younger than you.

But my intent here is not to tell you exactly *what to do;* it's to help you learn *how to think* about doing these things. For me, that has been the journey, an obsessive process of study and iteration that began that day on the rocky shore of Catalina Island.

More broadly, longevity demands a paradigm-shifting approach to medicine, one that directs our efforts toward preventing chronic diseases and improving our healthspan—and doing it now, rather than waiting until disease has taken hold or until our cognitive and physical function has already declined. It's not "preventive" medicine; it's *proactive* medicine, and I believe it has the potential not only to change the lives of individuals but also to relieve vast amounts of suffering in our society as a whole. This change is not coming from the medical establishment, either; it will happen only if and when patients and physicians demand it.

Only by altering our approach to medicine itself can we get to the rooftop and stop the eggs from falling. None of us should be satisfied racing around at the bottom to try to catch them.

Chapter 1 Summary:

Current medical philosophy is focused too late in the progression of "the big 4". Rather that working @ the diagnosis stage. Rather, preventing risk factors becoming acute symptoms becoming chronic conditions

"4 horsemen"
- heart disease
- cancer
- neurodegenerative disease
- type II diabetes & related metabolic disfunction

CHAPTER 2

Medicine 3.0

Rethinking Medicine for the Age of Chronic Disease

The time to repair the roof is when the sun is shining.

—John F. Kennedy

I don't remember what the last straw was in my growing frustration with medical training, but I do know that the beginning of the end came courtesy of a drug called gentamicin. Late in my second year of residency, I had a patient in the ICU with severe sepsis. He was basically being kept alive by this drug, which is a powerful IV antibiotic. The tricky thing about gentamicin is that it has a very narrow therapeutic window. If you give a patient too little, it won't do anything, but if you give him too much it could destroy his kidneys and hearing. The dosing is based on the patient's weight and the expected half-life of the drug in the body, and because I am a bit of a math geek (actually, more than a bit), one evening I came up with a mathematical model that predicted the precise time when this patient would need his next dose: 4:30 a.m.

Sure enough, when 4:30 rolled around we tested the patient and found that his blood levels of gentamicin had dropped to exactly the point where he needed another dose. I asked his nurse to give him the medication but found myself at odds with the ICU fellow, a trainee who was one level above us residents in the hospital pecking order. I wouldn't do that, she said. Just have them give it at seven, when the next nursing shift comes on. This puzzled me, because we knew that the patient would have to go for more than two hours basically unprotected from a massive infection that could kill him. Why wait? When the fellow left, I had the nurse give the medicine anyway.

Later that morning at rounds, I presented the patient to the attending physician and explained what I had done, and why. I thought she would appreciate my attention to patient care—getting the drug dosed just right—but instead, she turned and gave me a tongue-lashing like I'd never experienced. I'd been awake for more than twenty-four hours at this point, but I wasn't hallucinating. I was getting screamed at, even threatened with being fired, for trying to improve the way we delivered medication to a very sick patient. True, I had disregarded the suggestion (not a direct order) from the fellow, my immediate superior, and that was wrong, but the attending's tirade stunned me. Shouldn't we *always* be looking for better ways to do things?

Ultimately, I put my pride in check and apologized for my disobedience, but this was just one incident of many. As my residency progressed, my doubts about my chosen profession only mounted. Time and again, my colleagues and I found ourselves coming into conflict with a culture of resistance to change and innovation. There are some good reasons why medicine is conservative in nature, of course. But at times it seemed as if the whole edifice of modern medicine was so firmly rooted in its traditions that it was unable to change even slightly, even in ways that would potentially save the lives of people for whom we were supposed to be caring.

By my fifth year, tormented by doubts and frustration, I informed my superiors that I would be leaving that June. My colleagues and mentors thought I was insane; almost nobody leaves residency, certainly not at Hopkins with only two years to go. But there was no dissuading me. Throwing nine years of medical training out the window, or so it seemed, I took a job with McKinsey & Company, the well-known management consulting firm. My wife and I moved across the country to the posh playground of Palo Alto and San Francisco, where I had loved living while at Stanford. It was about as far away from medicine (and Baltimore) as it was possible to get, and I was glad. I felt as if I had wasted a decade of my life. But in the end, this seeming detour ended up reshaping the way I look at medicine—and more importantly, each of my patients.

The key word, it turned out, was *risk.*

McKinsey originally hired me into their healthcare practice, but because of my quantitative background (I had studied applied math and mechanical engineering in college, planning to pursue a PhD in aerospace engineering), they moved me over to credit risk. This was in 2006, during the runup to the global financial crisis, but before almost anyone besides the folks featured in Michael Lewis's *The Big Short* understood the magnitude of what was about to happen.

Our job was to help US banks comply with a new set of rules that required them to maintain enough reserves to cover their unexpected losses. The banks had done a good job of estimating their *expected* losses, but nobody really knew how to deal with the *unexpected* losses, which by definition were much more difficult to predict. Our task was to analyze the banks' internal data and come up with mathematical models to try to predict these unexpected losses on the basis of correlations among asset classes—which was just as tricky as it sounds, like a crapshoot on top of a crapshoot.

What started out as an exercise to help the biggest banks in the United States jump through some regulatory hoops uncovered a brewing disaster in what was considered to be one of their least risky, most stable portfolios: prime mortgages. By the late summer of 2007, we had arrived at the horrifying but inescapable conclusion that the big banks were about to lose more money on mortgages in the next two years than they had made in the previous decade.

In late 2007, after six months of round-the-clock work, we had a big meeting with the top brass of our client, a major US bank. Normally, my boss, as the senior partner on the project, would have handled the presentation. But instead he picked me. "Based on your previous career choice," he said, "I suspect you are better prepared to deliver truly horrible news to people."

This was not unlike delivering a terminal diagnosis. I stood up in a high-floor conference room and walked the bank's management team through the numbers that foretold their doom. As I went through my presentation, I watched the five stages of grief described by Elisabeth Kübler-Ross in her classic book *On Death and Dying*—denial, anger, bargaining, sadness, and acceptance—flash across the executives' faces. I had never seen that happen before outside of a hospital room.

My detour into the world of consulting came to an end, but it opened my eyes to a huge blind spot in medicine, and that is the understanding of risk. In finance and banking, understanding risk is key to survival. Great investors do not take on risk blindly; they do so with a thorough knowledge of both risk and reward. The study of credit risk is a science, albeit an imperfect one, as I learned with the banks. While risk is obviously also important in medicine, the medical profession often approaches risk more emotionally than analytically.

The trouble began with Hippocrates. Most people are familiar with the ancient Greek's famous dictum: "First, do no harm." It succinctly states the physician's primary responsibility, which is to not kill our patients or do anything that might make their condition worse instead of better. Makes sense. There are only three problems with this: (a) Hippocrates never actually said these words,* (b) it's sanctimonious bullshit, and (c) it's unhelpful on multiple levels.

"Do no harm"? Seriously? Many of the treatments deployed by our medi-

---

* The words "First, do no harm" do not appear in Hippocrates's actual writings. He urged physicians to "practice two things in your dealings with disease: either help or do not harm the patient." This was changed to "First, do no harm" by an aristocratic nineteenth-century British surgeon named Thomas Inman, whose other claim to fame was, well, nothing. Somehow it became the sacred motto of the medical profession for all of eternity.

cal forebears, from Hippocrates's time well into the twentieth century, were if anything *more* likely to do harm than to heal. Did your head hurt? You'd be a candidate for trepanation, or having a hole drilled in your skull. Strange sores on your private parts? Try not to scream while the Doktor of Physik dabs some toxic mercury on your genitals. And then, of course, there was the millennia-old standby of bloodletting, which was generally the very last thing that a sick or wounded person needed.

What bothers me most about "First, do no harm," though, is its implication that the best treatment option is always the one with the least immediate downside risk—and, very often, doing nothing at all. Every doctor worth their diploma has a story to disprove this nonsense. Here's one of mine: During one of the last trauma calls I took as a resident, a seventeen-year-old kid came in with a single stab wound in his upper abdomen, just below his xiphoid process, the little piece of cartilage at the bottom end of his sternum. He seemed to be stable when he rolled in, but then he started acting odd, becoming very anxious. A quick ultrasound suggested he might have some fluid in his pericardium, the tough fibrous sac around the heart. This was now a full-blown emergency, because if enough fluid collected in there, it would stop his heart and kill him within a minute or two.

There was no time to take him up to the OR; he could easily die on the elevator ride. As he lost consciousness, I had to make a split-second decision to cut into his chest right then and there and slice open his pericardium to relieve the pressure on his heart. It was stressful and bloody, but it worked, and his vital signs soon stabilized. No doubt the procedure was hugely risky and caused him great short-term harm, but had I not done it, he might have died waiting for a safer and more sterile procedure in the operating room. Fast death waits for no one.

The reason I had to act so dramatically in the moment was that the risk was so asymmetric: doing nothing—avoiding "harm"—would likely have resulted in his death. Conversely, even if I was wrong in my diagnosis, the hasty chest surgery we performed was quite survivable, though obviously not how one might wish to spend a Wednesday night. After we got him out of imminent danger, it became clear that the tip of the knife had just barely punctured his pulmonary artery, a simple wound that took two stitches to fix once he was stabilized and in the OR. He went home four nights later.

Risk is not something to be avoided at all costs; rather, it's something we need to understand, analyze, and work with. Every single thing we do, in medicine and in life, is based on some calculation of risk versus reward. Did you eat a salad from Whole Foods for lunch? There's a small chance there could have been *E. coli* on the greens. Did you drive to Whole Foods to get it? Also risky. But on balance, that salad is probably good for you (or at least less bad than some other things you could eat).

Sometimes, as in the case of my seventeen-year-old stab victim, you have to take the leap. In other, less rushed situations, you might have to choose more carefully between subjecting a patient to a colonoscopy, with its slight but real risk of injury, versus not doing the examination and potentially missing a cancer diagnosis. My point is that a physician who has never done *any* harm, or at least confronted the risk of harm, has probably never done much of anything to help a patient either. And as in the case of my teenage stabbing victim, sometimes doing nothing is the riskiest choice of all.

I actually kind of wish Hippocrates had been around to witness that operation on the kid who was stabbed—or any procedure in a modern hospital setting, really. He would have been blown away by all of it, from the precision steel instruments to the antibiotics and anesthesia, to the bright electric lights.

While it is true that we owe a lot to the ancients—such as the twenty thousand new words that medical school injected into my vocabulary, most derived from Greek or Latin—the notion of a continuous march of progress from Hippocrates's era to the present is a complete fiction. It seems to me that there have been two distinct eras in medical history, and that we may now be on the verge of a third.

The first era, exemplified by Hippocrates but lasting almost two thousand years after his death, is what I call *Medicine 1.0*. Its conclusions were based on direct observation and abetted more or less by pure guesswork, some of which was on target and some not so much. Hippocrates advocated walking for exercise, for example, and opined that "in food excellent medicine can be found; in food bad medicine can be found," which still holds up. But much of Medicine 1.0 missed the mark entirely, such as the notion of bodily "humors," to cite just one example of many. Hippocrates's major contribution was the insight that diseases are caused by nature and not by actions of the gods, as had previously been believed. That alone represented a huge step in the right direction. So it's hard to be too critical of him and his contemporaries. They did the best they could without an understanding of science or the scientific method. You can't use a tool that has not yet been invented.

Medicine 2.0 arrived in the mid-nineteenth century with the advent of the germ theory of disease, which supplanted the idea that most illness was spread by "miasmas," or bad air. This led to improved sanitary practices by physicians and ultimately the development of antibiotics. But it was far from a clean transition; it's not as though one day Louis Pasteur, Joseph Lister, and Robert Koch simply published their groundbreaking studies,* and the rest of the medical profession fell into line and changed the way they did everything overnight. In fact, the shift from Medicine 1.0 to Medicine 2.0 was a long, bloody slog that took centuries, meeting trench-warfare resistance from the establishment at many points along the way.

Consider the case of poor Ignaz Semmelweis, a Viennese obstetrician who was troubled by the fact that so many new mothers were dying in the hospital where he worked. He concluded that their strange "childbed fever" might somehow be linked to the autopsies that he and his colleagues performed in the mornings, before delivering babies in the afternoons—without washing their hands in between. The existence of germs had not yet been discovered, but Semmelweis nonetheless believed that the doctors were

* Pasteur discovered the existence of airborne pathogens and bacteria that caused food to rot; Lister developed antiseptic surgical techniques; and Koch identified the germs that caused tuberculosis and cholera.

transmitting *something* to these women that caused their illness. His observations were most unwelcome. His colleagues ostracized him, and Semmelweis died in an insane asylum in 1865.

That very same year, Joseph Lister first successfully demonstrated the principle of antiseptic surgery, using sterile techniques to operate on a young boy in a hospital in Glasgow. It was the first application of the germ theory of disease. Semmelweis had been right all along.

The shift from Medicine 1.0 to Medicine 2.0 was prompted in part by new technologies such as the microscope, but it was more about a *new way of thinking.* The foundation was laid back in 1628, when Sir Francis Bacon first articulated what we now know as the scientific method. This represented a major philosophical shift, from observing and guessing to observing, and then forming a hypothesis, which as Richard Feynman pointed out is basically a fancy word for a guess.

The next step is crucial: rigorously testing that hypothesis/guess to determine whether it is correct, also known as experimenting. Instead of using treatments that they *believed* might work, often despite ample anecdotal evidence to the contrary, scientists and physicians could systematically test and evaluate potential cures, then choose the ones that had performed best in experiments. Yet three centuries elapsed between Bacon's essay and the discovery of penicillin, the true game-changer of Medicine 2.0.

Medicine 2.0 was transformational. It is a defining feature of our civilization, a scientific war machine that has eradicated deadly diseases such as polio and smallpox. Its successes continued with the containment of HIV and AIDS in the 1990s and 2000s, turning what had seemed like a plague that threatened all humanity into a manageable chronic disease. I'd put the recent cure of hepatitis C right up there as well. I remember being told in medical school that hepatitis C was an unstoppable epidemic that was going to completely overwhelm the liver transplant infrastructure in the United States within twenty-five years. Today, most cases can be cured by a short course of drugs (albeit very expensive ones).

Perhaps even more amazing was the rapid development of not just one but several effective vaccines against COVID-19, not even a year after the pandemic took hold in early 2020. The virus genome was sequenced within weeks of the first deaths, allowing the speedy formulation of vaccines that specifically target its surface proteins. Progress with COVID treatments has also been remarkable, yielding multiple types of antiviral drugs within less than two years. This represents Medicine 2.0 at its absolute finest.

Yet Medicine 2.0 has proved far less successful against long-term diseases such as cancer. While books like this always trumpet the fact that lifespans have nearly doubled since the late 1800s, the lion's share of that progress may have resulted entirely from antibiotics and improved sanitation, as Steven Johnson points out in his book *Extra Life*. The Northwestern University economist Robert J. Gordon analyzed mortality data going back to 1900 (see figure 1) and found that if you subtract out deaths from the eight top infectious diseases, which were largely brought under control by the advent of antibiotics in the 1930s, overall mortality rates declined relatively little over the course of the twentieth century. That means that Medicine 2.0 has made scant progress against the Horsemen.

Figure 1. Change in Mortality Rates Since 1900

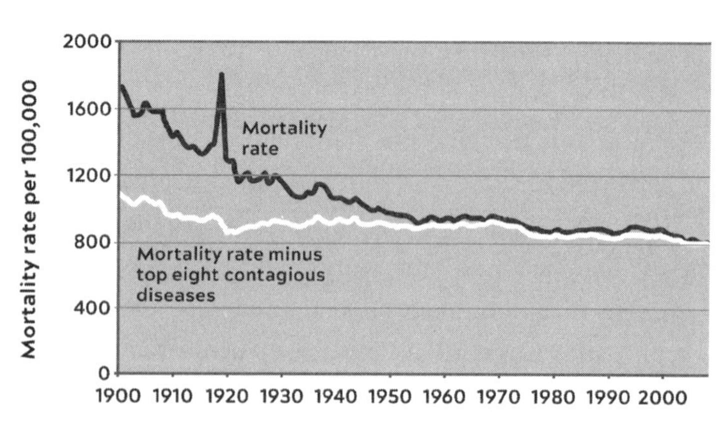

This graph shows how *little* real mortality rates have improved since 1900, once you remove the top eight contagious/infectious diseases, which were largely controlled by the advent of antibiotics in the early twentieth century.

Source: Gordon (2016).

Toward Medicine 3.0

During my stint away from medicine, I realized that my colleagues and I had been trained to solve the problems of an earlier era: the acute illnesses and injuries that Medicine 2.0 had evolved to treat. Those problems had a much shorter event horizon; for our cancer patients, time itself was the enemy. And we were always coming in too late.

This actually wasn't so obvious until I'd spent my little sabbatical immersed in the worlds of mathematics and finance, thinking every day about the nature of risk. The banks' problem was not all that different from the situation faced by some of my patients: their seemingly minor risk factors had, over time, compounded into an unstoppable, asymmetric catastrophe. Chronic diseases work in a similar fashion, building over years and decades—and once they become entrenched, it's hard to make them go away. Atherosclerosis, for example, begins many decades before the person has a coronary "event" that could result in their death. But that event, often a heart attack, too often marks the point where treatment begins.

This is why I believe we need a new way of thinking about chronic diseases, their treatment, and how to maintain long-term health. The goal of this new medicine—which I call *Medicine 3.0*—is not to patch people up and get them out the door, removing their tumors and hoping for the best, but rather to prevent the tumors from appearing and spreading in the first place. Or to avoid that first heart attack. Or to divert someone from the path to Alzheimer's disease. Our treatments, and our prevention and detection strategies, need to change to fit the nature of these diseases, with their long, slow prologues.

It is already obvious that medicine is changing rapidly in our era. Many pundits have been predicting a glorious new era of "personalized" or "precision" medicine, where our care will be tailored to our exact needs, down to our very genes. This is, obviously, a worthy goal; it is clear that no two patients are exactly alike, even when they are presenting with what appears to be an identical upper-respiratory illness. A treatment that works for one patient may prove useless in the other, either because her immune system is reacting differently or because her infection is viral rather than bacterial. Even now, it remains extremely difficult to tell the difference, resulting in millions of useless antibiotic prescriptions.

Many thinkers in this space believe that this new era will be driven by advances in technology, and they are likely right; at the same time, however, technology has (so far) been largely a limiting factor. Let me explain. On the one hand, improved technology enables us to collect much more data on patients than ever before, and patients themselves are better able to monitor their own biomarkers. This is good. Even better, artificial intelligence and machine learning are being harnessed to try to digest this massive profusion of data and come up with more definitive assessments of our risk of, say, heart disease than the rather simple risk factor–based calculators we have now. Others point to the possibilities of nanotechnology, which could enable doctors to diagnose and treat disease by means of microscopic bioactive particles injected into the bloodstream. But the nanobots aren't here yet, and barring a major public or private research push, it could be a while before they become reality.

The problem is that our *idea* of personalized or precision medicine remains some distance ahead of the technology necessary to realize its full promise. It's a bit like the concept of the self-driving car, which has been talked about for almost as long as automobiles have been crashing into each other and killing and injuring people. Clearly, removing human error from the equation as much as possible would be a good thing. But our technology is only today catching up to a vision we've held for decades.

If you had wanted to create a "self-driving" car in the 1950s, your best option might have been to strap a brick to the accelerator. Yes, the vehicle would have been able to move forward on its own, but it could not slow down, stop, or turn to avoid obstacles. Obviously not ideal. But does that mean the entire concept of the self-driving car is not worth pursuing? No, it only means that at the time we did not yet have the tools we now possess to help enable vehicles to operate both autonomously and safely: computers, sensors, artificial intelligence, machine learning, and so on. This once-distant dream now seems within our reach.

It is much the same story in medicine. Two decades ago, we were still taping bricks to gas pedals, metaphorically speaking. Today, we are approaching the point where we can begin to bring some appropriate technology to bear in ways that advance our understanding of patients as unique individuals. For example, doctors have traditionally relied on two tests to gauge their patients' metabolic health: a fasting glucose test, typically given once a year; or the HbA1c test we mentioned earlier, which gives us an estimate of their average blood glucose over the last 90 days. But those tests are of limited use because they are static and backward-looking. So instead, many of my patients have worn a device that monitors their blood glucose levels in real time, which allows me to talk to them about nutrition in a specific, nuanced, feedback-driven way that was not even possible a decade ago. This technology, known as continuous glucose monitoring (CGM), lets me observe how *their individual metabolism* responds to a certain eating pattern and make changes to their diet quickly. In time, we will have many more sensors like this that will allow us to tailor our therapies and interventions far more quickly and precisely. The self-driving car will do a better job of following the twists and turns of the road, staying out of the ditch.

But Medicine 3.0, in my opinion, is not really about technology; rather, it requires an evolution in our mindset, a shift in the way in which we approach medicine. I've broken it down into four main points.

First, *Medicine 3.0 places a far greater emphasis on prevention than treatment*. When did Noah build the ark? Long before it began to rain. Medicine 2.0 tries to figure out how to get dry after it starts raining. Medicine 3.0 studies meteorology and tries to determine whether we need to build a better roof, or a boat.

Second, *Medicine 3.0 considers the patient as a unique individual.* Medicine 2.0 treats everyone as basically the same, obeying the findings of the clinical trials that underlie evidence-based medicine. These trials take heterogeneous inputs (the people in the study or studies) and come up with homogeneous results (the average result across all those people). Evidence-based medicine then insists that we apply those average findings back to individuals. The problem is that no patient is strictly average. Medicine 3.0 takes the findings of evidence-based medicine and goes one step further, looking more deeply into the data to determine how our patient is similar or different from the "average" subject in the study, and how its findings might or might not be applicable to them. Think of it as "evidence-informed" medicine.

The third philosophical shift has to do with our attitude toward risk. *In Medicine 3.0, our starting point is the honest assessment, and acceptance, of risk—including the risk of doing nothing.*

There are many examples of how Medicine 2.0 gets risk wrong, but one of the most egregious has to do with hormone replacement therapy (HRT) for postmenopausal women, long entrenched as standard practice before the results of the Women's Health Initiative Study (WHI) were published in 2002. This large clinical trial, involving thousands of older women, compared a multitude of health outcomes in women taking HRT versus those who did not take it. The study reported a 24 percent relative increase in the risk of breast cancer among a subset of women taking HRT, and headlines all over the world condemned HRT as a dangerous, cancer-causing therapy. All of a sudden, on the basis of this one study, hormone replacement treatment became virtually taboo.

This reported 24 percent risk increase sounded scary indeed. But nobody seemed to care that the *absolute* risk increase of breast cancer for women in the study remained minuscule. Roughly five out of every one thousand women in the HRT group developed breast cancer, versus four out of every one thousand in the control group, who received no hormones. The absolute risk increase was just 0.1 percentage point. HRT was linked to, potentially, one additional case of breast cancer in every thousand patients. Yet this tiny increase in absolute risk was deemed to outweigh any benefits, meaning menopausal women would potentially be subject to hot flashes and night sweats, as well as loss of bone density and muscle mass, and other unpleasant symptoms of menopause—not to mention a potentially increased risk of Alzheimer's disease, as we'll see in chapter 9.

Medicine 2.0 would rather throw out this therapy entirely, on the basis of one clinical trial, than try to understand and address the nuances involved. Medicine 3.0 would take this study into account, while recognizing its inevitable limitations and built-in biases. The key question that Medicine 3.0 asks is whether this intervention, hormone replacement therapy, with its relatively small increase in *average* risk in a large group of women older than sixty-five, might still be net beneficial for our *individual* patient, with her own unique mix of symptoms and risk factors. How is she similar to or different from the population in the study? One huge difference: none of the women selected for the study were actually symptomatic, and most were many years out of menopause. So how applicable are the findings of this study to women who are in or just entering menopause (and are presumably younger)? Finally, is there some other possible explanation for the slight observed increase in risk with this specific HRT protocol?[*]

My broader point is that at the level of the individual patient, we should be willing to ask deeper questions of risk versus reward versus cost for this therapy—and for almost anything else we might do.

The fourth and perhaps largest shift is that where Medicine 2.0 focuses largely on lifespan, and is almost entirely geared toward staving off death, *Medicine 3.0 pays far more attention to maintaining healthspan, the quality of life.*

Healthspan was a concept that barely even existed when I went to medical school. My professors said little to nothing about how to help our patients maintain their physical and cognitive capacity as they aged. The word *exercise* was almost never uttered. Sleep was totally ignored, both in class and in residency, as we routinely worked twenty-four hours at a stretch. Our instruction in nutrition was also minimal to nonexistent.

Today, Medicine 2.0 at least acknowledges the importance of healthspan, but the standard definition—the period of life free of disease or disability—is totally insufficient, in my view. We want more out of life than simply the absence of sickness or disability. We want to be thriving, in every way, throughout the latter half of our lives.

Another, related issue is that longevity itself, and healthspan in particular, doesn't really fit into the business model of our current healthcare system.

---

* A deeper dive into the data suggests that the tiny increase in breast cancer risk was quite possibly due to the type of synthetic progesterone used in the study, and not the estrogen; the devil is always in the details.

There are few insurance reimbursement codes for most of the largely preventive interventions that I believe are necessary to extend lifespan and healthspan. Health insurance companies won't pay a doctor very much to tell a patient to change the way he eats, or to monitor his blood glucose levels in order to help prevent him from developing type 2 diabetes. Yet insurance will pay for this same patient's (very expensive) insulin *after* he has been diagnosed. Similarly, there's no billing code for putting a patient on a comprehensive exercise program designed to maintain her muscle mass and sense of balance while building her resistance to injury. But if she falls and breaks her hip, then her surgery and physical therapy will be covered. Nearly all the money flows to treatment rather than prevention—and when I say "prevention," I mean *prevention of human suffering*. Continuing to ignore healthspan, as we've been doing, not only condemns people to a sick and miserable older age but is guaranteed to bankrupt us eventually.

When I introduce my patients to this approach, I often talk about icebergs—specifically, the ones that ended the first and final voyage of the *Titanic*. At 9:30 p.m. on the fatal night, the massive steamship received an urgent message from another vessel that it was headed into an icefield. The message was ignored. More than an hour later, another ship telegraphed a warning of icebergs in the ship's path. The *Titanic*'s wireless operator, busy trying to communicate with Newfoundland over crowded airwaves, replied (via Morse code): "Keep out; shut up."

There were other problems. The ship was traveling at too fast a speed for a foggy night with poor visibility. The water was unusually calm, giving the crew a false sense of security. And although there was a set of binoculars on board, they were locked away and no one had a key, meaning the ship's lookout was relying on his naked eyes alone. Forty-five minutes after that last radio call, the lookout spotted the fatal iceberg just five hundred yards ahead. Everyone knows how that ended.

But what if the *Titanic* had had radar and sonar (which were not developed until World War II, more than fifteen years later)? Or better yet, GPS and satellite imaging? Rather than trying to dodge through the maze of deadly icebergs, hoping for the best, the captain could have made a slight course correction a day or two before and steered clear of the entire mess. This is exactly what ship captains do now, thanks to improved technology that has made *Titanic*-style sinkings largely a thing of the past, relegated to sappy, nostalgic movies with overwrought soundtracks.

The problem is that in medicine our tools do not allow us to see very far over the horizon. Our "radar," if you will, is not powerful enough. The longest randomized clinical trials of statin drugs for primary prevention of heart disease, for example, might last five to seven years. Our longest risk prediction time frame is ten years. But cardiovascular disease can take decades to develop.

Medicine 3.0 looks at the situation through a longer lens. A forty-year-old should be concerned with her thirty- or forty-year cardiovascular risk profile, not merely her ten-year risk. We therefore need tools with a much longer reach than relatively brief clinical trials. We need long-range radar and GPS, and satellite imaging, and all the rest. Not just a snapshot.

As I tell my patients, I'd like to be the navigator of your ship. My job, as I see it, is to steer you through the icefield. I'm on iceberg duty, 24-7. How many icebergs are out there? Which ones are closest? If we steer away from those, will that bring us into the path of other hazards? Are there bigger, more dangerous icebergs lurking over the horizon, out of sight?

Which brings us to perhaps the most important difference between Medicine 2.0 and Medicine 3.0. In Medicine 2.0, you are a passenger on the ship, being carried along somewhat passively. Medicine 3.0 demands much more from you, the patient: You must be well informed, medically literate to a reasonable degree, clear-eyed about your goals, and cognizant of the true nature of risk. You must be willing to change ingrained habits, accept new challenges, and venture outside of your comfort zone if necessary. You are always participating, never passive. You confront problems, even uncomfortable or scary ones, rather than ignoring them until it's too late. You have skin in the game, in a very literal sense. And you make important decisions.

Because in this scenario, you are no longer a passenger on the ship; you are its captain.

CHAPTER 3

Objective, Strategy, Tactics

A Road Map for Reading This Book

Strategy without tactics is the slowest route to victory.
Tactics without strategy is the noise before defeat.

—Sun Tzu

Several years ago, I flew to San Francisco to attend the funeral of the mother of a good friend from college, whom I'll call Becky. Because Becky's parents lived near Palo Alto, where I went to medical school, they invited me to dinner many times. We often ate in their garden, which had been beautifully planned and meticulously maintained by Becky's mother, whose name was Sophie.

I remembered Sophie as a vibrant, athletic woman who had seemed ageless. But I hadn't seen her since my wedding nearly fifteen years earlier. Becky filled me in on what I had missed. Beginning in her early seventies, Sophie had undergone a steep physical decline that began when she slipped and fell while gardening, tearing a muscle in her shoulder. That soon escalated into back and neck pain so severe that she could no longer work in the garden or play golf at all, her two primary passions in retirement. She simply sat around the house, feeling depressed. This was followed by a descent into dementia in the last couple of years of her life, before she died of a respiratory infection at age eighty-three.

At her memorial service, everyone agreed that it was a "blessing" that Sophie hadn't had to linger in that demented state for very long, but as I sat in the pew, I reflected on the fact that she had spent the last decade of her life being unable to participate in any of the activities that had given her pleasure. Instead, she had been in considerable pain. Nobody mentioned that. We were gathered to mourn Sophie's biological death, but it saddened me even more deeply that she had been robbed of the joy of her final years.

I often talk about Sophie with my patients, not because her tale is unusual but because it is so sadly typical. We have all watched our parents, grandparents, spouses, or friends undergo similar ordeals. The sad thing is that we almost expect this to happen to our elders; and even with this knowledge, relatively few of us take measures that might help ourselves avoid that fate. Even for Becky, who had cared for her mother during her difficult final years, the idea that she might end up in the same condition was probably the furthest thing from her mind. The future, for most of us, remains a hazy abstraction.

I tell Sophie's story to help illustrate a fundamental concept in my approach to longevity, which is the need to think about and plan for the later decades of our lives—our seventies, eighties, nineties, or beyond. For many people, like Sophie, the last ten years of life are not a particularly happy time. They typically suffer from one or more of the Horsemen diseases and the effects of the requisite treatments. Their cognitive and physical abilities may be weakening or gone. Generally, they are unable to participate in the activities they once loved, whether that means gardening, or playing chess, or riding a bicycle, or whatever else in their life gave them joy. I call this the Marginal Decade, and for many, if not most, it is a period of diminishment and limitation.

I ask all my patients to sketch out an alternative future for themselves.

What do you *want* to be doing in your later decades? What is your plan for the rest of your life?

Everyone has a slightly different answer—they might want to travel, or continue playing golf or hiking in nature, or simply be able to play with their grandkids and great-grandkids (top of my own list). The point of this exercise is twofold. First, it forces people to focus on their own endgame, which most of us might prefer to avoid thinking about. Economists call this "hyperbolic discounting," the natural tendency for people to choose immediate gratification over potential future gains, especially if those gains entail hard work. Second, it drives home the importance of healthspan. If Becky wants to enjoy a healthy, rewarding life in her later years, and not repeat her mother's fate, she will have to maintain and hopefully improve her physical and cognitive function every decade between now and then. Otherwise, the gravitational pull of aging will do its thing, and she will decline, just as her mother did.

Because I am a math guy, I like to visualize lifespan and healthspan in terms of a mathematical function, as in figure 2 on the following page—one of many graphs that I draw for my patients. The horizontal or x-axis of the graph represents your lifespan, how long you will live, while the vertical or y-axis represents a kind of sum total of your physical and cognitive function, the two age-dependent dimensions of healthspan. (Obviously, healthspan is not really quantifiable, but bear with my oversimplification.)

The black line represents the natural trajectory of your life: You are born at time zero, and for purposes of our diagram, we'll say your physical and cognitive health start out at 100 percent. You remain relatively robust until about the fifth decade of life, at which point your cognitive and physical health will likely begin a gradual but steady decline, until you die (healthspan = zero) sometime in your sixties or early seventies. This would have been a not untypical lifespan for someone born into a hunter-gatherer or primitive agrarian tribe, provided they managed to avoid early death thanks to infectious disease or another calamity.

Now look at the typical modern life course, represented by the short-dashed line on the graph, marked "Med 2.0." You will live a bit longer, thanks

Figure 2. Lifespan vs. Healthspan in Medicine 2.0 vs. Medicine 3.0

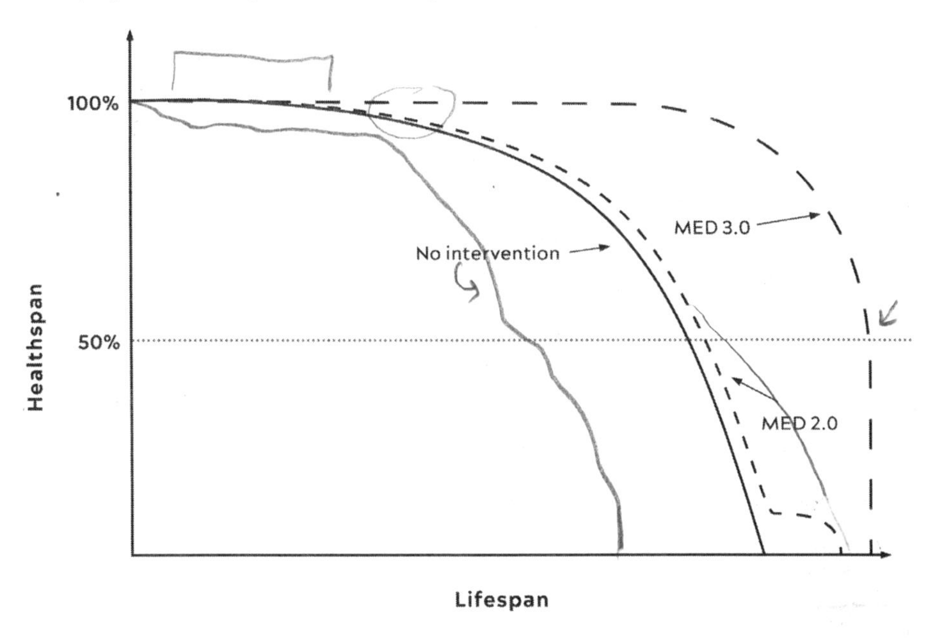

to the relative comfort and safety of our lives. But in midlife, you will gradually begin to feel some changes. You will lose a bit of your youthful strength and stamina. You might notice that you occasionally forget passwords, or the names of people you meet, or the names of actors in movies you watched long ago. Your friends and peers will begin to be diagnosed with cancer, cardiovascular disease and related conditions like high blood pressure, and diabetes or prediabetes. You will attend memorial services for friends from school.

At a certain point, the decline begins to steepen. Eventually, sometime around age seventy or seventy-five, give or take, your cognitive and physical capacities will diminish to roughly their halfway point (represented by the horizontal dotted line), which I sort of arbitrarily define as the point below which you are no longer able to do the things that you want to do with ease. You're constrained, and bad stuff starts to happen more frequently and with greater consequence. It's one thing to break your femur in a skiing accident when you're forty and still strong and resilient; it's quite another to break it falling off a curb when you're seventy-five and functioning at 25 percent of your capacity. At the same time, your own risk of chronic disease is rising exponentially.

This is where Medicine 2.0 steps in. We treat your heart disease, or cancer, or whatever else afflicts you, prolonging your life by a few months, or years if you're lucky. This is when the lifespan/healthspan curve flattens out horizontally to the right, representing this postponement of death. But now look at *where* this occurs: when your healthspan is already compromised. This means that we have delayed your death without significantly improving your quality of life—something at which Medicine 2.0 is quite adept. This is the Marginal Decade that most of us can expect, in our current system.

Now look at the long-dashed line on the graph. This represents your ideal trajectory. This is what you want. Instead of beginning a slow decline in midlife, your overall healthspan stays the same *or even improves* into your fifties and beyond. You will be fitter and healthier at fifty-five and even sixty-five than you were at forty-five and will remain physically fit and cognitively sharp well into your seventies and eighties, and possibly beyond. You will seem like someone a decade younger than the age on your passport, or possibly two. There is much more space under this curve, and all that space represents your longer, better life: more time being with your family, pursuing your passions, traveling, or continuing to do meaningful work. Moreover, when you do begin to decline, the descent is steep but relatively brief. This is called squaring the longevity curve.

In this scenario, we live longer, and we live better for longer. We outlive our life expectancy, and we also exceed society's expectations of what our later life is supposed to look like. Instead of a lousy Marginal Decade, we get to enjoy what feels more like a "Bonus Decade"—or decade*s*—when we are thriving in every dimension. This is our objective: to delay death, and to get the most out of our extra years. The rest of our lives becomes a time to relish rather than to dread.

The next obvious question is: How do we accomplish this? How do we lengthen our lifespan while simultaneously extending our healthspan? How do we stave off death via the Horsemen while slowing or even reversing physical, cognitive, and emotional decline?

What's our plan?

This is where most people make a wrong turn. They want to take a shortcut, right to the tactics: *this* is what to eat (and not eat), *that* is how you should exercise, *these* are the supplements or medications you need, and so on. There are warehouses full of books that purport to have the answers, but the one you are reading now is not one of them. Instead, I believe this is exactly where we need to hit pause and take a step back, lest we skip the most important step in the process: the strategy.

Take another look at the Sun Tzu quote that opens this chapter: "Tactics without strategy is the noise before defeat." He was talking about war, but it applies here as well. To achieve our objectives, we first need to have a *strategy:* an overall approach, a conceptual scaffolding or mental model that is informed by science, is tailored to our goals, and gives us options. Our specific tactics flow from our strategy, and the strategy derives from our objective. We know what the objective is by now, but the strategy is the key to victory.

The big mistake people often make is to conflate strategy and tactics, thinking they are the same. They are not. I like to explain this distinction using one of the more memorable boxing matches of all time: Muhammad Ali versus George Foreman, the famed "Rumble in the Jungle" that took place in Kinshasa, Zaire, in 1974. Ali's *objective,* obviously, was to win the match against Foreman and regain his heavyweight title. The problem Ali faced was that Foreman was younger, stronger, meaner, and favored to win in devastating fashion. It's hard to reconcile with the jovial guy who sells countertop grills now, but back in the day George Foreman was considered the meanest SOB who ever laced on boxing gloves. He was viewed as literally invincible. The pundits all agreed that colorful and beloved as he was, Ali didn't stand a chance—which is why he needed a strategy.

Ali knew he had certain slight advantages over Foreman in that he was faster, more experienced, and mentally tougher. He also knew that Foreman was hotheaded and prone to anger. Rather than try to counter Foreman punch for punch, Ali decided that he would attempt to induce the younger, less seasoned fighter to wear himself out, leaving him frustrated and tired, and thus vulnerable. If he could accomplish that, he knew it would be a more even match. This was his strategy: make Foreman angry, and then let him flail away until he had exhausted himself and Ali could mount an offensive.

From this strategy flowed the tactics that are now legendary: first, come at Foreman with a series of lead straight rights, an obvious, even disrespectful punch that was guaranteed to make Foreman mad. Nobody hits the heavyweight champion of the world like that. Ali then let an enraged Foreman chase him around the ring and press him up against the ropes, wasting energy, while he concentrated on trying to minimize the damage he absorbed—the famous "rope-a-dope."

For the first few rounds, everyone thought Foreman was absolutely crushing Ali, including Foreman. But because Ali's strategy was to try to outlast Foreman, he had trained himself to endure the abuse. By about the fifth round, you can almost see Foreman realizing, *Damn, I'm already gassed.* Meanwhile, Ali's superior physical conditioning meant he had much more left in the tank. He went on to win the match via a knockout in the eighth round.

The point is that the tactics are what you do when you are actually in the ring. The strategy is the harder part, because it requires careful study of one's opponent, identifying his strengths and weaknesses, and figuring out how to use both to your advantage, well before actually stepping in the ring. In this book, we will apply this three-part approach to longevity: **objective → strategy → tactics**.

Our Strategy

Going into the fight with Foreman, Ali knew that time was on his side. The longer he could keep his opponent riled up and wasting energy, while avoiding getting knocked out himself, the better his chances of winning in the long run. Unfortunately for us, time is definitely not on our side. Every moment we are alive, our risk of disease and death is tugging at us, the way gravity pulls a long jumper toward earth.

Of course, not every problem you face requires a strategy. In fact, many don't. You don't need a strategy if your objective is, say, to avoid getting a sunburn. Your straightforward tactical options are to put on sunblock, long sleeves and pants, and perhaps a big hat, or to stay out of the sun altogether. But we need a strategy in order to live longer and better, because longevity is a far more complex problem than sunburn.*

Living longer means delaying death from *all four* of the Horsemen. The Horsemen do have one powerful risk factor in common, and that is age. As you grow older, the risk grows exponentially that one or more of these diseases has begun to take hold in your body. Unfortunately, there's not much we can do about our chronological age—but what do we mean by "aging," exactly? It's not merely the passage of time, but what is happening inside us, beneath the surface, in our organs and our cells, as time passes. Entropy is working on us every single day.

"Aging is characterized by a progressive loss of physiological integrity, leading to impaired function and increased vulnerability to death," wrote the authors of an influential 2013 paper describing what they termed the "hallmarks of aging." They continued: "This deterioration is the primary risk factor for major human pathologies, including cancer, diabetes, cardiovascular disorders, and neurodegenerative diseases."

The very process of aging itself is what makes us vulnerable to these diseases, while also affecting our healthspan. Someone who drops dead of a heart attack did not just get sick an hour earlier. The disease was working inside them, silently and invisibly, for decades. As they grew older, their own internal defense mechanisms weakened, and disease gained the upper hand. We saw something similar in the COVID-19 pandemic. The virus infected people across all age groups, but it killed older people in vastly dispropor-

---

* Although avoiding sunburn, which contributes to the aged appearance of skin, not to mention melanoma risk, is unquestionably a good idea.

tionate numbers, precisely because it exposed and exploited their existing vulnerability to disease and death: their weakened immune systems, their cardiovascular and respiratory issues, and so on. Thus, our strategy must account for the effects of aging, just as Ali took his own advancing years into account as he searched for a way to defeat Foreman. Without the right strategy, Ali would have almost certainly lost the fight.

This is why we can't just skip ahead to the tactics, where I tell you what to do. If you are tempted, my advice is to pause, take a breath, and settle in. Without an understanding of the strategy, and the science that informs it, our tactics will not mean much, and you'll forever ride the merry-go-round of fad diets and trendy workouts and miracle supplements. You'll be stuck in a Medicine 2.0 mentality, seeking a quick fix to your problems. The only way to become an adroit tactician is to shift your mindset to Medicine 3.0, which requires becoming a master strategist first.

In the chapters to come, we will be taking a deep dive into some of the mechanisms underlying the aging process, and we will also be taking a very close look at the inner workings of each of the Horsemen diseases. How and when do they begin? What forces drive them—internal and external? How are they sustained? Most importantly, how can they be delayed or even prevented entirely? As we'll see in the next chapter, this is how the centenarians achieve their extraordinarily long lifespans: they delay or prevent the onset of chronic disease, by decades compared to the average.

We will also be taking a more detailed look at healthspan—another one of those overused buzzwords that has lost all meaning. The standard definition, as the period of life when we are free from disease and disability, sets far too low of a bar. If we're not sick and housebound, then we're "healthy"? I prefer to use more pointed language—so pointed that it often makes my patients uncomfortable.

Here's another way to think of it. Lifespan deals with death, which is binary: you're alive, and then you're dead. It's final. But before that happens, sometimes long before, most people suffer through a period of decline that, I would argue, is like dying in slow motion. Certainly, that was the case for

Sophie, Becky's mom. This can happen quickly, such as after a bad accident, but usually it goes so slowly that we barely perceive the change.

I think about healthspan and its deterioration in terms of three categories, or vectors. The first vector of deterioration is cognitive decline. Our processing speed slows down. We can't solve complex problems with the quickness and ease that we once did. Our memory begins to fade. Our executive function is less reliable. Our personality changes, and if it goes on for long enough, even our sentient self is lost. Fortunately, most people don't progress all the way to frank dementia, but many people experience some decline in their cognitive capacity as they age. Our objective is to minimize this.

The second vector of deterioration is the decline and eventual loss of function of our physical body. This may precede or follow cognitive decline; there is no predetermined order. But as we grow older, frailty stalks us. We lose muscle mass and strength, along with bone density, stamina, stability, and balance, until it becomes almost impossible to carry a bag of groceries into the house. Chronic pains prevent us from doing things we once did with ease. At the same time, the inexorable progression of atherosclerotic disease might leave us gasping for breath when we walk to the end of the driveway to fetch the newspaper (if newspapers still exist when we are old). Or we could be living a relatively active and healthy life until we fall or suffer some unexpected injury, as Sophie did, that tips us into a downward spiral from which we never recover.

My patients rarely expect this decline to affect *them*. I ask them to be very specific about their ideal future. What do they want to be doing when they are older? It's striking how rosy their predictions tend to be. They feel supremely confident that they will still be snowboarding or kickboxing, or whatever else it is they enjoy doing now, when they're in their seventies and eighties.

Then I stop them and explain: Look, in order to do that, you will need to have a certain level of muscular strength and aerobic fitness at that age. But even right now, at age fifty-two (for example), your strength and your maximum volume of oxygen uptake (VO_2 max) are already barely sufficient to do those things, and they are virtually certain to decline from here. So your choices are (a) surrender to the decline, or (b) come up with a plan, starting *now.*

No matter how ambitious your goals are for your later years, I suggest that you familiarize yourself with something called the "activities of daily living," a checklist used to assess the health and functionality of elderly people. The list includes such basic tasks as preparing a meal for oneself, walking without assistance, bathing and grooming, using a phone, going to the grocery store, handling personal finances, and so on. Now imagine living your life without the ability to feed or bathe yourself or walk a few blocks to meet friends for coffee. We take these for granted now, but to continue to live actively as we age, retaining even these minimal abilities, requires us to begin building a foundation of fitness and to maintain it diligently.

The third and final category of deterioration, I believe, has to do with emotional health. Unlike the others, this one is largely independent of age; it can afflict outwardly healthy young people in their twenties, or it can creep up on you in middle age, as it did with me. Or it can descend later in life. Surveys show that happiness tends to reach its nadir in our forties (age forty-seven, to be exact), but as I learned through painful experience, middle-aged distress often has its roots much earlier, in adolescence or childhood. And we may not recognize that we are in danger until we reach a crisis point, as I did. How we deal with it has a huge bearing on our physical health, our happiness, and our very survival.

To me, longevity as a concept is really only meaningful to the extent that we are defying or avoiding *all* these vectors of decline, simultaneously. And none of these individual components of longevity is worth much without all the others. To live to the age of one hundred without our mind *and* our body intact is not something that anyone would willingly choose. Similarly, to have the greatest quality of life, only to have it cut short at a young age, is also undesirable. And to retain good health as we age, but without love and friendship and purpose, is a purgatory I would not wish on my worst enemy.

The important distinction here is that while actual death is inevitable, this deterioration that we're talking about is less so. Not everyone who dies in their eighties or nineties passes through the valleys of cognitive, physical, or emotional destruction on the way there. They are preventable—and I believe that they are largely optional, despite their ever-increasing gravitational pull over time. As we will see in later chapters, cognitive, physical, and even emotional deterioration can all be slowed and even reversed in some cases with the application of the proper tactics.

The other key point is that lifespan and healthspan are not independent variables; they are tightly intertwined. If you increase your muscle strength and improve your cardiorespiratory fitness, you have also reduced your risk of dying from all causes by a far greater magnitude than you could achieve by taking any cocktail of medications. The same goes for better cognitive and emotional health. The actions we take to improve our healthspan will almost always result in a longer lifespan. This is why our tactics are largely aimed at improving healthspan first; the lifespan benefits will follow.

Tactics

The key difference between Medicine 2.0 and Medicine 3.0 has to do with how and *when* we apply our tactics. Typically, Medicine 2.0 steps in only when something is acutely wrong, like an infection or a broken bone, with short-term fixes for the immediate problem. In Medicine 3.0, our tactics must become interwoven into our daily lives. We eat, breathe, and sleep them—literally.

Medicine 2.0 relies on two types of tactics, broadly speaking: procedures (e.g., surgery) and medications. Our tactics in Medicine 3.0 fall into five broad domains: exercise, nutrition, sleep, emotional health, and exogenous molecules, meaning drugs, hormones, or supplements. I will not be talking much about molecules, because that would make this book twice as long as it already is, but one thing that I will say is that I do not shy away from pharmaceutical drugs because they are not "natural." I consider many drugs and supplements, including lipid-lowering medications, to be essential items in our longevity tool kit, and I hope that in the not-too-distant future we will have many even more effective tools at our disposal.

Drugs and supplements aside, our first tactical domain is exercise. Like "healthspan," exercise is another one of those overly broad blanket terms that annoy me, because it can encompass everything from a walk in the park to a hard bike ride up a mountain pass, a set of tennis, or a session in the gym lifting heavy weights. These all count as "exercise," but they obviously have very different effects (and risks, by the way). So we will break down this thing called exercise into its most important components: strength, stability, aerobic efficiency, and peak aerobic capacity. Increasing your limits in each of these areas is necessary if you are hoping to reach your limit of lifespan and healthspan. Again, my goal is not to tell you how to lose weight fast or improve the aesthetic quality of your midsection. We want to maintain physical strength, stamina, stability across a broad range of movements, while remaining free from pain and disability.

This is another area where my thinking has changed over time. I used to prioritize nutrition over everything else, but I now consider exercise to be the most potent longevity "drug" in our arsenal, in terms of lifespan and healthspan. The data are unambiguous: exercise not only delays actual death but also prevents both cognitive and physical decline, better than any other intervention. We also tend to feel better when we exercise, so it probably has some harder-to-measure effect on emotional health as well. My hope is that you will understand not only the *how* but the *why* of various types of exercise, so you will be able to formulate a program that fits your own personal goals.

Our second domain is nutrition. I won't be telling you to eat this, not that, or prescribing a specific diet that everyone should follow, and I'm definitely not taking sides in the pointless, never-ending diet wars pitting low carb versus paleo versus vegan, and so on. We will avoid such religious discussions in favor of biochemical evidence. The best science out there says that what you eat matters, but the first-order term is how *much* you eat: how many calories you take into your body.

How you go about achieving the Goldilocks zone here—not too much, not too little, but just right—will vary depending on numerous factors. My goal is to enable you to determine the best eating pattern for yourself. But please keep in mind that none of the tactics we will discuss are set in stone; we seek feedback from as many sources as possible to try to determine what works and what doesn't. A good strategy allows us to adopt new tactics and discard old ones in service of our objectives.

Next is sleep, which I and many others had ignored for far too long. Fortunately, over the past decade or so sleep has finally received the attention it deserves. Today, we have a far better understanding of its importance, and what goes wrong in the short and long term when our sleep is compromised (spoiler: a lot). There is not much that can compare to the feeling of waking up from a great night of sleep, feeling completely refreshed and totally primed for the day. Good sleep is critical to our innate physiological repair processes, especially in the brain, while poor sleep triggers a cascade of negative downstream consequences, from insulin resistance to cognitive decline, as well as mental health issues. I too used to be one of those people who enjoyed pulling all-nighters and thought sleep was for people who had nothing better to do. Long story short, I found out how wrong I was in very dramatic fashion. I am now convinced that Not-Thin Peter's biggest problem was less what he ate than how little he slept.

Finally, we will explore the importance of emotional health, which I believe is every bit as important a component of healthspan as the others. This is an area in which I have very little professional expertise but a great deal of personal experience. So while I do not have much hard experimental data and studies to point to, as in the other chapters, I will be sharing my own very long and painful journey to come to terms with things that happened to me in the past and to correct my own behavior and heal the relationships that I have damaged. If nothing else, it may serve as a cautionary tale—and a prod to get you to consider the state of your own emotional house, if warranted.

I will discuss my journey in much more detail in chapter 17, but one phrase from that period has stuck with me, almost like a mantra. It is something that one of my therapists, Esther Perel, said to me early in our work together.

"Isn't it ironic that your entire professional life is predicated around trying to make people live longer," she mused, "yet you're putting *no* energy into being less miserable, into suffering less emotionally?"

She continued: "Why would you want to live longer if you're so unhappy?"

Her logic was undeniable, and it changed my whole approach to longevity.

From Evidence Based to Evidence Informed

It's important, obviously, that our strategy be based on evidence. Unfortunately, the pursuit of longevity is where the most powerful tool of Medicine 2.0, the randomized clinical trial in humans, runs into a brick wall. Randomized controlled trials are used to determine cause and effect in relatively simple, short-term situations. It's fairly easy, for example, to run a study showing that sunscreen prevents sunburn. But such studies are of limited use in our quest for longevity.

This is where my approach may ruffle some people's feathers. The purists of evidence-based medicine demand data from randomized controlled trials (RCTs) before doing *anything*. Those trials are the gold standard of medical evidence, yet they also reinforce some major limitations of Medicine 2.0, beginning with its short time horizon. In general, the types of clinical questions that are best resolved by RCTs are those involving simple interventions such as a vaccine, or a medication to lower cholesterol. We give this treatment over a relatively short period, from six months up to maybe five or six years at the longest and look for its effect on a specified outcome. Does the vaccine reduce the rate of serious illness and death? Does this drug lower cholesterol and prevent cardiac death, or at least heart attacks, in highly susceptible individuals?

This type of study is the foundation of evidence-based medicine. But if our goal is longevity, the situation becomes more complicated. A one-year clinical trial, or even a five-year study, will not tell us everything we need to know about disease processes that take decades to unfold. There will never be a clinical trial to guide a cardiovascular prevention strategy for a healthy forty-year-old. It would simply take too long to do the study. Furthermore, outside of pharmacology, the interventions are very complex, particularly if they involve exercise, nutrition, and sleep. Studying longevity itself in this way is almost impossible—unless we could somehow take a hundred thousand babies, randomize them to four or five different interventions, and follow them throughout their lifetimes. That would (hopefully) yield a rock-solid, evidence-based prescription for maximizing lifespan and healthspan. But the obstacles to doing this are insurmountable, not least because it would require a century to complete.

Option B is to look at the different types of data that we do have and then develop a strategy that triangulates between them. This might not *definitively* solve the problem, but it can at least point us in the right direction. Our Option B strategy is based on combining insights from five different sources of data that, viewed separately, probably aren't strong enough to act on. When taken together, however, they can provide a solid foundation for our tactics. But our supporting framework must shift, from exclusively evidence-based to evidence-informed, risk-adjusted precision medicine.

meta-analysis?

Our first source of data comes from studies of centenarians, people who have lived to the age of one hundred and beyond, often in good health. These are the extreme outliers, the tiny sliver of the population who have outlived our usual life expectancy by two decades or more. By and large, they have delayed or evaded the diseases that kill most of the rest of us, and many of them have remained in fairly good shape. We would like to know how they accomplished this feat. What do centenarians have in common? What genes do they share that might give them an advantage over noncentenarians? What explains their survival and their apparently slower rate of aging? And most of all, what can the rest of us do to emulate their good fortune?

This evidence is made stronger by the fact that centenarians represent our "species of interest"—that is, they are human. Unfortunately, centenarian data are almost entirely observational rather than experimental, so we can't truly infer cause and effect. Centenarians' life histories and habits tend to be idiosyncratic, to say the least, and the fact that their numbers are relatively small means that it can be difficult to draw firm conclusions at all. (We will discuss centenarians in more detail in the next chapter.)

Next, we turn to lifespan data from animal "models," such as laboratory mice. It is obviously much easier, ethically and logistically, to test lifespan-altering tactics in mice, which typically only live about two or three years, than in humans. We have a huge amount of data about how different sorts of interventions, both dietary and in the form of exogenous molecules, affect mouse lifespan. The limitation, obviously, is that mice are not human; many drugs have succeeded in mice only to fail spectacularly in human studies. There are other types of animal models, including a tiny species of nematode worm called *C. elegans* that is often used in research, as well as fruit flies, dogs, primates, and even lowly yeast cells. All of these have strengths and weaknesses. My rule of thumb is that if a given intervention can be shown to extend lifespan or healthspan in multiple species spanning a billion years of evolution, for example, from worms to monkeys, then I am inclined to take it seriously.

A third and important source of information to support our strategy comes from human studies of the Horsemen: cardiovascular and cerebrovascular disease, cancer, Alzheimer's disease and related neurodegenerative conditions, and type 2 diabetes and related metabolic dysfunction. How do these diseases begin? How do they progress? What risk factors help to cause them, or fuel them? What underlying factors do they share? What are the cutting-edge treatment modalities for those with "advanced" disease—and what do they tell us about developing a strategy for prevention? We want to know each of these diseases inside and out, understanding their weaknesses and vulnerabilities, just as Ali scrutinized Foreman before their match.

Fourth, we consider the molecular and mechanistic insights derived from the study of aging in both humans and animal models. We have learned an enormous amount about the cellular changes that occur during the aging process and in specific diseases. From this, we also have developed some ideas for how to manipulate these changes, via exogenous molecules (e.g., drugs) or behavioral changes (e.g., exercise).

Our final source of insights is a very clever method of analysis called Mendelian randomization, or MR for short. MR helps bridge the gap between randomized controlled trials, which can establish causality, and pure epidemiology, which often cannot. We'll talk about epidemiology in more detail later, but while it has proved useful in certain situations, such as determining the link between smoking and lung cancer, it has been less useful in more complex scenarios. Mendelian randomization helps tease out causal relationships between modifiable risk factors (e.g., LDL cholesterol) and an outcome of interest (e.g., cancer) in situations where actual randomized experiments cannot easily be done. It accomplishes this by letting nature do the randomization.* By considering the random variation in relevant genes and comparing them against the observed results, it eliminates many of the biases and confounders that limit the usefulness of pure epidemiology.

For example, some epidemiologic studies have suggested an inverse relationship between LDL cholesterol and cancer risk. That is, people with lower LDL cholesterol appear to have a higher risk of cancer. But is the relationship *causal*? That's a tricky but important question. If true, it would imply that lowering LDL cholesterol, such as with statins, increases the risk of cancer, which would obviously be bad news. Epidemiology does not tell us the direction of causality, so we turn to MR.

With MR, we can look at genetic variations that result in low, medium, and high levels of LDL cholesterol. These genes are randomly occurring, so they serve as a proxy for a randomized natural experiment. By examining the relationship between the resulting LDL cholesterol levels and cancer in-

* For MR to work properly, certain conditions must be met. First, the genetic variant(s) being considered must associate with the risk factor of interest (this is called the relevance assumption); second, the genetic variant does not share a common cause with the outcome (this is called the independence assumption); and third, the genetic variant does not affect the outcome except through the risk factor (this is called the exclusion restriction assumption).

cidence, we can answer the question without the usual confounders that plague traditional epidemiology. And lo and behold, it turns out that low LDL cholesterol does not cause cancer or increase its risk. If we use the same technique to look at the effect of LDL levels on cardiovascular disease (our dependent variable), it turns out that higher LDL cholesterol *is* causally linked to the development of cardiovascular disease (as we'll discuss in chapter 7).

An astute reader will notice that one concept has been conspicuously absent from this chapter so far: absolute certainty. This took me a little while to grasp when I transitioned from mathematics to medicine, but in biology we can rarely "prove" anything definitively the way we can in mathematics. Living systems are messy, and confounding, and complex, and our understanding of even fairly simple things is constantly evolving. The best we can hope for is reducing our uncertainty. A good experiment in biology only increases or decreases our confidence in the probability that our hypothesis is true or false. (Although we can feel fairly certain about some things, such as the evidence supporting the idea that your doctor should wash her hands and put on sterile gloves before operating on you.)

In the absence of multiple, repeated, decades-long randomized clinical trials that might answer our questions with certainty, we are forced to think in terms of probabilities and risk. In a sense it's a bit like charting an investment strategy: we are seeking the tactics that are likeliest, based on what we know now, to deliver a better-than-average return on our capital, while operating within our own individual tolerance for risk. On Wall Street, gaining an advantage like this is called alpha, and we're going to borrow the idea and apply it to health. I propose that with some unorthodox but very reasonable lifestyle changes, you can minimize the most serious threats to your lifespan and healthspan and achieve your own measure of longevity alpha.

My aim here is to equip you with a set of tools that you can apply to your own specific situation—whether you need to pay attention to your glucose regulation, your weight, your physical condition, your Alzheimer's disease risk, and so on. Your personal tactics should never be static, but will evolve as needed, as you journey through life with all its uncertainties—and as we learn more about the science of aging and the workings of diseases like cancer. As your own situation changes, your tactics can (and must) change, because as the great philosopher Mike Tyson once put it, "Everyone has a plan until they get punched in the mouth."

Advice George Foreman could have used.

Chapter 3 Summary:

- Strategy is necessary if you are to use strategy successfully
- Healthspan deterioration is defined in 3 vectors
  1. cognitive decline
  2. decline + loss of physical functioning } no order
  3. emotional health

Medicine 2.0
↓
Procedures
Medications

Medicine 3.0
↓
Exercise
Sleep
Nutrition
Emotional health
Extrogenos Molecules

- MR (Mendelian Randomization) combines epidemiology + randomized trial ?

epidemiology ↓ helpful when looking @ variables like smoking ↓ A + B are connected

randomized trial ↓ establishes causality ↓ A causes B

PART II

CHAPTER 4

Centenarians

The Older You Get, the Healthier You Have Been

Whiskey's a good medicine.
It keeps your muscles tender.

—Richard Overton, 1906–2018

In his later years, Richard Overton liked to take the edge off his days with a shot of bourbon and a few puffs of a Tampa Sweet cigar, lit directly from the gas stove in his home in Austin, Texas. He insisted that he never inhaled—word to the wise. Mr. Overton, as he was known, was born during the Theodore Roosevelt administration and died in late 2018 at the age of 112.

Not to be outdone, British World War I veteran Henry Allingham attributed his own 113-year lifespan to "cigarettes, whiskey, and wild, wild women." It's a shame he never met the adventurous Frenchwoman Jeanne Calment, who once joked, "I've only ever had one wrinkle, and I'm sitting on it." She rode her bicycle until she was 100 and kept smoking until the age of 117.

Perhaps she shouldn't have quit, because she died five years later at 122, making her the oldest person ever to have lived.

Mildred Bowers, at a comparatively youthful 106, preferred beer, cracking open a cold one every day at 4 p.m. sharp—it's five o'clock somewhere, right? Theresa Rowley of Grand Rapids, Michigan, credited her daily Diet Coke for helping her live to the age of 104, while Ruth Benjamin of Illinois said the key to reaching her 109th birthday was her daily dose of bacon. "And potatoes, some way," she added. They were all youngsters compared with Emma Morano of Italy, who consumed three eggs a day, two of them raw, up until her death at age 117.

If we were epidemiologists from Saturn and all we had to go on was articles about centenarians in publications like *USA Today* and *Good Housekeeping*, we might conclude that the secret to extreme longevity is the breakfast special at Denny's, washed down with Jim Beam and a good cigar. And perhaps this is so. Another possibility is that these celebrity centenarians are messing with the rest of us. We cannot be certain, because the relevant experiment cannot be done, as much as I'd like to open *JAMA* and see the title "Do Cream-Filled Chocolate Doughnuts Extend Lifespan? A Randomized Clinical Trial."

We yearn for there to be some sort of "secret" to living a longer, healthier, happier life. That desire drives our obsession with knowing the special habits and rituals of those who live longest. We are fascinated by people like Madame Calment, who seem to have escaped the gravitational pull of mortality, despite having smoked or done other naughty things throughout their lives. Was it the bike riding that saved her? Or was it something else, such as the pound of chocolate that she purportedly consumed each week?

More broadly, it's worth asking: What do healthy centenarians actually have in common? And, more importantly, what can we learn from them—if anything? Do they really live longer because of their idiosyncratic behaviors, like drinking whiskey, or despite them? Is there some other common factor that explains their extreme longevity, or is it simply luck?

More rigorous research into large groups of centenarians has cast (further) doubt on the notion that "healthy" behaviors, which I can't resist putting into scare quotes, are required to attain extreme longevity. According to results from a large study of Ashkenazi Jewish centenarians, run by Nir Barzilai at the Albert Einstein College of Medicine in the Bronx, centenarians are no more health-conscious than the rest of us. They may actually be worse: a large proportion of the nearly five hundred subjects in the Einstein study drank alcohol and smoked, in some cases for decades. If anything, the centenarian males in the study were *less* likely to have exercised regularly at age seventy than age-matched controls. And many were overweight. So much for healthy lifestyles.

Could the centenarians merely be lucky? Certainly, their age alone makes them extreme statistical outliers. As of 2021, there were just under 100,000 centenarians in the United States, according to the Census Bureau. And although their number has increased by nearly 50 percent in just two decades, the over-one-hundred age group still represents only about 0.03 percent of the population, or about 1 out of every 3,333 of us.

After ten decades of age, the air gets pretty thin, pretty quickly. Those who live to their 110th birthday qualify for the ultra-elite cadre of "supercentenarians," the world's smallest age group, with only about three hundred members worldwide at any given time (although the number fluctuates). Just to give you a sense of how exclusive this club is, for every supercentenarian in the world at this writing, there are about nine billionaires.

Yet nobody has come close to Madame Calment's record. The next-longest-lived person ever recorded, Pennsylvania native Sarah Knauss, was a mere 119 when she died in 1999. Since then, the world's oldest person has rarely exceeded the age of 117, and she is almost always female. While some individuals have claimed extremely lengthy lifespans, of 140 years or more, Calment remains the only person ever to be verified as having lived past 120, leading some researchers to speculate that that may represent the upper limit of human lifespan, programmed into our genes.

We are interested in a slightly different question: Why are some people able to just blow past the eighty-year mark, which represents the finish line for most of the rest of us? Could their exceptional longevity—and exceptional healthspan—be primarily a function of their genes?

Studies of Scandinavian twins have found that genes may be responsible for only about 20 to 30 percent of the overall variation in human lifespan. The catch is that the older you get, the more genes start to matter. For centenarians, they seem to matter a lot. Being the sister of a centenarian makes you eight times more likely to reach that age yourself, while brothers of centenarians are seventeen times as likely to celebrate their hundredth birthday, according to data from the one-thousand-subject New England Centenarian Study, which has been tracking extremely long-lived individuals since 1995 (although because these subjects grew up in the same families, with presumably similar lifestyles and habits, this finding could be due to some environmental factors as well). If you don't happen to have centenarian siblings, the next best option is to choose long-lived parents.

This is part of why I place so much importance on taking a detailed family history from my patients: I need to know when your relatives died and why. What are your likely "icebergs," genetically speaking? And if you do happen to have centenarians in your family tree, let me offer my congratulations. Such genes are, after all, a form of inherited luck. But in my family, you were doing well if you made it to retirement age. So if you're like me, and most people reading this book, your genes aren't likely to take you very far. Why should we even bother with this line of inquiry?

Because we are probing a more relevant question: Can we, through our behaviors, somehow reap the same benefits that centenarians get for "free" via their genes? Or to put it more technically, can we mimic the centenarians' *phenotype*, the physical traits that enable them to resist disease and survive for so long, even if we lack their *genotype*? Is it possible to outlive our own life expectancy if we are smart and strategic and deliberate about it?

If the answer to this question is yes, as I believe it is, then understanding the inner workings of these actuarial lottery winners—how they achieve their extreme longevity—is a worthwhile endeavor that can inform our strategy.

When I first became interested in longevity, my greatest fear was that we would somehow figure out how to delay death without also extending the healthy period of people's lives, à la Tithonus (and à la Medicine 2.0). My mistake was to assume that this was *already* the fate of the very long-lived, and that all of them are essentially condemned to spend their extra years in a nursing home or under other long-term care.

A deeper look at the data from multiple large centenarian studies worldwide reveals a more hopeful picture. It's true that many centenarians exist in a somewhat fragile state: The overall mortality rate for Americans ages 100 and older is a staggering 36 percent, meaning that if Grandma is 101, she has about a one-in-three chance of dying in the next twelve months. Death is knocking at her door. Digging further, we find that many of the oldest old die from pneumonia and other opportunistic infections, and that a few centenarians, such as Madame Calment, really do die of what used to be called old age. But the vast majority still succumb to diseases of aging—the Horsemen—just like the rest of us.

The crucial distinction, the essential distinction, is that they tend to develop these diseases *much later in life* than the rest of us—if they develop them at all. We're not talking about two or three or even five years later; we're talking decades. According to research by Thomas Perls of Boston University and his colleagues, who run the New England Centenarian Study, one in five people in the general population will have received some type of cancer diagnosis by age seventy-two. Among centenarians, that one-in-five threshold is not reached until age one hundred, nearly *three decades* later. Similarly, one-quarter of the general population will have been diagnosed with clinically apparent cardiovascular disease by age seventy-five; among centenarians, that prevalence is reached only at age ninety-two. The same pattern holds for bone loss, or osteoporosis, which strikes centenarians sixteen years later than average, as well as for stroke, dementia, and hypertension: centenarians succumb to these conditions much later, if at all.

Their longevity is not merely a function of delaying disease. These people also often defy the stereotype of old age as a period of misery and decline. Perls, Barzilai, and other researchers have observed that centenarians tend to be in pretty good health overall—which, again, is not what most people expect. This doesn't mean that everyone who lives that long will be playing golf and jumping out of airplanes, but Perls's ninety-five-and-older study subjects scored very well on standard assessments of cognitive function and ability to perform those tasks of daily living we mentioned in chapter 3, such as cooking meals and clipping their own toenails, a seemingly simple job that becomes monumentally challenging in older age.

Curiously, despite the fact that female centenarians outnumber males by at least four to one, the men generally scored higher on both cognitive and functional tests. This might seem paradoxical at first, since women clearly live longer than men, on average. Perls believes there is a kind of selection process at work, because men are more susceptible to heart attacks and strokes beginning in middle age, while women delay their vulnerability by a decade or two and die less often from these conditions.

This tends to weed the frailer individuals out of the male population, so that *only* those men who are in relatively robust health even make it to their hundredth birthday, while women tend to be able to survive for longer *with* age-related disease and disability. Perls describes this as "a double-edged sword," in that women live longer but tend to be in poorer health. "The men tend to be in better shape," he has said. (The authors didn't measure this, but my hunch is that it may have something to do with men having more muscle mass, on average, which is highly correlated to longer lifespan and better function, as we'll discuss further in the chapters on exercise.)

But even if they are not in such great shape in their eleventh decade, these individuals have already enjoyed many extra years of healthy life compared with the rest of the population. Their healthspan, as well as their lifespan, has been extraordinarily long. What's even more surprising is that Perls's group has also found that the supercentenarians and the "semisupercentenarians" (ages 105 to 109) actually tend to be in even *better* health than garden-variety hundred-year-olds. These are the super survivors, and at those advanced ages, lifespan and healthspan are pretty much the same. As Perls and his colleagues put it in a paper title, "The Older You Get, the Healthier You Have Been."

In mathematical terms, the centenarians' genes have bought them a *phase shift* in time—that is, their entire lifespan and healthspan curve has been shifted a decade or two (or three!) to the right. Not only do they live longer, but these are people who have been healthier than their peers, and biologically younger than them, for virtually their entire lives. When they were sixty, their coronary arteries were as healthy as those of thirty-five-year-olds. At eighty-five, they likely looked and felt and functioned as if they were in their sixties. They seemed like people a generation younger than the age on their driver's license. *This* is the effect that we are seeking to mimic.

Think back to the notion of the Marginal Decade and Bonus Decade that we introduced in chapter 3, and the graph of lifespan versus healthspan. Because Medicine 2.0 often drags out lifespan in the context of low healthspan, it *lengthens* the window of morbidity, the period of disease and disability at the end of life. People are sicker for longer before they die. Their Marginal Decade is spent largely as a patient. When centenarians die, in contrast, they have generally (though not always) been sick and/or disabled for a much shorter period of time than people who die two or three decades earlier. This is called *compression of morbidity,* and it basically means shrinking or shortening the period of decline at the end of life and lengthening the period of healthy life, or healthspan.

One goal of Medicine 3.0 is to help people live a life course more like the centenarians—only better. The centenarians not only live longer but live longer in a healthier state, meaning many of them get to enjoy one, or two, or even three Bonus Decades. They are often healthier at ninety than the average person in their sixties. And when they do decline, their decline is typically brief. This is what we want for ourselves: to live longer with good function and without chronic disease, and with a briefer period of morbidity at the end of our lives.

The difference is that while most centenarians seem to get their longevity and good health almost accidentally, thanks to genes and/or good luck, the rest of us must try to achieve this intentionally. Which brings us to our next two questions: *How* do centenarians delay or avoid chronic disease? And how can we do the same?

This is where the genes likely come in—the longevity genes that most of us don't have because we failed to pick the right parents. But if we can identify specific genes that give centenarians their edge, perhaps we can reverse-engineer their phenotype, their effect.

This would seem to be a relatively straightforward task: sequence the genomes of a few thousand centenarians and see which individual genes or gene variants stand out as being more prevalent among this population than in the general population. Those would be your candidate genes. But when researchers did this, examining thousands of individuals via genome-wide association studies, they came up almost empty-handed. These individuals appeared to have very little in common with one another genetically. And their longevity may be due to dumb luck after all.

Why are longevity genes so elusive? And why are centenarians so rare in the first place? It comes down to natural selection.

Hold on a second, you might be saying. We've been taught all our lives that evolution and natural selection have relentlessly optimized us for a billion years, favoring beneficial genes and eliminating harmful ones—survival of the fittest, and all that. So why don't we *all* share these pro-longevity centenarian genes, whatever they are? Why aren't we all "fit" enough to live to be one hundred?

The short answer is that evolution doesn't really care if we live that long. Natural selection has endowed us with genes that work beautifully to help us develop, reproduce, and then raise our offspring, and perhaps help raise our offspring's offspring. Thus, most of us can coast into our fifth decade in relatively good shape. After that, however, things start to go sideways. The evolutionary reason for this is that after the age of reproduction, natural selection loses much of its force. Genes that prove unfavorable or even harmful in midlife and beyond are not weeded out because they have already been passed on. To pick one obvious example: the gene (or genes) responsible for male pattern baldness. When we are young, our hair is full and glorious, helping us attract mates. But natural selection does not really care whether a man in his fifties (or a woman, for that matter) has a full head of hair.

Hair loss is not terribly relevant to longevity, luckily for me. But this general phenomenon also explains why genes that might predispose someone to Alzheimer's disease or some other illness, later in life, have not vanished from our gene pool. In short, natural selection doesn't care if we develop Alzheimer's disease (or baldness) in old age. It doesn't affect our reproductive fitness. By the time dementia would appear, we have likely already handed down our genes. The same is true of genes that would accelerate our risk of heart disease or cancer in midlife. Most of us still carry these lousy genes—including some centenarians, by the way. Indeed, there is a chance that those same genes may have conferred some sort of advantage earlier in life, a phenomenon known as "antagonistic pleiotropy."

One plausible theory holds that centenarians live so long because they also possess certain other genes that protect them from the flaws in our typical genome, by preventing or delaying cardiovascular disease and cancer and maintaining their cognitive function decades after others lose it. But even as natural selection allows harmful genes to flourish in older age, it does almost nothing to promote these more helpful longevity-promoting genes, for the reasons discussed above. Thus it appears that no two centenarians follow the exact same genetic path to reaching extreme old age. There are many ways to achieve longevity; not just one or two.

That said, a handful of potential longevity genes have emerged in various studies, and it turns out that some of them are possibly relevant to our strategy. One of the most potent individual genes yet discovered is related to cholesterol metabolism, glucose metabolism—and Alzheimer's disease risk.

You may have heard of this gene, which is called *APOE,* because of its known effect on Alzheimer's disease risk. It codes for a protein called APOE (apolipoprotein E) that is involved in cholesterol transport and processing, and it has three variants: *e2, e3,* and *e4.* Of these, *e3* is the most common by far, but having one or two copies of the *e4* variant seems to multiply one's risk of developing Alzheimer's disease by a factor of between two and twelve. This is why I test all my patients for their *APOE* genotype, as we'll discuss in chapter 9.

The *e2* variant of *APOE,* on the other hand, seems to protect its carriers against dementia—and it also turns out to be very highly associated with longevity. According to a large 2019 meta-analysis of seven separate longevity studies, with a total of nearly thirty thousand participants, people who carried at least one copy of *APOE e2* (and no *e4*) were about 30 percent more likely to reach extreme old age (defined as ninety-seven for men, one hundred for women) than people with the standard *e3/e3* combination. Meanwhile, those with two copies of *e4,* one from each parent, were 81 percent *less* likely to live that long, according to the analysis. That's a pretty big swing.

We will explore the function of APOE in more detail in chapter 9, but it is likely relevant to our strategy on multiple levels. First and most obviously, it appears to play a role in delaying (or not delaying) the onset of Alzheimer's disease, depending on the variant. This is likely not a coincidence, because as we'll see, APOE plays an important role in shuttling cholesterol around the body, particularly in the brain; one's *APOE* variant also has a large influence on glucose metabolism. Its potent correlation with longevity suggests that we should focus our efforts on cognitive health and pay special attention to issues around cholesterol and lipoproteins (the particles that carry cholesterol, which we'll discuss in chapter 7), as well as glucose metabolism (chapter 6).

Researchers have identified two other cholesterol-related genes, known as *CETP* and *APOC3,* that are also correlated with extreme longevity (and may explain why centenarians rarely die from heart disease). But one individual gene, or even three dozen genes, is unlikely to be responsible for centenarians' extreme longevity and healthspan. Broader genetic studies suggest that hundreds, if not thousands, of genes could be involved, each making its own small contribution—and that there is no such thing as a "perfect" centenarian genome.

This is actually good news for those of us without centenarians in our family tree, because it suggests that even on this genetic level there may be no magic bullet; even for centenarians, longevity may be a game of inches, where relatively small interventions, with cumulative effect, could help us replicate the centenarians' longer lifespan and healthspan. Put another way, if we want to outlive our life expectancy and live better longer, we will have to work hard to earn it—through small, incremental changes.

One other possible longevity gene that has emerged, in multiple studies of centenarians worldwide, also provides some possible clues to inform our strategy. These are variants in a particular gene called *FOXO3* that seem to be directly relevant to human longevity.

In 2008, Bradley Willcox of the University of Hawaii and colleagues reported that a genetic analysis of participants in a long-running study of health and longevity in Hawaiian men of Japanese ancestry had identified three SNPs (or variants) in *FOXO3* that were strongly associated with healthy aging and longevity. Since then, several other studies have found that various other long-lived populations also appear to have *FOXO3* mutations, including Californians, New Englanders, Danes, Germans, Italians, French, Chinese, and American Ashkenazi Jews—making *FOXO3* one of the very few longevity-related genes to be found across multiple different ethnic groups and geographical locations.

FOXO3 belongs to a family of "transcription factors," which regulate how other genes are expressed—meaning whether they are activated or "silenced." I think of it as rather like the cellular maintenance department. Its responsibilities are vast, encompassing a variety of cellular repair tasks, regulating metabolism, caring for stem cells, and various other kinds of housekeeping, including helping with disposal of cellular waste or junk. But it doesn't do the heavy lifting itself, like the mopping, the scrubbing, the minor drywall repairs, and so on. Rather, it delegates the work to other, more specialized genes—its subcontractors, if you will. When *FOXO3* is activated, it in turn activates genes that generally keep our cells healthier. It seems to play an important role in preventing cells from becoming cancerous as well.

Here's where we start to see some hope, because *FOXO3* can be activated or suppressed by our own behaviors. For example, when we are slightly deprived of nutrients, or when we are exercising, *FOXO3* tends to be more activated, which is what we want.

Beyond *FOXO3*, gene expression itself seems to play an important but still poorly understood role in longevity. A genetic analysis of Spanish centenarians found that they displayed extremely youthful patterns of gene expression, more closely resembling a control group of people in their twenties than an older control group of octogenarians. Precisely how these centenarians achieved this is not clear, but it may have something to do with *FOXO3*—or some other, as yet unknown, governor of gene expression.

We still have more questions than answers when it comes to the genetics behind extreme longevity, but this at least points in a more hopeful direction. While your *genome* is immutable, at least for the near future, gene *expression* can be influenced by your environment and your behaviors. For example, a 2007 study found that older people who were put on a regular exercise program shifted to a more youthful pattern of gene expression after six months. This suggests that genetics *and* environment both play a role in longevity and that it may be possible to implement interventions that replicate at least some of the centenarians' good genetic luck.

I find it useful to think of centenarians as the results of a natural experiment that tells us something important about living longer and living better. Only in this case, Darwin and Mendel, the Russian geneticist, are the scientists. The experiment entails taking a random collection of human genomes and exposing them to a variety of environments and behaviors. The centenarians possess the correct combination of genome X required to survive in environment Y (perhaps with help from behaviors Z). The experiment is not simple; there are likely many pathways to longevity, genetic and otherwise.

Most of us, obviously, cannot expect to get away with some of the centenarians' naughty behaviors, such as smoking and drinking for decades. But even if we don't (and in many cases, shouldn't) imitate their "tactics," the centenarians can nevertheless help inform our strategy. Their superpower is their ability to resist or delay the onset of chronic disease by one or two or even three decades, while also maintaining relatively good healthspan.

It's this phase shift that we want to emulate. But Medicine 2.0, which is almost solely focused on helping us live longer *with* disease, is not going to get us there. Its interventions almost always come too late, when disease is already established. We have to look at the other end of the time line, trying to slow or stop diseases before they start. We must focus on delaying the *onset* rather than extending the *duration* of disease—and not just one disease but *all* chronic diseases. Our goal is to live longer *without* disease.

This points to another flaw of Medicine 2.0, which is that it generally looks at these diseases as entirely separate from one another. We treat diabetes as if it were unrelated to cancer and Alzheimer's, for example, even though it is a major risk factor for both. This disease-by-disease approach is reflected in the "silo" structure of the National Institutes of Health, with separate institutions dedicated to cancer, heart disease, and so on. We treat them as distinct when we should be looking for their commonalities.

"We're trying to attack heart disease, cancer, stroke, and Alzheimer's one disease at a time, as if somehow these diseases are all unrelated to each other," says S. Jay Olshansky, who studies the demography of aging at the University of Illinois–Chicago, "when in fact the underlying risk factor for almost everything that goes wrong with us as we grow older, both in terms of diseases we experience, and of the frailty and disability associated with it, is related to the underlying biological process of aging."

In the next chapter, we will look at one particular intervention, a drug that likely slows or delays that underlying biological process of aging at a mechanistic level. It may become relevant to our strategy as well, but for now this means pursuing two approaches in parallel. We need to think about very early *disease-specific* prevention, which we will explore in detail in the next few chapters dedicated to the Horsemen diseases. And we need to think about very early *general* prevention, targeting all the Horsemen at once, via common drivers and risk factors.

These approaches overlap, as we'll see: reducing cardiovascular risk by targeting specific lipoproteins (cholesterol) may also reduce Alzheimer's disease risk, for example, though not cancer. The steps we take to improve metabolic health and prevent type 2 diabetes almost certainly reduce the risk of cardiovascular disease, cancer, and Alzheimer's simultaneously. Some types of exercise reduce risk for all chronic diseases, while others help maintain the physical and cognitive resilience that centenarians largely get via their genes. This level of prevention and intervention may seem excessive by the standards of Medicine 2.0, but I would argue that it is necessary.

In the end, I think that the centenarians' secret comes down to one word: *resilience*. They are able to resist and avoid cancer and cardiovascular disease, even when they have smoked for decades. They are able to maintain ideal metabolic health, often despite a lousy diet. And they resist cognitive and physical decline long after their peers succumb. It is this resilience that we want to cultivate, just as Ali prepared himself to withstand and ultimately outlast Foreman. He prepared intelligently and thoroughly, he trained for a long time before the match, and he deployed his tactics from the opening bell. He could not have lasted forever, but he made it through enough rounds that he was able to fulfill his objective and win the fight.

Chapter 4 Summary

-Centenarians live healthier for longer

-Longevity could be] cummulative small things

-FOXO3 which is a transcription gene affecting longevity somehow

-Genes & environment can impact gene expression/ longevity

CHAPTER 5

Eat Less, Live Longer?

The Science of Hunger and Health

Scientists who play by someone else's rules
don't have much chance of making discoveries.

—Jack Horner

In the fall of 2016, I met three friends at George Bush Intercontinental Airport in Houston to embark on a somewhat unusual vacation. We flew eleven hours overnight to Santiago, Chile, where we drank coffee and ate breakfast before boarding another plane to fly six more hours to the west, across 2,500 miles of open ocean, to Easter Island, the world's most isolated body of land that is inhabited by humans. We were all men in our forties, but this was not your typical guys' weekend.

Most people know about Easter Island because of the thousand or so mysterious giant stone heads, called *moai,* dotting its shoreline, but there's a lot more to it. The island was named by European explorers who landed there on

Easter Sunday in 1722, but the natives call it Rapa Nui. It is an extreme, isolated, spectacular place. The triangle-shaped island of roughly sixty-three square miles is what's left of a trio of ancient volcanoes that surged up more than two miles from the seabed millions of years ago. One end of the island is ringed by very high cliffs that plunge down into the gorgeous blue ocean. The nearest human settlement is more than one thousand miles away.

We were not there as tourists. We were on a pilgrimage to the source of one of the most intriguing molecules in all of medicine, one that most people have never even heard of. The story of how this molecule was discovered, and how it revolutionized the study of longevity, is one of the most incredible sagas in biology. This molecule, which came to be known as rapamycin, had also transformed transplant medicine, giving millions of patients a second chance at life. But that was not why we had traveled ten thousand miles to this remote spot. We had come because rapamycin had been demonstrated to do something that no other drug had ever done before: extend maximum lifespan in a mammal.

This discovery came about at least in part thanks to the work of one member of our group, David Sabatini, who was then a professor of biology at MIT's Whitehead Institute. David had helped discover the key cellular pathway that rapamycin acts upon. Also on the trip was another biologist named Navdeep Chandel (Nav to his friends), a friend of David's who studies metabolism and mitochondria, the little organelles that produce power (and do much more) in our cells, at Northwestern University. Completing our foursome was my close friend Tim Ferriss. Tim is an entrepreneur and author, not a scientist, but he has a knack for asking the right questions and bringing a fresh perspective to something. Plus, I knew that he would be willing to swim in the ocean with me every day, reducing my chances of being eaten by a shark by approximately 50 percent.

One purpose of our trip was to scout out the location for a scientific conference that would be entirely devoted to research about this amazing substance. But mostly, we wanted to make a pilgrimage to the place where this extraordinary molecule had come from and to pay homage to its almost accidental discovery.

After we dropped off our luggage at our thirty-room tourist hotel, our first stop was Rano Kau, the one-thousand-foot-tall extinct volcano that dominates the southwest corner of the island. Our destination was the center of the crater, where there is a large swampy lake, nearly a mile across, that had a certain mystique among the locals. According to a local legend that we had heard, when people were feeling sick or unwell, they would make their way down into the crater, perhaps spending a night in the belly of the volcano, which was believed to have special healing powers.

This is where the story of rapamycin begins. In late 1964, a Canadian scientific and medical expedition arrived on Easter Island, having sailed all the way from Halifax aboard a naval vessel. They spent several weeks conducting research and dispensing much-needed medical care to the local inhabitants, and they brought home numerous specimens of the island's unusual flora and fauna, including soil samples from the area of the crater. The scientists might have heard the same legend about its healing properties that we did.

A few years later, a jar of Easter Island dirt ended up on the lab bench of a biochemist in Montreal named Suren Sehgal, who worked for a Canadian pharmaceutical company then called Ayerst. Sehgal found that this soil sample was saturated with a strange and potent antifungal agent that was seemingly produced by a soil bacterium called *Streptomyces hygroscopicus*. Curious, Sehgal isolated the bacterium and grew it in culture, then began testing this mysterious compound in his lab. He named it rapamycin, after Rapa Nui, the native name for Easter Island (*mycin* is the suffix typically applied to antimicrobial agents). But then Ayerst abruptly closed its Montreal lab, and Sehgal's bosses ordered him to destroy all the compounds he was researching.

Sehgal disobeyed the order. One day, he smuggled a jar of rapamycin home from work. His son Ajai, who was originally supposed to be the fifth member of our pilgrimage, remembers opening the family freezer to get ice cream when he was a kid and seeing a well-wrapped container in there marked DO NOT EAT. The jar survived the family's move to Princeton, New

Jersey, where Sehgal was ultimately transferred, and when the pharmaceutical giant Wyeth acquired Ayerst in 1987, his new bosses asked Sehgal if he had any interesting projects he'd like to pursue. He pulled the jar of rapamycin out of the freezer and went back to work.

Sehgal believed he had found a cure for athlete's foot, which would have been a big enough deal. At one point, his son Ajai recalls, he prepared a homemade ointment containing rapamycin for a neighbor who had developed some sort of weird body rash; her rash cleared up almost immediately. But rapamycin turned out to be so much more than the next Dr. Scholl's foot spray. It proved to have powerful effects on the immune system, and in 1999 it was approved by the US Food and Drug Administration (FDA) to help transplant patients accept their new organs. As a surgical resident, I used to give it out like Tic Tacs to kidney and liver transplant patients. Sometimes referred to as sirolimus, rapamycin is also used as a coating on arterial stents because it prevents the stented blood vessels from reoccluding. The hits kept coming, even after Sehgal died in 2003: In 2007, a rapamycin analog* called everolimus was approved for use against a type of kidney cancer.

The compound was deemed so important that in the early 2000s Wyeth-Ayerst placed a plaque on Easter Island, not far from the volcano crater, honoring the place where rapamycin had been discovered. But when we went looking for the plaque, we found to our dismay that it had been stolen.

The reason rapamycin has so many diverse applications is thanks to a property that Sehgal had observed, but never explored, which is that it tends to slow down the process of cellular growth and division. David Sabatini was one of a handful of scientists who picked up the baton from Sehgal, seeking to explain this phenomenon. Understanding rapamycin became his life's work. Beginning when he was a graduate student, working from a sheaf of

---

* A drug analog is a compound with similar but not identical molecular structure; e.g., oxycodone is an analog of codeine.

papers that Sehgal himself had photocopied, Sabatini helped to elucidate how this unique compound worked on the cell. Ultimately, he and others discovered that rapamycin acted directly on a very important intracellular protein complex called mTOR (pronounced "em-tor"), for "mechanistic target of rapamycin."*

Why do we care about mTOR? Because this mechanism turns out to be one of the most important mediators of longevity at the cellular level. Not only that, but it is highly "conserved," meaning it is found in virtually all forms of life, ranging from yeast to flies to worms and right on up to us humans. In biology, "conserved" means that something has been passed on via natural selection, across multiple species and classes of organisms—a sign that evolution has deemed it to be very important.

It was uncanny: this exotic molecule, found only on an isolated scrap of land in the middle of the ocean, acts almost like a switch that inhibits a very specific cellular mechanism that exists in nearly everything that lives. It was a perfect fit, and this fact still blows my mind every time I think about it.

The job of mTOR is basically to balance an organism's need to grow and reproduce against the availability of nutrients. When food is plentiful, mTOR is activated and the cell (or the organism) goes into growth mode, producing new proteins and undergoing cell division, as with the ultimate goal of reproduction. When nutrients are scarce, mTOR is suppressed and cells go into a kind of "recycling" mode, breaking down cellular components and generally cleaning house. Cell division and growth slow down or stop, and reproduction is put on hold to allow the organism to conserve energy.

"To some extent, mTOR is like the general contractor for the cell," Sabatini explains. It lies at the nexus of a long and complicated chain of upstream

---

* This is where the nomenclature gets a bit confusing. Briefly, the drug rapamycin blocks or inhibits the activity of mTOR, mechanistic target of rapamycin, the protein complex found in cells. Adding to the confusion, mTOR was originally called *mammalian target of rapamycin,* to distinguish it from a version of *target of rapamycin,* TOR, that had first been discovered in yeast. TOR and mTOR are essentially the same, meaning this same basic mechanism is found up and down the tree of life, across a billion years of evolution.

and downstream pathways that basically work together to regulate metabolism. It senses the presence of nutrients, especially certain amino acids, and it helps assemble proteins, the essential cellular building blocks. As he put it, "mTOR basically has a finger in every major process in the cell."

On July 9, 2009, a brief but important science story appeared in *The New York Times:* "Antibiotic Delayed Aging in Experiments with Mice," the headline read. Yawn. The "antibiotic" was rapamycin (which is not really an antibiotic), and according to the study, mice that had been given the drug lived significantly longer on average than controls: 13 percent longer for females, 9 percent for males.

The story was buried on page A20, but it was a stunning result. Even though the drug had been given late in life, when the mice were already "old" (six hundred days, roughly the equivalent of humans in their sixties), it had still boosted the animals' remaining life expectancy by 28 percent for males and 38 percent for females. It was the equivalent of a pill that could make a sixty-year-old woman live to the age of ninety-five. The authors of the study, published in *Nature,* speculated that rapamycin might extend lifespan "by postponing death from cancer, by retarding mechanisms of aging, or both." The real headline here, however, was that no other molecule had been shown to extend lifespan in a mammal. Ever.

The results were especially convincing because the experiment had been run by three different teams of researchers in three separate labs, using a total of 1,901 genetically diverse animals, and the results had been consistent across the board. Even better, other labs quickly and readily reproduced these results, which is a relative rarity, even with much-ballyhooed findings.

You might find this surprising, but many of the most headline-grabbing studies, the ones you read about in the newspaper or see reported on the news, are never repeated. Case in point: the well-publicized finding from 2006 that a substance found in the skins of grapes (and in red wine), resveratrol, extended lifespan in overweight mice. This generated countless news articles and even a long segment on *60 Minutes* about the benefits of this amazing molecule (and, by extension, red wine). Resveratrol supplement sales shot through the roof. But other labs could not reproduce the initial findings. When resveratrol was subjected to the same sort of rigorous testing as rapamycin, as part of a National Institute on Aging program to test potential antiaging interventions, it did *not* extend lifespan in a similar diverse population of normal mice.

The same is true of other well-hyped supplements such as nicotinamide riboside, or NR: it, too, failed to extend lifespan consistently in mice. Of course, there are no data showing that any of these supplements lengthen life or improve health in humans. But study after study since 2009 has confirmed that rapamycin can extend mouse lifespans pretty reliably. It has also been shown to do so in yeast and fruit flies, sometimes alongside genetic manipulations that reduced mTOR activity. Thus, a reasonable person could conclude that there was something good about turning down mTOR, at least temporarily—and that rapamycin may have potential as a longevity-enhancing drug.

To scientists who study aging, the life-extending effect of rapamycin was hugely exciting, but it also wasn't exactly a surprise. It appeared to represent the culmination of decades, if not centuries, of observations that how much food we eat correlates somehow with how long we live. This idea goes all the way back to Hippocrates, but more modern experiments have demonstrated, over and over, that reducing the food intake of lab animals could lengthen their lives.

The first person to really put the idea of *eating less* into practice, in a rigorous, documented way, was not an ancient Greek or a modern scientist but a sixteenth-century Italian businessman named Alvise Cornaro. A self-made real estate developer who had become tremendously wealthy by draining swamps and turning them into productive farmland, Cornaro (whose friends called him "Luigi") had a beautiful young wife and a villa outside Venice with its own theater. He loved to throw parties. But as he neared forty, he found himself suffering from "a train of infirmities," as he put it—stomach pains, weight gain, and continual thirst, a classic symptom of incipient diabetes.

The cause was obvious: too much feasting. The cure was also obvious: knock off the huge meals and parties, his doctors advised him. Not-Thin Luigi balked. He didn't want to give up his lavish lifestyle. But as his symptoms became more and more unbearable, he realized that he had to make a hard course correction or he would never get to see his young daughter grow up. Summoning all his willpower, he cut himself back to a Spartan diet that consisted of about twelve ounces of food per day, typically in the form of some sort of chicken-based stew. It was nourishing, but not overly filling. "[I] constantly rise from the table with a disposition to eat and drink still more," he wrote later.

After a year on this regimen, Cornaro's health had improved dramatically. As he put it, "I found myself . . . entirely freed from all my complaints." He stuck to the diet, and by the time he reached his eighties he was so thrilled to have lived so long in such good health that he felt compelled to share his secret with the world. He penned an autobiographical tract that he called "Discourses on the Sober Life," although it was emphatically not a teetotaler's screed, for he washed down his longevity stew with two generous glasses of wine each day.

Cornaro's prescriptions lived on long after he died in 1565. His book was reprinted in several languages over the next few centuries, lauded by Benjamin Franklin, Thomas Edison, and other luminaries, making it perhaps the first bestselling diet book in history. But it was not until the mid-twentieth century that scientists would begin rigorously testing Cornaro's notion that eating less can lengthen one's life (or at least, the lives of laboratory animals).

We're not talking about simply putting animals on Weight Watchers. Caloric restriction without malnutrition, commonly abbreviated as CR, is a precise experimental method where one group of animals (the controls) are fed *ad libitum*, meaning they eat as much as they want, while the experimental group or groups are given a similar diet containing all the necessary nutrients but 25 or 30 percent fewer total calories (more or less). The restricted animals are then compared against the controls.

The results have been remarkably consistent. Studies dating back to the 1930s have found that limiting caloric intake can lengthen the lifespan of a mouse or a rat by anywhere from 15 to 45 percent, depending on the age of onset and degree of restriction. Not only that, but the underfed animals also seem to be markedly healthier for their age, developing fewer spontaneous tumors than normally fed mice. CR seems to improve their healthspan in addition to their lifespan. You'd think that hunger might be unhealthy, but the scientists have actually found that the less they feed the animals, the longer they live. Its effects seem to be dose dependent, up to a point, almost like a drug.

The life-extending effect of CR seems to be almost universal. Numerous labs have found that restricting caloric intake lengthens lifespan not only in rats and mice (usually) but also in yeast, worms, flies, fish, hamsters, dogs, and even, weirdly, spiders. It has been found to extend lifespan in just about every model organism on which it has been tried, with the odd exception of houseflies. It seems that, across the board, hungry animals become more resilient and better able to survive, at least inside a well-controlled, germ-free laboratory.

That doesn't mean that I will be recommending this kind of radical caloric restriction as a tactic for my patients, however. For one, CR's usefulness remains doubtful outside of the lab; very lean animals may be more susceptible to death from infection or cold temperatures. And while eating a bit less worked for Luigi Cornaro, as well as for some of my own patients, long-term severe caloric restriction is difficult if not impossible for most humans to sustain. Furthermore, there is no evidence that extreme CR would truly maximize the longevity function in an organism as complex as we humans, who live in a more variable environment than the animals described above. While it seems likely that it would reduce the risk of succumbing to at least some of the Horsemen, it seems equally likely that the uptick in mortality due to infections, trauma, and frailty might offset those gains.

The real value of caloric restriction research lies in the insights it has contributed to our understanding of the aging process itself. CR studies have helped to uncover critical cellular mechanisms related to nutrients and longevity. Reducing the amount of nutrients available to a cell seems to trigger a group of innate pathways that enhance the cell's stress resistance and metabolic efficiency—all of them related, in some way, to mTOR.

The first of these is an enzyme called AMP-activated protein kinase, or AMPK for short. AMPK is like the low-fuel light on the dashboard of your car: when it senses low levels of nutrients (fuel), it activates, triggering a cascade of actions. While this typically happens as a response to lack of nutrients, AMPK is also activated when we exercise, responding to the transient drop in nutrient levels. Just as you would change your itinerary if your fuel light came on, heading for the nearest gas station rather than Grandma's house, AMPK prompts the cell to conserve and seek alternative sources of energy.

It does this first by stimulating the production of new mitochondria, the tiny organelles that produce energy in the cell, via a process called mitochondrial biogenesis. Over time—or with disuse—our mitochondria become vulnerable to oxidative stress and genomic damage, leading to dysfunction and failure. Restricting the amount of nutrients that are available, via dietary restriction or exercise, triggers the production of newer, more efficient mitochondria to replace old and damaged ones. These fresh mitochondria help the cell produce more ATP, the cellular energy currency, with the fuel it does have. AMPK also prompts the body to provide more fuel for these new mitochondria, by producing glucose in the liver (which we'll talk about in the next chapter) and releasing energy stored in fat cells.

More importantly, AMPK works to inhibit the activity of mTOR, the cellular growth regulator. Specifically, it seems to be a drop in amino acids that induces mTOR to shut down, and with it all the anabolic (growth) processes that mTOR controls. Instead of making new proteins and undergoing cell division, the cell goes into a more fuel-efficient and stress-resistant mode, activating an important cellular recycling process called *autophagy,* which means "self-eating" (or better yet, "self-devouring").

Autophagy represents the catabolic side of metabolism, when the cell stops producing new proteins and instead begins to break down old proteins and other cellular structures into their amino acid components, using the scavenged materials to build new ones. It's a form of cellular recycling, cleaning out the accumulated junk in the cell and repurposing it or disposing of it. Instead of going to Home Depot to buy more lumber and drywall and screws, the cellular "contractor" scavenges through the debris from the house he just tore down for spare materials that he can reuse, either to build and repair the cell or to burn to produce energy.

Autophagy is essential to life. If it shuts down completely, the organism dies. Imagine if you stopped taking out the garbage (or the recycling); your house would soon become uninhabitable. Except instead of trash bags, this cellular cleanup is carried out by specialized organelles called lysosomes, which package up the old proteins and other detritus, including pathogens, and grind them down (via enzymes) for reuse. In addition, the lysosomes also break up and destroy things called aggregates, which are clumps of damaged proteins that accumulate over time. Protein aggregates have been implicated in diseases such as Parkinson's and Alzheimer's disease, so getting rid of them is good; impaired autophagy has been linked to Alzheimer's disease–related pathology and also to amyotrophic lateral sclerosis (ALS), Parkinson's disease, and other neurodegenerative disorders. Mice who lack one specific autophagy gene succumb to neurodegeneration within two to three months.

By cleansing our cells of damaged proteins and other cellular junk, autophagy allows cells to run more cleanly and efficiently and helps make them more resistant to stress. But as we get older, autophagy declines. Impaired autophagy is thought to be an important driver of numerous aging-related phenotypes and ailments, such as neurodegeneration and osteoarthritis. Thus, I find it fascinating that this very important cellular mechanism can be triggered by certain kinds of interventions, such as a temporary reduction in nutrients (as when we are exercising or fasting)—and the drug rapamycin. (The Nobel Committee shares this fascination, having awarded the 2016 Nobel Prize in Physiology or Medicine to Japanese scientist Yoshinori Ohsumi for his work in elucidating the genetic regulation of autophagy.)

Yet its autophagy-promoting effect is only one reason why rapamycin may have a future as a longevity drug, according to Matt Kaeberlein, a researcher at the University of Washington. Kaeberlein, who has been studying rapamycin and mTOR for a couple of decades, believes that the drug's benefits are much more wide-ranging and that rapamycin and its derivatives have huge potential for use in humans, for the purpose of extending lifespan and healthspan.

Even though rapamycin is already approved for use in humans for multiple indications, there are formidable obstacles to launching a clinical trial to look at its possible impact on human aging—mainly, its potential side effects in healthy people, most notably the risk of immunosuppression.

Historically, rapamycin was approved to treat patients indefinitely following organ transplantation, as part of a cocktail of three or four drugs meant to suppress the part of their immune system that would otherwise attack and destroy their new organ. This immune-suppressing effect explains why there has been some reluctance to consider using (or even studying) rapamycin in the context of delaying aging in healthy people, despite ample animal data suggesting that it might lengthen lifespan and healthspan. Its purported immune-suppressing effects just seemed to be too daunting to overcome. Thus, it has seemed unlikely that rapamycin could ever realize its promise as a longevity-promoting drug for humans.

But all that started to change in late December 2014 with the publication of a study showing that the rapamycin analog everolimus actually *enhanced* the adaptive immune response to a vaccine in a group of older patients. In the study, led by scientists Joan Mannick and Lloyd Klickstein, who then worked at Novartis, the group of patients on a moderate weekly dose of everolimus seemed to have the best response to the flu vaccine, with the fewest reported side effects. This study suggested that rapamycin (and its derivatives) might actually be more of an immune *modulator* than an "immunosuppressor," as it had almost always been described before this study: that is, under some dosing regimens it can enhance immunity, while under completely different dosing regimens it may inhibit immunity.

Until this study appeared, I (like many others) had largely given up on the possibility that rapamycin could ever be used as a preventive therapy in healthy people. I had assumed that its apparent immunosuppressive effects were too serious. But this very well-done and well-controlled study actually suggested the opposite. It appeared that the immune suppression resulted from daily use of rapamycin at low to moderate doses. The study subjects had been given moderate to high doses followed by a rest period, and this cyclical administration had had an opposite, immune-enhancing effect.

It seems odd that giving different doses of the same drug could have such disparate effects, but it makes sense if you understand the structure of mTOR, which is actually composed of two separate complexes, called mTOR complex 1 (mTORC1) and mTOR complex 2 (mTORC2). The two complexes have different jobs, but (at risk of oversimplifying) the longevity-related benefits seem to result from inhibiting complex 1. Giving the drug daily, as is typically done with transplant patients, appears to inhibit both complexes, while dosing the drug briefly or cyclically inhibits mainly mTORC1, unlocking its longevity-related benefits, with fewer unwanted side effects. (A rapamycin analog or "rapalog" that selectively inhibited mTORC1 but not mTORC2 would thus be more ideal for longevity purposes, but no one has successfully developed one yet.)

As it is, its known side effects remain an obstacle to any clinical trial of rapamycin for geroprotection (delaying aging) in healthy people. To get around these objections, Kaeberlein is doing a large clinical trial of rapamycin in companion (pet) dogs, which are not a bad proxy for humans—they're large, they're mammals, they share our environment, and they age in ways similar to us. In a preliminary phase of this study, which he calls the Dog Aging Project, Kaeberlein found that rapamycin actually seemed to improve cardiac function in older animals. "One thing that's been surprising to me," he says, "is the different ways that rapamycin not only seems to delay the decline but seems to make things better. There clearly seems to be, at least in some organs, a rejuvenating function."

Kaeberlein has also observed that rapamycin seems to reduce systemic inflammation, perhaps by tamping down the activity of so-called senescent cells, which are "older" cells that have stopped dividing but have not died;

these cells secrete a toxic cocktail of inflammatory cytokines, chemicals that can harm surrounding cells. Rapamycin seems to reduce these inflammatory cytokines. It also improves cancer surveillance, the ways in which our body, most likely the immune system, detects and eliminates cancer cells. In another recent study, Kaeberlein's group found that rapamycin appeared to improve periodontal (gum) health in older dogs.

The main phase of the Dog Aging Project, involving some 600 pet dogs, is now under way; results from this larger clinical trial are expected in 2026. (Disclosure: I am a partial funder of this research.) The dogs in this study are also following a weekly, cyclical dosing schedule with rapamycin, similar to the protocol in the 2014 immune study in humans. If the results are positive, it would not surprise me if the use of rapamycin for longevity purposes becomes more common. A small but growing number of people, including me and a handful of my patients, already take rapamycin off-label for its potential geroprotective benefits. I can't speak for everyone, but taking it cyclically does appear to reduce unwanted side effects, in my experience.

Even so, the hurdles it would have to clear to gain approval for broader human use remain daunting. The vast majority of people who currently take rapamycin comprise transplant patients who already have serious health issues and multiple comorbidities. In populations like this, rapamycin's side effects seem less significant than they might in healthier people.

"There is a very low tolerance for side effects, by the public and by regulatory agencies, if you're talking about treating a healthy person," says Kaeberlein. "The intent is to slow aging in people before they get sick, to keep them healthy longer, so in many ways it is the opposite of the traditional biomedical approach, where normally we wait until people are sick and then we try to cure their diseases."

The real obstacle here is a regulatory framework rooted in Medicine 2.0, which does not (yet) recognize "slowing aging" and "delaying disease" as fully legitimate end points. This would represent a Medicine 3.0 use for this drug, where we would be using a drug to help healthy people stay healthy, rather than to cure or relieve a specific ailment. Thus, it would face much more scrutiny and skepticism. But if we're talking about preventing the diseases of aging, which kill 80 percent of us, then it's certainly worth having a serious conversation about what level of risk is and isn't acceptable in order to achieve that goal. Part of my aim in writing this book is to move that conversation forward.

This may already be starting to happen. The FDA has given the green light for a clinical trial of another drug with potential longevity benefits, the diabetes medication metformin. This trial is called TAME (Targeting Aging with Metformin), and it came about in a very different way. Metformin has been taken by millions of people for years. Over time, researchers noticed (and studies appeared to confirm) that patients on metformin appeared to have a lower incidence of cancer than the general population. One large 2014 analysis seemed to show that diabetics on metformin actually lived longer than nondiabetics, which is striking. But none of these observations "prove" that metformin is geroprotective—hence the need for a clinical trial.

But aging itself is difficult—if not impossible—to measure with any accuracy. Instead, TAME lead investigator Nir Barzilai, whom we met in the previous chapter, decided to look at a different endpoint: whether giving metformin to healthy subjects delays the onset of aging-related diseases, as a proxy for its effect on aging. I'm hopeful that someday, maybe in the near future, we could attempt a similar human trial of rapamycin, which I believe has even greater potential as a longevity-promoting agent.*

For the moment, though, let's think about the fact that *all* of what we've talked about in this chapter, from mTOR and rapamycin to caloric restriction, points in one direction: that what we eat and how we metabolize it appear to play an outsize role in longevity. In the next chapter, we will take a much more detailed look at how metabolic disorders help to instigate and promote chronic disease.

* Before leaving Rapa Nui, the four of us vowed to replace the missing plaque honoring the discovery of rapamycin with a new one saluting the island's unique contribution to molecular biology and the role of Suren Sehgal in preserving and elucidating the importance of this molecule.

CHAPTER 6

The Crisis of Abundance

Can Our Ancient Genes Cope with Our Modern Diet?

Avoidable human misery is more often caused not so much by stupidity as by ignorance, particularly our ignorance about ourselves.

—Carl Sagan

When it comes to managing junior surgical residents, there is a sort of unwritten rule that Hippocrates might have stated as follows: *First, let them do no harm.* That rule was in full effect during my early months at Johns Hopkins, in 2001, on the surgical oncology service. We were removing part of a patient's cancerous ascending colon, and one of my jobs was to "pre-op" him, which was basically like a briefing/semi-interrogation the day before surgery to be sure we knew everything that we needed to know about his medical history.

I met with this patient and outlined the procedure he was about to undergo, reminded him not to eat anything after 8 p.m., and asked him a series of routine questions, including whether or not he smoked and how much alcohol he drank. I had practiced asking this last one in a disarming, seemingly offhand way, but I knew it was among the most important items on my checklist. If we believed that a patient consumed significant amounts of alcohol (typically more than four or five drinks per day), we had to make sure that the anesthesiologists knew this, so they could administer specific drugs during recovery, typically benzodiazepines such as Valium, in order to ward off alcohol withdrawal. Otherwise, the patient could be at risk for delirium tremens, or the DTs, a potentially fatal condition.

I was relieved when he told me that he drank minimally. One less thing to worry about. The next day, I wheeled the patient into the OR and ran my checklist of mundane intern-level stuff. It would take a few minutes for the anesthesiologists to put him to sleep, after which I could place the Foley catheter into his bladder, swab his skin with Betadine, place the surgical drapes, and then step aside while the chief resident and attending surgeon made the first incision. If I was lucky, I would get to assist with the opening and closing of the abdomen. Otherwise, I was there to retract the liver, holding it out of the way so the senior surgeons could have an unobstructed view of the organ they needed to remove, which was sort of tucked in underneath the liver.

As the surgery got under way, nothing seemed out of the ordinary. The surgeons had to make their way through a bit of abdominal fat before they could get to the peritoneal cavity, but nothing we didn't see most days. There is an incredible rush of anticipation one feels just before cutting through the last of several membranes separating the outside world from the inner abdominal cavity. One of the first things you see, as the incision grows, is the tip of the liver, which I've always considered to be a really underappreciated organ. The "cool kids" in medicine specialize in the brain or the heart, but the liver is the body's true workhorse—and also, it's simply breathtaking to behold. Normally, a healthy liver is a deep, dark purple color, with a gorgeous silky-smooth texture. Hannibal Lecter was not too far off: it really does look as if it might be delicious with some fava beans and a nice Chianti.

This patient's liver appeared rather less appetizing as it emerged from beneath the omental fat. Instead of a healthy, rich purple, it was mottled and sort of orangish, with protruding nodules of yellow fat. It looked like foie gras gone bad. The attending looked up at me sharply. "You said this guy was not a drinker!" he barked.

Clearly, this man was a very heavy drinker; his liver showed all the signs of it. And because I had failed to elicit that information, I had potentially placed his life in danger.

But it turned out that I hadn't made a mistake. When the patient awoke after surgery, he confirmed that he rarely drank alcohol, if ever. In my experience, patients confronting cancer surgery rarely lied about drinking or anything else, especially when fessing up meant getting some Valium or even better, a couple of beers with their hospital dinner. But he definitely had the liver of an alcoholic, which struck everyone as odd.

This would happen numerous times during my residency. Every time, we would scratch our heads. Little did we know that we were witnessing the beginning, or perhaps the flowering, of a silent epidemic.

Five decades earlier, a surgeon in Topeka, Kansas, named Samuel Zelman had encountered a similar situation: he was operating on a patient whom he knew personally, because the man was an aide in the hospital where he worked. He knew for a fact that the man did not drink any alcohol, so he was surprised to find out that his liver was packed with fat, just like that of my patient, decades later.

This man did, in fact, drink a lot—of Coca-Cola. Zelman knew that he consumed a staggering quantity of soda, as many as twenty bottles (or more) in a single day. These were the older, smaller Coke bottles, not the supersizes we have now, but still, Zelman estimated that his patient was taking in an extra 1,600 calories per day on top of his already ample meals. Among his colleagues, Zelman noted, he was "distinguished for his appetite."

His curiosity piqued, Zelman recruited nineteen other obese but nonalcoholic subjects for a clinical study. He tested their blood and urine and conducted liver biopsies on them, a serious procedure performed with a serious needle. All of the subjects bore some sign or signs of impaired liver function, in a way eerily similar to the well-known stages of liver damage seen in alcoholics.

This syndrome was often noticed but little understood. It was typically attributed to alcoholism or hepatitis. When it began to be seen in teenagers, in the 1970s and 1980s, worried doctors warned of a hidden epidemic of teenage binge drinking. But alcohol was not to blame. In 1980, a team at the Mayo Clinic dubbed this "hitherto unnamed disease" nonalcoholic steatohepatitis, or NASH. Since then, it has blossomed into a global plague. More than one in four people on this planet have some degree of NASH or its precursor, known as nonalcoholic fatty liver disease, or NAFLD, which is what we had observed in our patient that day in the operating room.

NAFLD is highly correlated with both obesity and hyperlipidemia (excessive cholesterol), yet it often flies under the radar, especially in its early stages. Most patients are unaware that they have it—and so are their doctors, because NAFLD and NASH have no obvious symptoms. The first signs would generally show up only on a blood test for the liver enzyme alanine aminotransferase (ALT for short). Rising levels of ALT are often the first clue that something is wrong with the liver, although they could also be a symptom of something else, such as a recent viral infection or a reaction to a medication. But there are many people walking around whose physicians have no idea that they are in the early stages of this disease, because their ALT levels are still "normal."

Next question: What is normal? According to Labcorp, a leading testing company, the acceptable range for ALT is below 33 IU/L for women and below 45 IU/L for men (although the ranges can vary from lab to lab). But "normal" is not the same as "healthy." The reference ranges for these tests are based on current percentiles,* but as the population in general becomes less healthy, the average may diverge from optimal levels. It's similar to what has happened with weight. In the late 1970s, the average American adult male weighed 173 pounds. Now the average American man tips the scale at nearly 200 pounds.

* Typically, "normal" means between the 2.5th and 97.5th percentiles, a very wide range.

In the 1970s, a 200-pound man would have been considered very overweight; today he is merely average. So you can see how in the twenty-first century, "average" is not necessarily optimal.

With regard to ALT liver values, the American College of Gastroenterology recently revised its guidelines to recommend clinical evaluation for liver disease in men with ALT above 33 and women with ALT above 25—significantly below the current "normal" ranges. Even that may not be low enough: a 2002 study that excluded people who *already* had fatty liver suggested upper limits of 30 for men, and 19 for women. So even if your liver function tests land within the reference range, that does not imply that your liver is actually healthy.

NAFLD and NASH are basically two stages of the same disease. NAFLD is the first stage, caused by (in short) more fat entering the liver or being produced there than exiting it. The next step down the metabolic gangplank is NASH, which is basically NAFLD plus inflammation, similar to hepatitis but without a viral infection. This inflammation causes scarring in the liver, but again, there are no obvious symptoms. This may sound scary, but all is not yet lost. Both NAFLD and NASH are still reversible. If you can somehow remove the fat from the liver (most commonly via weight loss), the inflammation will resolve, and liver function returns to normal. The liver is a highly resilient organ, almost miraculously so. It may be the most regenerative organ in the human body. When a healthy person donates a portion of their liver, both donor and recipient end up with an almost full-sized, fully functional liver within about eight weeks of the surgery, and the majority of that growth takes place in just the first two weeks.

In other words, your liver can recover from fairly extensive damage, up to and including partial removal. But if NASH is not kept in check or reversed, the damage and the scarring may progress into cirrhosis. This happens in about 11 percent of patients with NASH and is obviously far more serious. It now begins to affect the cellular architecture of the organ, making it much more difficult to reverse. A patient with cirrhosis is likely to die from various complications of their failing liver unless they receive a liver transplant. In 2001, when we did the operation on the man with the fatty liver, NASH officially accounted for just over 1 percent of liver transplants in the United States; by 2025, NASH with cirrhosis is expected to be the leading indication for liver transplantation.

As devastating as it is, cirrhosis is not the only end point I'm worried about here. I care about NAFLD and NASH—and you should too—because they represent the tip of the iceberg of a global epidemic of metabolic disorders, ranging from insulin resistance to type 2 diabetes. Type 2 diabetes is technically a distinct disease, defined very clearly by glucose metrics, but I view it as simply the last stop on a railway line passing through several other stations, including hyperinsulinemia, prediabetes, and NAFLD/NASH. If you find yourself anywhere on this train line, even in the early stages of NAFLD, you are likely also en route to one or more of the other three Horsemen diseases (cardiovascular disease, cancer, and Alzheimer's disease). As we will see in the next few chapters, metabolic dysfunction vastly increases your risk for all of these. So you can't fight the Horsemen without taking on metabolic dysfunction first.

Notice that I said "metabolic dysfunction" and not "obesity," everybody's favorite public health bogeyman. It's an important distinction. According to the Centers for Disease Control (CDC), more than 40 percent of the US population is obese (defined as having a BMI* greater than 30), while roughly another third is overweight (BMI of 25 to 30). Statistically, being obese means someone is at greater risk of chronic disease, so a lot of attention is focused on the "obesity problem," but I take a broader view: obesity is merely one symptom of an underlying metabolic derangement, such as hyperinsulinemia, that also happens to cause us to gain weight. But not everyone who is obese is metabolically unhealthy, and not everyone who is metabolically unhealthy is obese. There's more to metabolic health than meets the eye.

As far back as the 1960s, before obesity had become a widespread problem, a Stanford endocrinologist named Gerald Reaven had observed that excess weight often traveled in company with certain other markers of poor

* BMI is far from perfect, as it does not capture the proportion of fat to muscle, but is good enough for our purposes here.

health. He and his colleagues noticed that heart attack patients often had both high fasting glucose levels and high triglycerides, as well as elevated blood pressure and abdominal obesity. The more of these boxes a patient checked, the greater their risk of cardiovascular disease.

In the 1980s, Reaven labeled this collection of related disorders "Syndrome X"—where the X factor, he eventually determined, was insulin resistance. Today we call this cluster of problems "metabolic syndrome" (or MetSyn), and it is defined in terms of the following five criteria:

1. high blood pressure (>130/85)
2. high triglycerides (>150 mg/dL)
3. low HDL cholesterol (<40 mg/dL in men or <50 mg/dL in women)
4. central adiposity (waist circumference >40 inches in men or >35 in women)
5. elevated fasting glucose (>110 mg/dL)

If you meet three or more of these criteria, then you have the metabolic syndrome—along with as many as 120 million other Americans, according to a 2020 article in *JAMA*. About 90 percent of the US population ticks at least one of these boxes. But notice that obesity is merely one of the criteria; it is *not* required for the metabolic syndrome to be diagnosed. Clearly the problem runs deeper than simply unwanted weight gain. This tends to support my view that obesity itself is not the issue but is merely a symptom of other problems.

Studies have found that approximately one-third of those folks who are obese by BMI are actually metabolically *healthy*, by many of the same parameters used to define the metabolic syndrome (blood pressure, triglycerides, cholesterol, and fasting glucose, among others). At the same time, some studies have found that between 20 and 40 percent of nonobese adults may be metabolically unhealthy, by those same measures. A high percentage of obese people are also metabolically sick, of course—but as figure 3 illustrates, many normal-weight folks are in the same boat, which should be a wake-up call to all. This is *not* about how much you weigh. Even if you happen to be thin, you still need to read this chapter.

Figure 3. Uncoupling Obesity from Metabolic Health

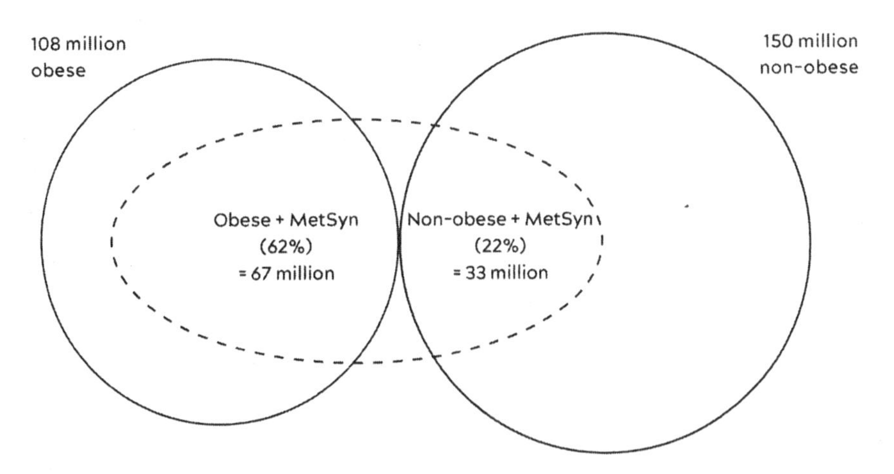

Relative prevalence of metabolic dysfunction ("MetSyn") across the obese and nonobese segments of the population.

Source: Internal analysis based on data from National Institute of Diabetes and Digestive and Kidney Diseases (2021).

This figure (based on NIH data and not the *JAMA* article just mentioned) shows quite dramatically how obesity and metabolic dysfunction are not the same thing—far from it, in fact. Some 42 percent of the US population is obese (BMI>30). Out of a conservatively estimated 100 million Americans who meet the criteria for the metabolic syndrome (i.e., metabolically unhealthy), almost exactly one-third are *not* obese. Many of these folks are overweight by BMI (25-29.9), but nearly 10 million Americans are normal weight (BMI 19-24.9) but metabolically unhealthy.

Some research suggests that these people might be in the most serious danger. A large meta-analysis of studies with a mean follow-up time of 11.5 years showed that people in this category have more than triple the risk of all-cause mortality and/or cardiovascular events than metabolically healthy normal-weight individuals. Meanwhile, the metabolically healthy but obese subjects in these studies were *not* at significantly increased risk. The upshot is that it's not only obesity that drives bad health outcomes; it's metabolic dysfunction. That's what we're concerned with here.

Metabolism is the process by which we take in nutrients and break them down for use in the body. In someone who is metabolically healthy, those nutrients are processed and sent to their proper destinations. But when someone is metabolically unhealthy, many of the calories they consume end up where they are not needed, at best—or outright harmful, at worst.

If you eat a doughnut, for example, the body has to decide what to do with the calories in that doughnut. At the risk of oversimplifying a bit, the carbohydrate from our doughnut has two possible fates. First, it can be converted into glycogen, the storage form of glucose, suitable for use in the near term. About 75 percent of this glycogen ends up in skeletal muscle and the other 25 percent goes to the liver, although this ratio can vary. An adult male can typically store a total of about 1,600 calories worth of glycogen between these two sites, or about enough energy for two hours of vigorous endurance exercise. This is why if you are running a marathon or doing a long bike ride, and do not replenish your fuel stores in some way, you are likely to "bonk," or run out of energy, which is not a pleasant experience.

One of the liver's many important jobs is to convert this stored glycogen back to glucose and then to release it as needed to maintain blood glucose levels at a steady state, known as *glucose homeostasis.* This is an incredibly delicate task: an average adult male will have about five grams of glucose circulating in his bloodstream at any given time, or about a teaspoon. That teaspoon won't last more than a few minutes, as glucose is taken up by the muscles and especially the brain, so the liver has to continually feed in more, titrating it precisely to maintain a more or less constant level. Consider that five grams of glucose, spread out across one's entire circulatory system, is normal, while seven grams—a teaspoon and a half—means you have diabetes. As I said, the liver is an amazing organ.

We have a far greater capacity, almost unlimited, for storing energy as fat—the second possible destination for the calories in that doughnut. Even a relatively lean adult may carry ten kilograms of fat in their body, representing a whopping ninety thousand calories of stored energy.

That decision—where to put the energy from the doughnut—is made via hormones, chief among them insulin, which is secreted by the pancreas when the body senses the presence of glucose, the final breakdown product of most carbohydrates (such as those in the doughnut). Insulin helps shuttle the glucose to where it's needed, while maintaining glucose homeostasis. If you happen to be riding a stage of the Tour de France while you eat the doughnut, or are engaged in other intense exercise, those calories will be consumed almost instantly in the muscles. But in a typical sedentary person, who is not depleting muscle glycogen rapidly, the excess energy from the doughnut will largely end up in fat cells (or more specifically, as triglycerides contained within fat cells).

The twist here is that fat—that is, subcutaneous fat, the layer of fat just beneath our skin—is actually the *safest* place to store excess energy. Fat in and of itself is not bad. It's where we should put surplus calories. That's how we evolved. While fat might not be culturally or aesthetically desirable in our modern world, subcutaneous fat actually plays an important role in maintaining metabolic health. The Yale University endocrinologist Gerald Shulman, one of the leading researchers in diabetes, once published an elegant experiment demonstrating the necessity of fat: when he surgically implanted fat tissue into insulin-resistant mice, thereby making them *more* fat, he found that their metabolic dysfunction was cured almost instantly. Their new fat cells sucked up their excess blood glucose and stored it safely.

Think of fat as acting like a kind of metabolic buffer zone, absorbing excess energy and storing it safely until it is needed. If we eat extra doughnuts, those calories are stored in our subcutaneous fat; when we go on, say, a long hike or swim, some of that fat is then released for use by the muscles. This fat flux goes on continually, and as long as you haven't exceeded your own fat storage capacity, things are pretty much fine.

But if you continue to consume energy in excess of your needs, those subcutaneous fat cells will slowly fill up, particularly if little of that stored energy is being utilized. When someone reaches the limit of their capacity to store energy in their subcutaneous fat, yet they continue to take on excess calories, all that energy still has to go somewhere. The doughnuts or whatever they might be eating are probably still getting converted into fat, but now the body has to find other places to store it.

It's almost as if you have a bathtub, and you're filling it up from the faucet. If you keep the faucet running even after the tub is full and the drain is closed (i.e., you're sedentary), water begins spilling over the rim of the tub, flowing into places where it's not wanted or needed, like onto the bathroom floor, into the heating vents or down the stairs. It's the same with excess fat. As more calories flood into your subcutaneous fat tissue, it eventually reaches capacity and the surplus begins spilling over into other areas of your body: into your blood, as excess triglycerides; into your liver, contributing to NAFLD; into your muscle tissue, contributing directly to insulin resistance in the muscle (as we'll see); and even around your heart and your pancreas (figure 4). None of these, obviously, are ideal places for fat to collect; NAFLD is just one of many undesirable consequences of this fat spillover.

Fat also begins to infiltrate your abdomen, accumulating in between your organs. Where subcutaneous fat is thought to be relatively harmless, this "visceral fat" is anything but. These fat cells secrete inflammatory cytokines such as TNF-alpha and IL-6, key markers and drivers of inflammation, in close proximity to your most important bodily organs. This may be why visceral fat is linked to increased risk of both cancer and cardiovascular disease.

Fat storage capacity varies widely among individuals. Going back to our tub analogy, some people have subcutaneous fat-storage capacity equivalent to a regular bathtub, while others may be closer to a full-sized Jacuzzi or hot tub. Still others may have only the equivalent of a five-gallon bucket. It also matters, obviously, how much "water"is flowing *into* the tub via the faucet (as calories in food) and how much is flowing *out* via the drain (or being consumed via exercise or other means).

Individual fat-storage capacity seems to be influenced by genetic factors. This is a generalization, but people of Asian descent (for example), tend to have much lower capacity to store fat, on average, than Caucasians. There are other factors at play here as well, but this explains in part why some people can be obese but metabolically healthy, while others can appear "skinny" while still walking around with three or more markers of metabolic syndrome. It's these

Figure 4. How Excess Fat Increases Cardiometabolic Risk

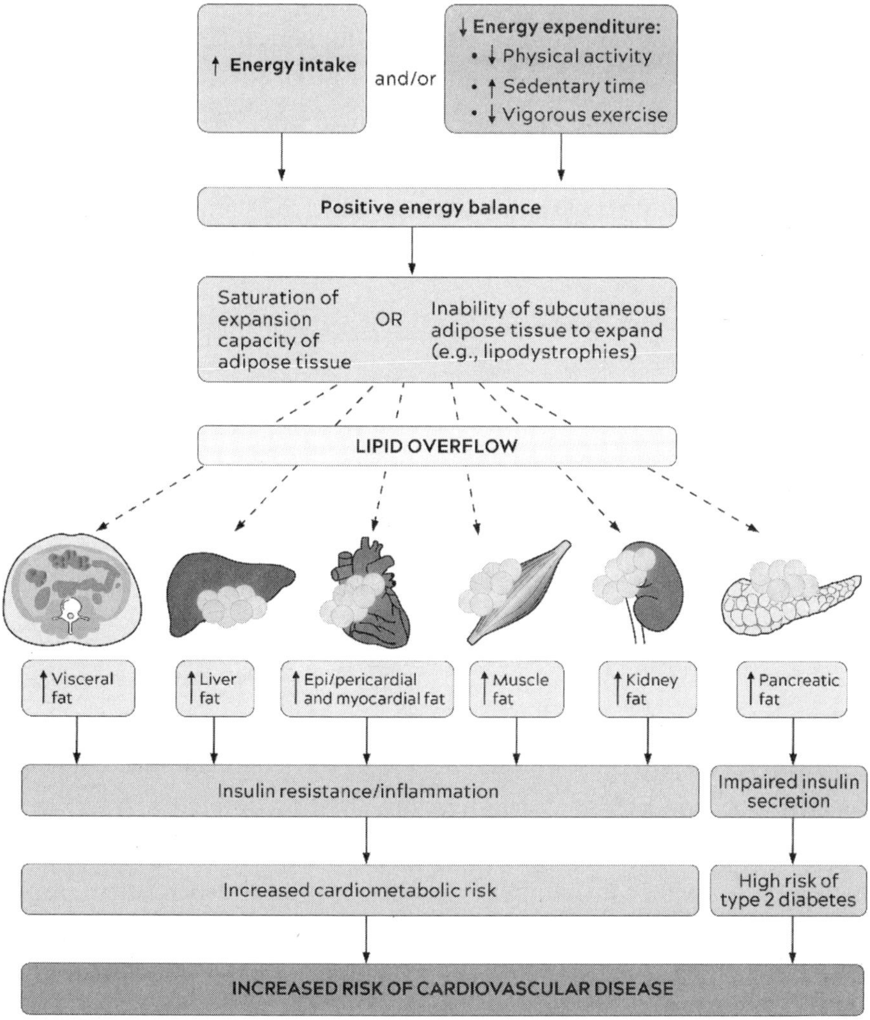

Source: Tchernof and Després (2013).

people who are most at risk, according to research by Mitch Lazar at the University of Pennsylvania, because a "thin" person may simply have a much lower capacity to safely store fat. All other things being equal, someone who carries a bit of body fat may also have greater fat-storage capacity, and thus more metabolic leeway than someone who appears to be more lean.

It doesn't take much visceral fat to cause problems. Let's say you are a forty-year-old man who weighs two hundred pounds. If you have 20 percent body fat, making you more or less average (50th percentile) for your age and sex, that means you are carrying 40 pounds of fat throughout your body. Even if just 4.5 pounds of that is visceral fat, you would be considered at exceptionally high risk for cardiovascular disease and type 2 diabetes, in the top 5 percent of risk for your age and sex. This is why I insist my patients undergo a DEXA scan annually—and I am far more interested in their visceral fat than their total body fat.

It may have taken you a long time to get there, but now you are in trouble—even if you, and your doctor, may not yet realize it. You have fat accumulating in many places where it should not be, such as in your liver, between your abdominal organs, even around your heart—regardless of your actual weight. But one of the first places where this overflowing fat will cause problems is in your muscle, as it worms its way in between your muscle fibers, like marbling on a steak. As this continues, microscopic fat droplets even appear *inside* your muscle cells.

This is where insulin resistance likely begins, Gerald Shulman concludes from three decades' worth of investigation. These fat droplets may be among the first destinations of excess energy/fat spillover, and as they accumulate they begin to disrupt the complex network of insulin-dependent transport mechanisms that normally bring glucose in to fuel the muscle cell. When these mechanisms lose their function, the cell becomes "deaf" to insulin's signals. Eventually, this insulin resistance will progress to other tissues, such as the liver, but Shulman believes that it originates in muscle. It's worth noting that one key ingredient in this process seems to be inactivity. If a person is not physically active, and they are not consuming energy via their muscles, then this fat-spillover-driven insulin resistance develops much more quickly. (This is why Shulman requires his study subjects, mostly young college students, to refrain from physical activity, in order to push them towards insulin resistance.)

Insulin resistance is a term that we hear a lot, but what does it really mean? Technically, it means that cells, initially muscle cells, have stopped listening to insulin's signals, but another way to visualize it is to imagine the cell as a balloon being blown up with air. Eventually, the balloon expands to the point where it gets more difficult to force more air inside. You have to blow harder and harder. This is where insulin comes in, to help facilitate the process of blowing air into the balloon. The pancreas begins to secrete even more insulin, to try to remove excess glucose from the bloodstream and cram it into cells. For the time being it works, and blood glucose levels remain normal, but eventually you reach a limit where the "balloon" (cells) cannot accept any more "air" (glucose).

This is when the trouble shows up on a standard blood test, as fasting blood glucose begins to rise. This means you have high insulin levels *and* high blood glucose, and your cells are shutting the gates to glucose entry. If things continue in this way, then the pancreas becomes fatigued and less able to mount an insulin response. This is made worse by, you guessed it, the fat now residing in the pancreas itself. You can see the vicious spiral forming here: fat spillover helps initiate insulin resistance, which results in the accumulation of still more fat, eventually impairing our ability to store calories as anything *other* than fat. There are many other hormones involved in the production and distribution of fat, including testosterone, estrogen, hormone-sensitive lipase* and cortisol. Cortisol is especially potent, with a double-edged effect of depleting subcutaneous fat (which is generally beneficial) and replacing it with more harmful visceral fat. This is one reason why stress levels and sleep, both of which affect cortisol release, are pertinent to metabolism. But insulin seems to be the most potent as far as promoting fat accumulation because it acts as kind of a one-way gate, allowing fat to enter the cell while impairing the release of energy from fat cells (via a process called lipolysis). Insulin is all about fat storage, not fat utilization.

When insulin is chronically elevated, more problems arise. Fat gain and

* An enzyme expressed in fat cells that helps convert stored triglycerides into free fatty acids.

ultimately obesity are merely one symptom of this condition, known as hyperinsulinemia. I would argue that they are hardly even the most serious symptoms: as we'll see in the coming chapters, insulin is also a potent growth-signaling hormone that helps foster both atherosclerosis and cancer. And when insulin resistance begins to develop, the train is already well down the track toward type 2 diabetes, which brings a multitude of unpleasant consequences.

Our slowly dawning awareness of NAFLD and NASH mirrors the emergence of the global epidemic of type 2 diabetes a century ago. Like cancer, Alzheimer's, and heart disease, type 2 diabetes is known as a "disease of civilization," meaning it has only come to prominence in the modern era. Among primitive tribes and in prior times, it was largely unknown. Its symptoms had been recognized for thousands of years, going back to ancient Egypt (as well as ancient India), but it was the Greek physician Aretaeus of Cappadocia who named it *diabetes,* describing it as "a melting down of the flesh and limbs into urine."

Back then, it was vanishingly rare, observed only occasionally. As type 2 diabetes emerged, beginning in the early 1700s, it was at first largely a disease of the superelite, popes and artists and wealthy merchants and nobles who could afford this newly fashionable luxury food known as sugar. The composer Johann Sebastian Bach is thought to have been afflicted, among other notable personages. It also overlapped with gout, a more commonly recognized complaint of the decadent upper classes. This, as we'll soon see, was not a coincidence.

By the early twentieth century, diabetes was becoming a disease of the masses. In 1940 the famed diabetologist Elliott Joslin estimated that about one person in every three to four hundred was diabetic, representing an enormous increase from just a few decades earlier, but it was still relatively uncommon. By 1970, around the time I was born, its prevalence was up to one in every fifty people. Today over 11 percent of the US adult population, one in nine, has clinical type 2 diabetes, according to a 2022 CDC report, including more than 29 percent of adults over age sixty-five. Another 38 percent of US adults—more than one in three—meet at least one of the criteria for prediabetes. That means that nearly half of the population is either on the road to type 2 diabetes or already there.

One quick note: diabetes ranks as only the seventh or eighth leading cause of death in the United States, behind things like kidney disease, accidents, and Alzheimer's disease. In 2020, a little more than one hundred thousand deaths were attributed to type 2 diabetes, a fraction of the number due to either cardiovascular disease or cancer. By the numbers, it barely qualifies as a Horseman. But I believe that the actual death toll due to type 2 diabetes is much greater and that we undercount its true impact. Patients with diabetes have a much greater risk of cardiovascular disease, as well as cancer and Alzheimer's disease and other dementias; one could argue that diabetes with related metabolic dysfunction is one thing that all these conditions have in common. This is why I place such emphasis on metabolic health, and why I have long been concerned about the epidemic of metabolic disease not only in the United States but around the world.

Why is this epidemic happening now?

The simplest explanation is likely that our metabolism, as it has evolved over millennia, is not equipped to cope with our ultramodern diet, which has appeared only within the last century or so. Evolution is no longer our friend, because our environment has changed much faster than our genome ever could. Evolution wants us to get fat when nutrients are abundant: the more energy we could store, in our ancestral past, the greater our chances of survival and successful reproduction. We needed to be able to endure periods of time without much food, and natural selection obliged, endowing us with genes that helped us conserve and store energy in the form of fat. That enabled our distant ancestors to survive periods of famine, cold climates, and physiologic stressors such as illness and pregnancy. But these genes have proved less advantageous in our present environment, where many people in the developed world have access to almost unlimited calories.

Another problem is that not all of these calories are created equal, and not all of them are metabolized in the same way. One abundant source of calories in our present diet, fructose, also turns out to be a very powerful driver of metabolic dysfunction if consumed to excess. Fructose is not a novel nutrient, obviously. It's the form of sugar found in nearly all fruits, and as such it is essential in the diets of many species, from bats and hummingbirds up to bears and monkeys and humans. But as it turns out, we humans have a unique capacity for turning calories from fructose into fat.

Lots of people like to demonize fructose, especially in the form of high-fructose corn syrup, without really understanding why it's supposed to be so harmful. The story is complicated but fascinating. The key factor here is that fructose is metabolized in a manner different from other sugars. When we metabolize fructose, along with certain other types of foods, it produces large amounts of uric acid, which is best known as a cause of gout but which has also been associated with elevated blood pressure.

More than two decades ago, a University of Colorado nephrologist named Rick Johnson noticed that fructose consumption appeared to be an especially powerful driver not only of high blood pressure but also of fat gain. "We realized fructose was having effects that could not be explained by its calorie content," Johnson says. The culprit seemed to be uric acid. Other mammals, and even some other primates, possess an enzyme called uricase, which helps them clear uric acid. But we humans lack this important and apparently beneficial enzyme, so uric acid builds up, with all its negative consequences.

Johnson and his team began investigating our evolutionary history, in collaboration with a British anthropologist named Peter Andrews, a retired researcher at the Natural History Museum in London and an expert on primate evolution. Others had observed that our species had lost this uricase enzyme because of some sort of random genetic mutation, far back in our evolutionary past, but the reason why had remained mysterious. Johnson and Andrews scoured the evolutionary and fossil record and came up with an intriguing theory: that this mutation may have been essential to the very emergence of the human species.

The story they uncovered was that, millions of years ago, our primate ancestors migrated north from Africa into what is now Europe. Back then, Europe was lush and semitropical, but as the climate slowly cooled, the forest changed. Deciduous trees and open meadows replaced the tropical forest, and the fruit trees on which the apes depended for food began to disappear, especially the fig trees, a staple of their diets. Even worse, the apes now had to endure a new and uncomfortably cold season, which we know as "winter." In order to survive, these apes now needed to be able to store some of the calories they did eat as fat. But storing fat did not come naturally to them because they had evolved in Africa, where food was always available. Thus, their metabolism did not prioritize fat storage.

At some point, our primate ancestors underwent a random genetic mutation that effectively switched on their ability to turn fructose into fat: the gene for the uricase enzyme was "silenced," or lost. Now, when these apes consumed fructose, they generated lots of uric acid, which caused them to store many more of those fructose calories as fat. This newfound ability to store fat enabled them to survive in the colder climate. They could spend the summer gorging themselves on fruit, fattening up for the winter.

These same ape species, or their evolutionary successors, migrated back down into Africa, where over time they evolved into hominids and then *Homo sapiens*—while also passing their uricase-silencing mutation down to us humans. This, in turn, helped enable humans to spread far and wide across the globe, because we could store energy to help us survive cold weather and seasons without abundant food.

But in our modern world, this fat-storage mechanism has outlived its usefulness. We no longer need to worry about foraging for fruit or putting on fat to survive a cold winter. Thanks to the miracles of modern food technology, we are almost literally swimming in a sea of fructose, especially in the form of soft drinks, but also hidden in more innocent-seeming foods like bottled salad dressing and yogurt cups.*

Whatever form it takes, fructose does not pose a problem when consumed the way that our ancestors did, before sugar became a ubiquitous

---

* While it may be in vogue to vilify high-fructose corn syrup, which is 55 percent fructose and 45 percent glucose, it's worth pointing out that good old table sugar (sucrose) is about the same, consisting of 50 percent fructose and 50 percent glucose. So there's really not much of a difference between the two.

commodity: mostly in the form of actual fruit. It is very difficult to get fat from eating too many apples, for example, because the fructose in the apple enters our system relatively slowly, mixed with fiber and water, and our gut and our metabolism can handle it normally. But if we are drinking quarts of apple juice, it's a different story, as I'll explain in a moment.

Fructose isn't the only thing that creates uric acid; foods high in chemicals called purines, such as certain meats, cheeses, anchovies, and beer, also generate uric acid. This is why gout, a condition of excess uric acid, was so common among gluttonous aristocrats in the olden days (and still today). I test my patients' levels of uric acid, not only because high levels may promote fat storage but also because it is linked to high blood pressure. High uric acid is an early warning sign that we need to address a patient's metabolic health, their diet, or both.

Another issue is that glucose and fructose are metabolized very differently at the cellular level. When a brain cell, muscle cell, gut cell, or any other type of cell breaks down glucose, it will almost instantly have more ATP (adenosine triphosphate), the cellular energy "currency," at its disposal. But this energy is not free: the cell must expend a small amount of ATP in order to make more ATP, in the same way that you sometimes have to spend money to make money. In glucose metabolism, this energy expenditure is regulated by a specific enzyme that prevents the cell from "spending" too much of its ATP on metabolism.

But when we metabolize fructose in large quantities, a different enzyme takes over, and this enzyme does *not* put the brakes on ATP "spending." Instead, energy (ATP) levels inside the cell drop rapidly and dramatically. This rapid drop in energy levels makes the cell think that we are still hungry. The mechanisms are a bit complicated, but the bottom line is that even though it is rich in energy, fructose basically tricks our metabolism into thinking that we are depleting energy—and need to take in still more food and store more energy as fat.*

---

* This drop in cellular ATP triggers an enzyme called *AMP deaminase*, or AMPD, which is sort of like the evil twin to AMPK, the reverse-fuel gauge enzyme that we discussed in the

On a more macro level, consuming large quantities of liquid fructose simply overwhelms the ability of the gut to handle it; the excess is shunted to the liver, where many of those calories are likely to end up as fat. I've seen patients work themselves into NAFLD by drinking too many "healthy" fruit smoothies, for the same reason: they are taking in too much fructose, too quickly. Thus, the almost infinite availability of liquid fructose in our already high-calorie modern diet sets us up for metabolic failure if we're not careful (and especially if we are not physically active).

I sometimes think back to that patient who first introduced me to fatty liver disease. He and Samuel Zelman's Patient Zero, the man who drank a dozen Cokes a day, had the same problem: they consumed many more calories than they needed. In the end, I still think excess calories matter the most.

Of course, my patient was in the hospital not because of his NAFLD, but because of his colon cancer. His operation turned out beautifully: we removed the cancerous part of his colon and sent him off to a speedy recovery. His colon cancer was well established, but it had not metastasized or spread. I remember the attending surgeon feeling pretty good about the operation, as we had caught the cancer in time. This man was maybe forty or forty-five, with a long life still ahead of him.

But what became of him? He was obviously also in the early stages of metabolic disease. I keep wondering whether the two might have been connected in some way, his fatty liver and his cancer. What did he look like, metabolically, ten years before he came in for surgery that day? As we'll see in chapter 8, obesity and metabolic dysfunction are both powerful risk factors for cancer. Could this man's underlying metabolic issues have been fueling his cancer somehow? What would have happened if his underlying issues,

previous chapter. When AMPK is activated, it triggers all sorts of cellular survival programs, including the burning of stored fat, that help enable the organism to survive without food. When fructose triggers AMPD, on the other hand, it sends us down the path of fat storage. (This cascade also triggers hunger by blocking the satiety hormone leptin.)

which his fatty liver made plain as day, had been recognized a decade or more earlier? Would we have ever even met?

It seems unlikely that Medicine 2.0 would have addressed his situation at all. The standard playbook, as we touched on in chapter 1, is to wait until someone's HbA1c rises above the magic threshold of 6.5 percent before diagnosing them with type 2 diabetes. But by then, as we've seen in this chapter, the person may already be in a state of elevated risk. To address this rampant epidemic of metabolic disorders, of which NAFLD is merely a harbinger, we need to get a handle on the situation much earlier.

One reason I find value in the concept of metabolic syndrome is that it helps us see these disorders as part of a continuum and not a single, binary condition. Its five relatively simple criteria are useful for predicting risk at the population level. But I still feel that reliance on it means waiting too long to declare that there is a problem. Why wait until someone has three of the five markers? Any one of them is generally a bad sign. A Medicine 3.0 approach would be to look for the warning signs years earlier. We want to intervene *before* a patient actually develops metabolic syndrome.

This means keeping watch for the earliest signs of trouble. In my patients, I monitor several biomarkers related to metabolism, keeping a watchful eye for things like elevated uric acid, elevated homocysteine, chronic inflammation, and even mildly elevated ALT liver enzymes. Lipoproteins, which we will discuss in detail in the next chapter, are also important, especially triglycerides; I watch the ratio of triglycerides to HDL cholesterol (it should be less than 2:1 or better yet, less than 1:1), as well as levels of VLDL, a lipoprotein that carries triglycerides—all of which may show up many years before a patient would meet the textbook definition of metabolic syndrome. These biomarkers help give us a clearer picture of a patient's overall metabolic health than HbA1c, which is not very specific by itself.

But the first thing I look for, the canary in the coal mine of metabolic disorder, is elevated insulin. As we've seen, the body's first response to incipient insulin resistance is to produce *more* insulin. Think back to our analogy with the balloon: as it gets harder to get air (glucose) into the balloon (the cell), we have to blow harder and harder (i.e., produce more insulin). At first, this appears to be successful: the body is still able to maintain glucose homeostasis, a steady blood glucose level. But insulin, especially postprandial insulin, is already on the rise.

One test that I like to give patients is the oral glucose tolerance test, or OGTT, where the patient swallows ten ounces of a sickly-sweet, almost undrinkable beverage called Glucola that contains seventy-five grams of pure glucose, or about twice as much sugar as in a regular Coca-Cola.* We then measure the patient's glucose *and* their insulin, every thirty minutes over the next two hours. Typically, their blood glucose levels will rise, followed by a peak in insulin, but then the glucose will steadily decrease as insulin does its job and removes it from circulation.

On the surface, this is fine: insulin has done its job and brought glucose under control. But the insulin in someone at the early stages of insulin resistance will rise very dramatically in the first thirty minutes and then remain elevated, or even rise further, over the next hour. This postprandial insulin spike is one of the biggest early warning signs that all is not well.

Gerald Reaven, who died in 2018 at the age of eighty-nine, would have agreed. He had to fight for decades for insulin resistance to be recognized as a primary cause of type 2 diabetes, an idea that is now well accepted. Yet diabetes is only one danger: Studies have found that insulin resistance itself is associated with huge increases in one's risk of cancer (up to twelvefold), Alzheimer's disease (fivefold), and death from cardiovascular disease (almost sixfold)—all of which underscores why addressing, and ideally preventing, metabolic dysfunction is a cornerstone of my approach to longevity.

It seems at least plausible that my patient with the fatty liver had developed elevated insulin at some point, well before his surgery. But it is also extremely unlikely that Medicine 2.0 would have even considered treating him, which boggles my mind. If any other hormone got out of balance like this,

---

* For comparison, a regular twelve-ounce Coca-Cola contains thirty-nine grams of high-fructose corn syrup, about half of which is glucose and half is fructose.

such as thyroid hormone or even cortisol, doctors would act swiftly to rectify the situation. The latter might be a symptom of Cushing's disease, while the former could be a possible sign of Graves' disease or some other form of hyperthyroidism. Both of these endocrine (read: *hormone*) conditions require and receive treatment as soon as they are diagnosed. To do nothing would constitute malpractice. But with hyperinsulinemia, for some reason, we wait and do nothing. Only when type 2 diabetes has been diagnosed do we take any serious action. This is like waiting until Graves' disease has caused *exophthalmos*, the signature bulging eyeballs in people with untreated hyperthyroidism, before stepping in with treatment.

It is beyond backwards that we do not treat hyperinsulinemia like a bona fide endocrine disorder of its own. I would argue that doing so might have a greater impact on human health and longevity than any other target of therapy. In the next three chapters, we will explore the three other major diseases of aging—cardiovascular disease, cancer, and neurodegenerative diseases—all of which are fueled in some way by metabolic dysfunction. It will hopefully become clear to you, as it is to me, that the logical first step in our quest to delay death is to get our metabolic house in order.

The good news is that we have tremendous agency over this. Changing how we exercise, what we eat, and how we sleep (see Part III) can completely turn the tables in our favor. The bad news is that these things require effort to escape the default modern environment that has conspired against our ancient (and formerly helpful) fat-storing genes, by overfeeding, undermoving, and undersleeping us all.

CHAPTER 7

The Ticker

Confronting—and Preventing—Heart Disease, the Deadliest Killer on the Planet

There is some risk involved in action, there always is.
But there is far more risk in failure to act.

—Harry S. Truman

One downside of my profession is that too much knowledge can become its own kind of curse. As I delved back into the practice of medicine, after my hiatus in the business world, it dawned on me that I already know how I am likely to die: I seem destined to die from heart disease.

I wonder what took me so long to get the hint. When I was five, my father's older brother Francis—his favorite of eight siblings—died of a sudden heart attack at forty-six. When my own brother Paul was born two days later, my grief-stricken father chose Francis as his middle name. Just as certain names run in families, a propensity for early cardiovascular disease seems to run in mine. Another uncle suffered a fatal heart attack at forty-two, while a third made it to age sixty-nine before his heart killed him, which is perhaps more typical but still way too young.

My dad is lucky, because he has lived to the ripe old age of eighty-five (so far). But even he has a stent in one of his coronary arteries, a souvenir of his own minor event in his midsixties. He felt chest pain one day at the limestone quarry where he worked, and ended up in the ER, where it was determined that he had had a recent infarction. The stent, a little sleeve of metal wire, was put in a year or so later. I'm not actually convinced that the stent did anything—he was not experiencing symptoms at the time it was placed—but perhaps it scared him enough to be more diligent with his medications and his diet.

So even though my cholesterol profile is excellent, and I eat sensibly, never smoke, have normal blood pressure, and rarely drink alcohol, I'm still at risk. I feel like I'm trapped in that Charlie Munger anecdote, about the guy who just wants to know *where* he's going to die, so he'll make sure never to go there. Unfortunately, too often heart disease finds you.

When I was in medical school, my first-year pathology professor liked to ask a trick question: What is the most common "presentation" (or symptom) of heart disease? It wasn't chest pain, left arm pain, or shortness of breath, the most common answers; it was sudden death. You know the patient has heart disease because he or she has just died from it. This is why, he claimed, the only doctors who truly understand cardiovascular disease are pathologists. His point: by the time a pathologist sees your arterial tissue, you are dead.

While mortality rates from those first, surprise heart attacks have dropped significantly, thanks to improvements in basic cardiac life support and time-sensitive interventions such as cardiac catheterization and clot-obliterating drugs that can halt a heart attack almost in its tracks, they are still fatal roughly one-third of the time, according to Ron Krauss, senior scientist and director of atherosclerosis research at Children's Hospital Oakland Research Institute.

Globally, heart disease and stroke (or cerebrovascular disease), which I lump together under the single heading of atherosclerotic cardiovascular disease, or ASCVD, represent the leading cause of death, killing an estimated

2,300 people every day in the United States, according to the CDC—more than any other cause, including cancer. It's not just men who are at risk: American women are up to ten times more likely to die from atherosclerotic disease than from breast cancer (not a typo: one in three versus one in thirty). But pink ribbons for breast cancer far outnumber the American Heart Association's red ribbons for awareness of heart disease among women.

My uncles' deaths remain a mystery to me. They lived in Egypt, and I have no idea what their blood work looked like or, more importantly, what kind of shape their coronary arteries were in. I'm pretty sure they smoked, but perhaps if they had had access to better medical care they too might have survived their heart attacks, as my father did. Or perhaps their fates were inescapable, tied to their genes. All I know is that forty-two seems really young to keel over from a heart attack.

I'd known about my uncles all my life, but the implication of their stories only really hit me in my midthirties, when I became a father for the first time. All of a sudden, the awareness of my own mortality crashed over my head like a rogue wave, coming out of nowhere on one of my long swims. This book probably wouldn't have been written if not for that family history.

Like most thirty-six-year-olds, Not-Thin Peter scarcely gave a thought to heart disease. Why should I have? My heart was strong enough to have propelled me across the twenty-one-mile-wide Catalina Channel, working steadily for more than fourteen hours, like a Mercedes diesel engine purring along smoothly inside my chest. I was in great shape, I thought. But I was nevertheless worried, on account of my family history. So I insisted that my doctor order a CT scan of my heart, and it wound up changing my whole outlook on life.

The scan was calibrated to detect calcification in my coronary arteries, a sign of advanced atherosclerosis. The results showed that I had a calcium "score" of 6. That sounds low, and in absolute terms it was; someone with severe disease could return a score well over 1,000. But for someone age thirty-six, it should have been zero. My score of 6 meant that I had more calcium in my coronary arteries than 75 to 90 percent of people my age. As I dug deeper into the pathology of this disease, I was dismayed to learn that it was already fairly late in the game. A calcium score is treated as a predictor of future risk, which it is, but it is also a measure of historical and existing damage. And I was already off the charts. I was only in my midthirties, but I had the arteries of a fifty-five-year-old.

I was upset by this revelation, although knowing what I know now, it's not at all surprising. At the time, I was overweight and borderline insulin resistant, two huge risk factors that by themselves help create an environment that fosters and accelerates the development of atherosclerotic lesions. Yet because my calcium score was "only" 6, and my all-important LDL ("bad") cholesterol was "normal," the medical advice I received was—wait for it—to do nothing. Sound familiar?

Doing nothing is not my style, as you may have gathered by now. I sensed that I was not on a good trajectory, and I needed to figure out how to change it. My curiosity launched me on a years-long quest to truly understand atherosclerosis. The story I uncovered, with the generous help of my mentors Tom Dayspring, Allan Sniderman, and Ron Krauss (among others), all of whom are world-renowned experts in cardiac pathology and/or the study of lipids, was flabbergasting.

While heart disease is the most prevalent age-related condition, it is also more easily prevented than either cancer or Alzheimer's disease. We know a lot about how and why it begins and the manner in which it progresses. While it can't exactly be cured or reversed the way type 2 diabetes (sometimes) can, it is relatively easy to delay if you're smart and you get on the case early. It's also the rare example of a chronic disease where Medicine 2.0 already does focus on prevention, to some extent. We have a cabinet full of blood pressure–and cholesterol–lowering medications that really do reduce the risk of death for many patients, and we have blood tests and imaging tests (like my calcium scan) that can at least give us a snapshot, however blurry, of someone's cardiovascular health. It's a start.

In spite of how well we understand atherosclerotic disease and its progression, and how many tools we have to prevent it, it *still* kills more people than cancer in the United States each year, many of them completely out of the blue. We're losing the war. I don't claim to have all the answers, but I think this is at least partly due to the fact that we still have some major blind spots in our understanding of what truly drives our risk for the disease, how it develops, and most of all when we need to act to counter its momentum.

The fundamental problem, I believe, is classic Medicine 2.0: guidelines for managing cardiovascular risk are based on an overly short time horizon, compared to the time line of the disease. We need to begin treating it, and preventing it, much earlier. If we could get it right, the potential payoff would be huge: the high prevalence of male centenarians on the island of Sardinia, for example, has largely been attributed to their ability to avoid or delay circulatory disease. Fewer Sardinian men die from heart disease between the ages of eighty and one hundred than anywhere else in Italy.

But we're nowhere near that. Heart disease remains our deadliest killer, the worst of the Horsemen. In the next few pages, I hope to convince you that it need not be—that with the right strategy, and attention to the correct risk factors *at the correct time,* it should be possible to eliminate much of the morbidity and mortality that is still associated with atherosclerotic cardiovascular and cerebrovascular disease.

Bluntly put: this should be the tenth leading cause of death, not the first.

Scientists have been exploring the medical mysteries of the human heart for almost as long as poets have been probing its metaphorical depths. It is a wondrous organ, a tireless muscle that pumps blood around the body every moment of our lives. It pounds hard when we are exercising, slows down when we sleep, and even microadjusts its rate between beats, a hugely important phenomenon called heart rate variability. And when it stops, we stop.

Our vascular network is equally miraculous, a web of veins, arteries, and capillaries that, if stretched out and laid end to end, would wrap around the earth more than twice (about sixty thousand miles, if you're keeping score). Each individual blood vessel is a marvel of material science and engineering, capable of expanding and contracting dozens of times per minute, allowing vital substances to pass through its membranes, and accommodating huge swings in fluid pressure, with minimal fatigue. No material created by man can even come close to matching this. If one vessel is injured, others regrow to take its place, ensuring continuous blood flow throughout the body.

Incredible as it is, however, our circulatory system is far from perfect—in fact, it is almost perfectly designed to generate atherosclerotic disease, just in the course of daily living. This is in large part because of another important function of our vasculature. In addition to transporting oxygen and nutrients to our tissues and carrying away waste, our blood traffics cholesterol molecules between cells.

It's practically a dirty word, *cholesterol.* Your doctor will probably utter it with a frown, because as everyone knows, cholesterol is evil stuff. Well, some of it is—you know, the LDL or "bad" cholesterol, which is inevitably counterpoised against the HDL, or "good" cholesterol. I practically need to be restrained when I hear these terms, because they're so meaningless. And your "total cholesterol," the first number that people offer up when we're talking about heart disease, is only slightly more relevant to your cardiovascular risk than the color of your eyes. So let's hit rewind and look at what cholesterol really is, what it does, and how it contributes to heart disease.

Cholesterol is essential to life. It is required to produce some of the most important structures in the body, including cell membranes; hormones such as testosterone, progesterone, estrogen, and cortisol; and bile acids, which are necessary for digesting food. All cells can synthesize their own cholesterol, but some 20 percent of our body's (large) supply is found in the liver, which acts as a sort of cholesterol repository, shipping it out to cells that need it and receiving it back via the circulation.

Because cholesterol belongs to the lipid family (that is, fats), it is not water soluble and thus cannot dissolve in our plasma like glucose or sodium and travel freely through our circulation. So it must be carted around in tiny spherical particles called *lipoproteins*—the final "L" in LDL and HDL—which act like little cargo submarines. As their name suggests, these lipoproteins are part lipid (inside) and part protein (outside); the protein is essentially the vessel that allows them to travel in our plasma while carrying their water-insoluble cargo of lipids, including cholesterol, triglycerides, and phospholipids, plus vitamins and other proteins that need to be distributed to our distant tissues.

The reason they're called high- and low-density lipoproteins (HDL and LDL, respectively) has to do with the amount of fat relative to protein that each one carries. LDLs carry more lipids, while HDLs carry more protein in relation to fat, and are therefore more dense. Also, these particles (and other lipoproteins) frequently exchange cargo with one another, which is part of what drives me crazy about labeling them "good" and "bad." When an HDL transfers its "good cholesterol" to an LDL particle, does that cholesterol suddenly become "bad"?

The answer is no—because it's not the cholesterol per se that causes problems but the nature of the particle in which it's transported. Each lipoprotein particle is enwrapped by one or more large molecules, called apolipoproteins, that provide structure, stability, and, most importantly solubility to the particle. HDL particles are wrapped in a type of molecule called apolipoprotein A (or apoA), while LDL is encased in apolipoprotein B (or apoB). This distinction may seem trivial, but it goes to the very root cause of atherosclerotic disease: every single lipoprotein that contributes to atherosclerosis—not only LDL but several others*—carries this apoB protein signature.

Another major misconception about heart disease is that it is somehow caused by the cholesterol that we eat in our diet. According to this dated and simplistic view, eating cholesterol-rich foods causes the so-called bad cholesterol to accumulate in our blood and then build up on our artery walls, as if

* There are also very-low-density lipoproteins, or VLDLs, which we mentioned in the previous chapter, as well as intermediate-density lipoproteins, or IDLs. These carry even more fat than the LDLs, much of it in the form of triglycerides, and they are also marked with apoB. Also: While HDL particles have multiple apoAs, each LDL (or VLDL, or IDL) has only one apoB particle, making it relatively easy to measure their concentration.

you poured bacon grease down the kitchen drain every time you made breakfast. Sooner or later, your sink will back up.

The humble egg, in particular, was singled out in a 1968 proclamation by the American Heart Association, accused of causing heart disease because of its high cholesterol content. It has remained in nutritional purgatory for decades, even after reams of research papers showing that dietary cholesterol (and particularly egg consumption) may not have much to do with heart disease at all. Eating lots of saturated fat *can* increase levels of atherosclerosis-causing lipoproteins in blood, but most of the actual cholesterol that we consume in our food ends up being excreted out our backsides. The vast majority of the cholesterol in our circulation is actually produced by our own cells. Nevertheless, US dietary guidelines warned Americans away from consuming foods high in cholesterol for decades, and nutrition labels still inform American consumers about how much cholesterol is contained in each serving of packaged foods.

Even Ancel Keys, the famed nutrition scientist who was one of the founding fathers of the notion that saturated fat causes heart disease, knew this was nonsense. The problem he recognized was that much of the basic research into cholesterol and atherosclerosis had been conducted in rabbits, which have a unique ability to absorb cholesterol into their blood from their food and form atherosclerotic plaques from it; the mistake was to assume that humans also absorb dietary cholesterol as readily. "There's no connection whatsoever between cholesterol in food and cholesterol in blood," Keys said in a 1997 interview. "None. And we've known that all along. Cholesterol in the diet doesn't matter at all unless you happen to be a chicken or a rabbit."

It took nearly two more decades before the advisory committee responsible for the US government dietary guidelines finally conceded (in 2015) that "cholesterol is not a nutrient of concern for overconsumption." Glad we settled that.

The final myth that we need to confront is the notion that cardiovascular disease primarily strikes "old" people and that therefore we don't need to worry much about prevention in patients who are in their twenties and thirties and forties. Not true. I'll never forget the one-question pop quiz that Allan

Sniderman dropped on me over dinner at Dulles Airport, back in 2014: "What proportion of heart attacks occur in people younger than age sixty-five?" I guessed high, one in four, but I was way low. Fully *half* of all major adverse cardiovascular events in men (and a third of those in women), such as heart attack, stroke, or any procedure involving a stent or a graft, occur before the age of sixty-five. In men, one-quarter of all events occur before age fifty-four.

But while the events themselves may have seemed sudden, the problem was likely lurking for years. Atherosclerosis is a slow-moving, sneaky disease, which is why I take such a hard line on it. Our risk of these "events" rises steeply in the second half of our lifespan, but some scientists believe the underlying processes are set into motion in late adolescence, even as early as our teens. The risk builds throughout our lives, and the critical factor is time. Therefore it is critical that we understand how it develops, and progresses, so we can develop a strategy to try to slow or stop it.

Back when I had an office, pre-COVID, I kept a clutter-free desk, but one book in particular was always there: *Atlas of Atherosclerosis Progression and Regression,* by Herbert C. Stary. It will never be a bestseller, but in the field of cardiovascular pathology it is legendary. It also happens to be a highly effective tool for communicating the seriousness of this disease to my patients, thanks to its lavish and gruesome photographs of arterial lesions as they form, develop, and rupture—all taken of the arteries of dead people, many of them in their twenties and thirties. The story it lays out, in graphic detail, is equal parts fascinating and terrifying. By the time I finished, my patients would often have this kind of harrowed expression on their faces, as if they'd just leafed through a coffee-table book documenting their own death.

This isn't a perfect analogy, but I think of atherosclerosis as kind of like the scene of a crime—breaking and entering, more or less. Let's say we have a street, which represents the blood vessel, and the street is lined with houses, representing the arterial wall. The fence in front of each house is analogous to something called the endothelium, a delicate but critical layer of tissue that lines all our arteries and veins, as well as certain other tissues, such as the kidneys. Composed of just a single layer of cells, the endothelium acts as a semipermeable barrier between the vessel lumen (i.e., the street, where the blood flows) and the arterial wall proper, controlling the passage of materials and nutrients and white blood cells into and out of the bloodstream. It also helps maintain our electrolyte and fluid balance; endothelial problems can lead to edema and swelling. Another very important job it does is to dilate and contract to allow increased or decreased blood flow, a process modulated by nitric oxide. Last, the endothelium regulates blood-clotting mechanisms, which can be important if you accidentally cut yourself. It's a pretty important little structure.

The street is very busy, with a constant flow of blood cells and lipoproteins and plasma and everything else that our circulation carries, all brushing past the endothelium. Inevitably, some of these cholesterol-bearing lipoprotein particles will penetrate the barrier, into an area called the subendothelial space—or in our analogy, the front porch. Normally, this is fine, like guests stopping by for a visit. They enter, and then they leave. This is what HDL particles generally do: particles tagged with apoA (HDL) can cross the endothelial barrier easily in both directions, in and out. LDL particles and other particles with the apoB protein are far more prone to getting stuck inside.

This is what actually makes HDL particles potentially "good" and LDL particles potentially "bad"—not the cholesterol, but the particles that carry it. The trouble starts when LDL particles stick in the arterial wall and subsequently become oxidized, meaning the cholesterol (and phospholipid) molecules they contain come into contact with a highly reactive molecule known as a reactive oxygen species, or ROS, the cause of oxidative stress. It's the oxidation of the lipids on the LDL that kicks off the entire atherosclerotic cascade.

Now that it is lodged in the subendothelial space and oxidized, rendering it somewhat toxic, the LDL/apoB particle stops behaving like a polite guest, refusing to leave—and inviting its friends, other LDLs, to join the party. Many of these also are retained and oxidized. It is not an accident that the two biggest risk factors for heart disease, smoking and high blood pressure, cause damage to the endothelium. Smoking damages it chemically, while high blood pressure does so mechanically, but the end result is endothelial harm that, in turn, leads to greater retention of LDL. As oxidized LDL accumulates, it causes still more damage to the endothelium.

I've been saying LDL, but the key factor here is actually *exposure* to apoB-tagged particles, over time. The more of these particles that you have in your circulation, not only LDL but VLDL and some others, the greater the risk that some of them will penetrate the endothelium and get stuck. Going back to our street analogy, imagine that we have, say, one ton of cholesterol moving down the street, divided among four pickup trucks. The chance of an accident is fairly small. But if that same total amount of cholesterol is being carried on five hundred of those little rental scooters that swarm around cities like Austin, where I live, we are going to have absolute mayhem* on our hands. So to gauge the true extent of your risk, we *have* to know how many of these apoB particles are circulating in your bloodstream. That number is much more relevant than the total quantity of cholesterol that these particles are carrying.

If you take a healthy coronary artery and expose it to high enough concentrations of apoB particles, over a long enough time, a certain amount of LDL (and VLDL) will get stuck in that subendothelial space and become oxidized, which then leads to it sticking together in clumps or aggregates. In response to this incursion, the endothelium dials up the biochemical equivalent of 911, summoning specialized immune cells called monocytes to the scene to confront the intruders. Monocytes are large white blood cells that enter the subendothelial space and transform into macrophages, larger and hungrier immune cells that are sometimes compared to Pac-Man. The macrophage, whose name means "big eater," swallows up the aggregated or oxidized LDL, trying to remove it from the artery wall. But if it consumes too much cholesterol, then it blows up into a foam cell, a term that you may have heard—so named because under a microscope it looks foamy or soapy. When enough foam cells gather together, they form a "fatty streak"—literally a streak of fat that you can see with your naked eye during an autopsy of a splayed-open coronary artery.

---

* The scientific term for this is *stochastic*, meaning it's a largely random process.

The fatty streak is a precursor of an atherosclerotic plaque, and if you are reading this and are older than fifteen or so, there is a good chance you already have some of these lurking in your arteries. Yes, I said "fifteen" and not "fifty"—this is a lifelong process and it starts very early. Autopsy data from young people who died from accidents, homicides, or other noncardiovascular causes have revealed that as many as a third of sixteen- to twenty-year-olds *already* had actual atherosclerotic lesions or plaques in their coronary arteries when they died. As teenagers.

Figure 5. **Atherosclerotic Disease in a 23-Year-Old**

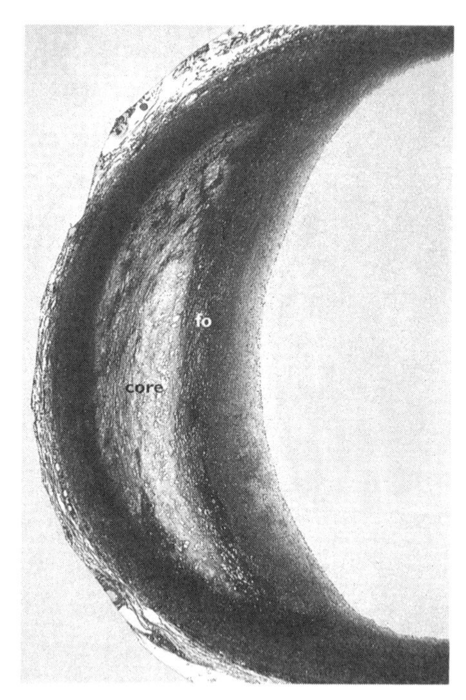

This is a cross-sectional view of the proximal left anterior descending artery, one of the key vessels supplying blood to the heart, from a twenty-three-year-old male homicide victim. Note that he already has extensive atherosclerotic damage in the wall of this artery: a significant core ("core") of accumulated lipids, and macrophages and foam cells ("fo") in the subendothelial space, beginning to encroach on the lumen, the passage where blood flows. He would likely not suffer a heart attack anytime soon, but this is very advanced disease nonetheless.

Source: Stary, (2003).

It's not as if they were about to have heart attacks. The atherosclerotic process moves very slowly. This may be partly because of the action of HDLs. If an HDL particle arrives at our crime scene, with the foam cells and fatty streaks, it can suck the cholesterol back *out* of the macrophages in a process called delipidation. It then slips back across the endothelial layer and into the bloodstream, to deliver the excess cholesterol back to the liver and other tissues (including fat cells and hormone-producing glands) for reuse.

Its role in this process of "cholesterol efflux" is one reason why HDL is considered "good," but it does more than that. Newer research suggests that HDL has multiple other atheroprotective functions that include helping maintain the integrity of the endothelium, lowering inflammation, and neutralizing or stopping the oxidation of LDL, like a kind of arterial antioxidant.

The role of HDL is far less well understood than that of LDL. The cholesterol content in your LDL particles, your "bad" cholesterol number (technically expressed as LDL-C),* is actually a decent if imperfect proxy for its biologic impact; lots of studies have shown a strong correlation between LDL-C and event risk. But the all-important "good cholesterol" number on your blood test, your HDL-C, doesn't actually tell me very much if anything about your overall risk profile. Risk does seem to decline as HDL-C rises to around the 80th percentile. But simply raising HDL cholesterol concentrations by brute force, with specialized drugs, has not been shown to reduce cardiovascular risk at all. The key seems to be to increase the *functionality* of the particles—but as yet we have no way to do that (or measure it).

HDL may or may not explain why centenarians develop heart disease two decades later than average, if at all; remember, three of the most prominent "longevity genes" discovered to date are involved with cholesterol transport and processing (APOE and two others, CETP and APOC3). And it's not just centenarians: I have patients walking around whose lipoprotein panels read like a death sentence, with sky-high LDL-C and apoB, but by every single

---

* A brief word about nomenclature: When we say LDL or HDL, we are typically referring to a type of *particle;* when we say LDL-C or HDL-C, we are talking about a laboratory measurement of the *concentration* of cholesterol within those particles.

measure that we have—calcium score, CT angiogram, you name it—they show no sign of disease. Yet as of now, we cannot give a satisfactory explanation for why. I feel strongly that if we are to make any further progress attacking cardiovascular disease with drugs, we must start by better understanding HDL and hopefully figuring out how to enhance its function.

But I digress. Back at the crime scene, the ever-growing number of foam cells begin to sort of ooze together into a mass of lipids, like the liquefying contents of a pile of trash bags that have been dumped on the front lawn. This is what becomes the core of our atherosclerotic plaque. And this is the point where breaking and entering tilts over into full-scale looting. In an attempt to control the damage, the "smooth muscle" cells in the artery wall then migrate to this toxic waste site and secrete a kind of matrix in an attempt to build a kind of barrier around it, like a scar. This matrix ends up as the fibrous cap on your brand-new arterial plaque.

More bad news: None of what's gone on so far is easily detectable in the various tests we typically use to assess cardiovascular risk in patients. We might expect to see evidence of inflammation, such as elevated levels of C-reactive protein, a popular (but poor) proxy of arterial inflammation. But it's still mostly flying below our medical radar. If you look at the coronary arteries with a CT scan at this very early stage, you will likely miss this if you're looking only for calcium buildup. (You have a better chance of spotting this level of damage if using a more advanced type of CT scan, called a CT angiogram, which I much prefer to a garden-variety calcium scan* because it can also identify the noncalcified or "soft" plaque that precedes calcification.)

As this maladaptive repair or remodeling process continues, the plaque

---

* While the CT angiogram costs a bit more, requires IV dye, and exposes the patient to slightly more radiation, I struggle to find credible arguments against its use. Approximately 15 percent of people who have a normal calcium score (0) are still found to have soft plaque or even small calcifications on CT angiograms, and as many as 2 to 3 percent of people with a zero calcium score are found on CT angiogram to have high-risk plaques. For this reason, I almost always prefer my patients to have a CT angiogram over a calcium scan if we opt to look for evidence of disease via imaging studies.

will continue to grow. At first, this expansion is directed toward the outer arterial wall, but as it continues it may encroach on the lumen, the passage through which blood flows—in our analogy, blocking traffic in the street itself. This luminal narrowing, known as *stenosis*, can also be seen in an angiogram.

At a certain point in this process, the plaque may start to become calcified. This is what (finally) shows up on a regular calcium scan. Calcification is merely another way in which the body is trying to repair the damage, by stabilizing the plaque to protect the all-important arteries. But it's like pouring concrete on the Chernobyl reactor: you're glad it's there, but you know there's been an awful lot of damage in the area to warrant such an intervention. A positive calcium score is really telling us that there are almost certainly other plaques around that may or may not be stabilized (calcified).

If the plaque does become unstable, eroding or even rupturing, you've really got problems. The damaged plaque may ultimately cause the formation of a clot, which can narrow and ultimately block the lumen of the blood vessel—or worse, break free and cause a heart attack or stroke. This is why we worry more about the noncalcified plaques than the calcified ones.

Normally, however, most atherosclerotic plaques are fairly undramatic. They grow silently and invisibly, gradually occluding the blood vessel until one day the obstruction, due to the plaque itself or a plaque-induced clot, becomes a problem. For example, a sedentary person may not notice that she has a partially blocked coronary artery until she goes outside to shovel snow. The sudden demands on her circulatory system can trigger ischemia (decreased blood delivery of oxygen) or infarction (tissue death from no blood flow)—or, in layman's terms, a heart attack or a stroke.

It may seem sudden, but the danger was lurking all along.

When I finally recognized my own cardiovascular risk, back in my thirties, I had no idea about how any of this complex disease process worked. Looking back, it's clear that I had quite a few of the major and lesser risk factors already checked off. I didn't smoke, which is perhaps the most potent environmental risk driver, and my blood pressure was normal, but I had other problems. And as my calcium score revealed, I already had a small, calcified plaque in the upper part of my left anterior descending (LAD) artery, one of the main arteries supplying my heart. There may have been other bad things happening there as well, but because I did not have a CT angiogram at this time, I had no sense of what kind of damage existed elsewhere in my coronary arteries. Anything shy of calcification is *not* identified by the calcium score.

Clearly, Not-Thin Peter was already on the road to heart disease. My waist size was on track to hit 40 by the time I turned forty, a clear sign of my metabolic dysfunction. Underneath my belt, I was likely accumulating visceral fat. I was also insulin resistant, an enormous risk driver for cardiovascular disease. Though my blood pressure was fine, I suspect it would have deteriorated fairly rapidly as I aged, as hypertension seems rampant in my family. I probably also had high levels of uric acid, which as we saw in the previous chapter is often found in the company of high blood pressure and other signs of metabolic dysfunction. All of these contribute to another necessary (but not sufficient) condition that is required for atherosclerosis to develop, and that is inflammation. The endothelial barrier, in particular, is uniquely vulnerable to damage from inflammation.

But no physician would likely have treated me for any of this. My blood panel did not signal any significant risk. My LDL-C tested at 110 to 120 mg/dL, just slightly higher than normal but not a cause for concern, particularly in a younger person. My triglycerides were on the higher side, slightly north of 150 mg/dL, but that did not set off any alarm bells either. I now know that these numbers almost certainly indicated a high concentration of atherogenic apoB particles—but no one bothered to test my apoB number, either.

Back then, nearly fifteen years ago, the apoB test (simply, measuring the concentration of apoB-tagged particles) was not commonly done. Since then, evidence has piled up pointing to apoB as far more predictive of cardiovascular disease than simply LDL-C, the standard "bad cholesterol" measure. According to an analysis published in *JAMA Cardiology* in 2021, each standard-deviation increase in apoB raises the risk of myocardial infarction by 38 percent in patients without a history of cardiac events or a diagnosis of cardiovascular disease (i.e., primary prevention). That's a powerful correlation. Yet even now, the American Heart Association guidelines still favor LDL-C testing instead of apoB. I have all my patients tested for apoB regularly, and you should ask for the same test the next time you see your doctor. (Don't be waved off by nonsensical arguments about "cost": It's about twenty to thirty dollars.)

I was still in my thirties, yet I likely already had all three of the major prerequisites for heart disease: significant lipoprotein burden or apoB, LDL oxidation or modification (leading to the plaques that my calcium scan revealed), and a high level of background inflammation. None of these is enough to *guarantee* that someone will develop heart disease, but all three are *necessary* to develop it. We are fortunate that many of these conditions can be modulated or nearly eliminated—including apoB, by the way—via lifestyle changes and medications. As we'll discuss in the final section, I take a very hard line on lowering apoB, the particle that causes all this trouble. (In short: get it as low as possible, as early as possible.)

But before we go there, I want to talk about another deadly but relatively little-known lipoprotein that is likely responsible for graveyards full of sudden cardiac arrest victims, people whose conventional cholesterol panels and risk factor profiles otherwise looked fine. I don't have this particular issue, thankfully, but a very good friend of mine does, and finding out about it in a timely manner likely saved his life.

I met Anahad O'Connor in 2012 on a junket to France, courtesy of the French-American Foundation and an award we had both won, and we immediately bonded. I think it was because we were the only two guys on the trip who skipped the *pain au chocolat* and spent our spare time in the gym. Also, he wrote about health and science for *The New York Times,* so we had plenty to talk about.

Because I am a cholesterol nerd, I badgered Anahad into doing a comprehensive lipoprotein panel when we got back to the United States. He looked at me funny—Why should he do that? He was only in his early thirties, an extremely fit vegetarian with maybe 6 or 7 percent body fat. He should have been fine in the lipid department. But you never know; his father had died of an aneurysm, which may have been a sign of circulatory problems.

As expected, his standard lipid numbers looked great. There was only one thing that seemed off, so I suggested that he should probably also get a calcium scan, as I had done, so we could get a better sense of the state of his arteries. That's where things got interesting. Recall, my calcium score had come back at 6, placing me at higher risk than 75 to 90 percent of people my age. Anahad's calcium score was 125—off the charts for someone so young and otherwise healthy. "Can this be real?" he said.

It was. It turned out that the culprit was a little-known but very deadly type of particle called Lp(a) (pronounced "el-pee-little-A"). This hot mess of a lipoprotein is formed when a garden-variety LDL particle is fused with another, rarer type of protein called apolipoprotein(a), or apo(a) for short (not to be confused with apolipoprotein A or apoA, the protein that marks HDL particles). The apo(a) wraps loosely around the LDL particle, with multiple looping amino acid segments called "kringles," so named because their structure resembles the ring-shaped Danish pastry by that name. The kringles are what make Lp(a) so dangerous: as the LDL particle passes through the bloodstream, they scoop up bits of oxidized lipid molecules and carry them along.

As my lipid guru Tom Dayspring points out, this isn't entirely bad. There is some evidence that Lp(a) may act as a sort of cleansing agent, like a street-sweeping truck that gathers up unpleasant and potentially harmful lipid junk and delivers it to the liver. But because Lp(a) is a member of the apoB particle family, it also has the potential to penetrate the endothelium and get lodged in an artery wall; because of its structure, Lp(a) may be even more likely than a normal LDL particle to get stuck, with its extra cargo of lipids gone bad. Even worse, once in there, it acts partly as a thrombotic or proclotting factor, which helps to speed the formation of arterial plaques.

Often, the way Lp(a) announces itself is via a sudden, seemingly premature heart attack. This is what happened to *Biggest Loser* host Bob Harper, who suffered cardiac arrest at a gym in New York in 2017, at age fifty-two.

Harper's life was saved by a bystander who performed CPR until the paramedics arrived. He woke up in the hospital two days later, wondering what had hit him. Turns out, his very high level of Lp(a) was what had hit him. He had no idea that he was at risk.

This is not an atypical scenario: when a patient comes to me and says their father or grandfather or aunt, or all three, died of "premature" heart disease, elevated Lp(a) is the first thing I look for. It is the most prevalent hereditary risk factor for heart disease, and its danger is amplified by the fact that it is still largely flying under the radar of Medicine 2.0, although that is beginning to change.

Most people have relatively small concentrations of this particle, but some individuals can have as much as one hundred times more than others. The variation is largely genetic, and an estimated 20 to 30 percent of the US population has levels high enough that they are at increased risk; also, people of African descent tend to have higher levels of Lp(a), on average, than Caucasians. This is why, if you have a history of premature heart attacks in your family, you should definitely ask for an Lp(a) test. We test every single patient for Lp(a) during their first blood draw. Because elevated Lp(a) is largely genetic, the test need only be done once (and cardiovascular disease guidelines are beginning to advise a once-a-lifetime test for it anyway).

Anahad was fortunate that he found out about his situation when he did. His calcium score meant that he had already suffered significant atherosclerotic damage due to his Lp(a). Beyond the harm it causes to coronary arteries, Lp(a) is particularly destructive to the aortic valve, one of the more important structures in the heart, by promoting the formation of tiny, bony particles in the valve leaflets, which leads to stenosis or narrowing of the aortic outlet.

There was no quick fix for Anahad, or anyone else with elevated Lp(a). It does not seem to respond to behavioral interventions such as exercise and dietary changes the way that, say, LDL-C does. A class of drug called PCSK9 inhibitors, aimed at lowering apoB concentrations, does seem to be able to reduce Lp(a) levels by approximately 30 percent, but as yet there are no data suggesting that they reduce the excess events (heart attacks) attributable to that particle. Thus, the only real treatment for elevated Lp(a) right now is aggressive management of apoB overall. Though we can't reduce Lp(a) directly, beyond what a PCSK9 inhibitor can do, we can lower the remaining apoB concentration sufficiently that we can reduce a patient's overall risk.* Because Anahad is relatively young, he also has more time to address his other risk factors.

Luckily, we found the trouble before the trouble found him.

How to Reduce Cardiovascular Risk

In a way, Not-Thin Peter and Anahad O'Connor were like two sides of the same coin. While our stories don't seem to have much in common, both underscore the insidious, almost sneaky nature of heart disease: my risk should have been obvious, based on my family history, while Anahad's disease remained all but invisible until he happened to have a calcium scan, which is not normally done on healthy-seeming people in their thirties. We only learned of our risk thanks to dumb luck, because few doctors would have considered screening either of us for heart disease at our ages.

Together, our stories illustrate three blind spots of Medicine 2.0 when it comes to dealing with atherosclerotic disease: first, an overly simplistic view of lipids that fails to understand the importance of total lipoprotein burden (apoB) and how much one needs to reduce it in order to truly reduce risk; second, a general lack of knowledge about other bad actors such as Lp(a); and third, a failure to fully grasp the lengthy time course of atherosclerotic disease, and the implications this carries if we seek true prevention.

When I look at a patient's blood panel for the first time, my eyes immediately dart to two numbers: apoB and Lp(a). I look at the other numbers, too, but these two tell me the most when it comes to predicting their risk of ASCVD. ApoB not only tells me the concentration of LDL particles (which, you'll recall, is more predictive of disease than the concentration of choles-

* There is a new class of drug called antisense oligonucleotides, or ASOs, which are currently in clinical trials to virtually eliminate Lp(a) in circulation. So far, these trials look promising in that they dramatically reduce Lp(a) concentration, but it's too soon to know if they are effective at doing what matters most: reducing cardiovascular events.

terol found *within* LDL particles, LDL-C), but it also captures the concentration of VLDL particles, which as members of the apoB family can also contribute to atherosclerosis. Furthermore, even someone whose apoB is low can still have a dangerously elevated Lp(a).*

Once you establish the central importance of apoB, the next question becomes, By how much does one need to lower it (or its proxy LDL-C) to achieve meaningful risk reduction? The various treatment guidelines specify target ranges for LDL-C, typically 100 mg/dL for patients at normal risk, or 70 mg/dL for high-risk individuals. In my view, this is still far too high. Simply put, I think you can't lower apoB and LDL-C too much, provided there are no side effects from treatment. You want it as low as possible.

As Peter Libby, one of the leading authorities on cardiovascular disease, and colleagues wrote in *Nature Reviews* in 2019, "Atherosclerosis *probably would not occur* [emphasis mine] in the absence of LDL-C concentrations in excess of physiological needs (on the order of 10 to 20 mg/dL)." Furthermore, the authors wrote: "If the entire population maintained LDL concentrations akin to those of a neonate (or to those of adults of most other animal species), atherosclerosis might well be an orphan disease."

Translation: if we all maintained the apoB levels we had when we were babies, there wouldn't be enough heart disease on the planet for people to know what it was. Kind of like 3-hydroxyisobutyric aciduria. What, you haven't heard of it? Well, that's because there have been only thirteen reported cases. Ever. That is an orphan disease. I'm being a bit facetious, but my point is that atherosclerotic disease shouldn't even be in the top ten causes of death, if we treated it more aggressively. Instead, we have over eighteen million cases of fatal atherosclerotic disease per year globally.

Many doctors, and in fact many of you reading this, might be shocked to see such a low LDL-C target: 10 to 20 mg/dL? Most guidelines consider lowering LDL-C to 70 mg/dL to be "aggressive," even for secondary prevention in

---

* This is because the total number of LDL particles is far higher than the number of Lp(a) particles, but Lp(a) still has an outsize ability to cause damage, even in relatively small numbers.

high-risk patients, such as those who have already had a heart attack. It's also natural to ask whether such extremely low levels of LDL-C and apoB are safe, given the ubiquity and importance of cholesterol in the human body. But consider the following: infants, who presumably require the *most* cholesterol, in order to meet the enormous demands of their rapidly growing central nervous system, have similarly low levels of circulating cholesterol, without any developmental impairment. Why? Because the total amount of cholesterol contained in all our lipoproteins—not just LDL, but also HDL and VLDL—represents only about 10 to 15 percent of our body's total pool of cholesterol. So the concern is unwarranted, as demonstrated by scores of studies showing no ill effects from extremely low LDL concentrations.

This is my starting point with any patient, whether they are like Anahad (with one prominent risk factor) or like me (lots of smaller risk factors). Our first order of business is to reduce the burden of apoB particles, primarily LDLs but also VLDLs, which can be dangerous in their own right. And do so dramatically, not marginally or incrementally. We want it as low as possible, sooner rather than later. We must also pay attention to other markers of risk, notably those associated with metabolic health, such as insulin, visceral fat, and homocysteine, an amino acid that in high concentrations* is strongly associated with increased risk of heart attack, stroke, and dementia.

You'll note that I don't pay much attention to HDL-C, because while having very low HDL-C is *associated* with higher risk, it does not appear to be *causal*. This is why drugs aimed at raising HDL-C have generally failed to reduce risk and events in clinical trials. The reasons why are suggested by two elegant Mendelian randomization studies that examined both sides of the HDL-C question: Does low HDL-C *causally* increase the risk of myocardial infarction? No. Does raising HDL-C *causally* lower the risk of myocardial infarction? No.

Why? Probably because whatever benefit HDLs provide in the battle for arterial supremacy, it (again) seems to be driven by their *function*—which

---

* Homocysteine is broken down by B vitamins, which is why deficiency in B vitamins or genetic mutations in enzymes involved in their metabolism *(e.g., MTHFR)* can raise homocysteine.

doesn't seem to be related to their cholesterol content. But we cannot test for HDL functionality, and until we have a better grasp on how HDL actually works, it will likely remain elusive as a target of therapy.

Lipoproteins aren't the only significant risk factors for cardiovascular disease; as noted earlier, smoking and high blood pressure both damage the endothelium directly. Smoking cessation and blood pressure control are thus non-negotiable first steps in reducing cardiovascular risk.

We'll talk about nutrition in much more detail, but my first step in controlling my own cardiovascular risk was to begin to change my own diet, so as to lower my triglycerides (a contributor to apoB when they are high, as mine were), but more importantly to manage my insulin levels. I needed to get my metabolic house in order. I should note that my own solution at the time, a ketogenic diet, might not work for everyone, nor is it a diet to which I continue to adhere. In my clinical experience, about a third to half of people who consume high amounts of saturated fats (which sometimes goes hand in hand with a ketogenic diet) will experience a dramatic *increase* in apoB particles, which we obviously don't want.* Monounsaturated fats, found in high quantities in extra virgin olive oil, macadamia nuts, and avocados (among other foods), do not have this effect, so I tend to push my patients to consume more of these, up to about 60 percent of total fat intake. The point is not necessarily to limit fat overall but to shift to fats that promote a better lipid profile.

But for many patients, if not for most, lowering apoB to the levels we aim for—the physiologic levels found in children—cannot be accomplished with diet alone, so we need to use nutritional interventions in tandem with drugs. Here we are fortunate because we have more preventive options in our armamentarium than we do for cancer or neurodegenerative disease. Statins are

---

* There are at least two reasons for this. First, it seems saturated fat contributes directly to the synthesis of excess cholesterol. Second, and more importantly, excess saturated fat causes the liver to reduce expression of LDL receptors, thereby reducing the amount of LDL removed from circulation.

far and away the most prescribed class of drugs for lipid management, but there are several other options that might be right for a given individual, and often we need to combine classes of drugs, so it's not uncommon for a patient to take two lipid-lowering drugs that operate via distinct mechanisms. These are typically thought of as "cholesterol-lowering" medications, but I think we are better served to think about them in terms of increasing apoB *clearance,* enhancing the body's ability to get apoBs out of circulation. That's really our goal. Mostly this is done by amplifying the activity of LDL receptors (LDLR) in the liver, which absorb cholesterol from the bloodstream.

Different drugs arrive at this effect via different paths. Typically our first line of defense (or attack), statins inhibit cholesterol synthesis, prompting the liver to increase the expression of LDLR, taking more LDL out of circulation. They may have other benefits too, including an apparent anti-inflammatory effect, so while I don't think statins should be dissolved into the drinking water, as some have suggested, I do think they are very helpful drugs for reducing apoB or LDL concentration in many patients. Not everyone can take statins comfortably; about 5 percent of patients experience deal-breaking side effects, most notably statin-related muscle pain. Also, a smaller but nonzero subset of patients taking statins experience a disruption in glucose homeostasis, which may explain why statins are associated with a small increase in the risk for type 2 diabetes. Another fraction of patients experience an asymptomatic rise in liver enzymes, which is even more common in patients also taking the drug ezetimibe. All these side effects are completely and rapidly reversible when the drug is discontinued. But for those who can tolerate them (i.e., most people), I deploy them early and often. (For more on specific statin and other apoB-lowering medications, see the sidebar on pages 137–138.)

This brings us to the final, and perhaps greatest, major blind spot of Medicine 2.0: time.

The process I've outlined in this chapter unfolds very slowly—not over two or three or even five years, but over many decades. The fact that younger people have been found to have lesions and plaques, without suffering many events, tells us that there is a considerable period of time when the disease is not harmful. Dying from cardiovascular disease is certainly not inevitable: the centenarians delay it for decades, and many avoid it altogether, their arteries remaining as clean as those of people a generation younger. Somehow, they manage to slow the process down.

Nearly all adults are coping with some degree of vascular damage, no matter how young and vital they may seem, or how pristine their arteries appear on scans. There is always damage, especially in regions of shear stress and elevated local blood pressure, such as curves and splits in the vasculature. Atherosclerosis is with us, in some form, throughout our life course. Yet most doctors consider it "overtreatment" to intervene if a patient's computed ten-year risk of a major adverse cardiac event (e.g., heart attack or stroke) is below 5 percent, arguing that the benefits are not greater than the risks, or that treatment costs too much. In my opinion, this betrays a broader ignorance about the inexorable, long-term unfolding of heart disease. Ten years is far too short a time horizon. If we want to reduce deaths from cardiovascular disease, we need to begin thinking about prevention in people in their forties and even thirties.

Another way to think of all this is that someone might be considered "low risk" at a given point—but on what time horizon? The standard is ten years. But what if our time horizon is "the rest of your life"?

Then nobody is at low risk.

When I had my first calcium scan in 2009, at the age of thirty-six, my ten-year risk was incalculably low—literally. The dominant mathematical models for risk assessment have a lower limit for age of forty or forty-five. My parameters could not even be entered into the models. So it's no wonder nobody was alarmed by my findings. Despite my calcium score of 6, my ten-year risk of a heart attack was far less than 5 percent.

In 2016, seven years after my initial calcium scan, I had a CT angiogram (the better, higher-res scan), which showed the same small speck of calcium but no evidence of additional soft plaque elsewhere. In 2022, I went back again for a repeat CT angiogram, and the result was the same. There was no indication whatsoever of soft plaque this time either, and only that tiny speck of calcium remained from 2009.* Thus, at least at the resolution of the sharpest CT scanner commercially available, there is no reason to believe that my atherosclerosis has progressed over thirteen years.

I have no idea if this means I'm free from risk—I frankly doubt it—but I no longer fear dying from cardiovascular disease the way I once did. My long, comprehensive program of prevention seems to have paid off. I feel a lot better now, at age fifty, than I did at age thirty-six, and my risk is a lot lower by any metric other than age. One major reason for this is that I started early, well before Medicine 2.0 would have suggested *any* intervention.

Yet most physicians and cardiology experts would still insist that one's thirties are too young to begin to focus on primary prevention of cardiac disease. This viewpoint is directly challenged by a 2018 *JAMA Cardiology* paper coauthored by Allan Sniderman, comparing ten-year versus thirty-year risk horizons in terms of prevention. Sniderman and colleagues' analysis found that looking at a thirty-year time frame rather than the standard ten years and taking aggressive precautionary measures *early*—like beginning statin treatment earlier in certain patients—could prevent hundreds of thousands more cardiac events, and by implication could save many lives.

For context, most studies of statins used in primary prevention (that is, prevention of a first cardiac event) last about five years and typically find a "number needed to treat" (or NNT, the number of patients who need to take a drug in order for it to save one life) of between about 33 and 130, depending on the baseline risk profile of the patients. (Amazingly, the *longest* statin trials to date have lasted just seven years.) But looking at their risk reduction potential over a thirty-year time frame, as the Sniderman study did, reduces the NNT down to less than 7: For every seven people who are put on a statin at this early stage, we could potentially save one life. The reason for this is sim-

* The only difference was that the 2016 test gave me a calcium score of 0; in 2022 my calcium score was 2, and the original CT scan scored this same tiny plaque as a 6. This reinforces my belief that while calcium scores are useful, they are by no means sufficient on their own.

ple math: risk is proportional to apoB exposure over time. The sooner we lower apoB exposure, thus lowering risk, the more the benefits compound over time—and the greater our overall risk reduction.

This encapsulates the fundamental difference between Medicine 2.0 and Medicine 3.0 when it comes to cardiovascular disease. The former views prevention largely as a matter of managing relatively short-term risk. Medicine 3.0 takes a much longer view—and more importantly seeks to identify and eliminate the *primary causative agent* in the disease process: apoB. This changes our approach to treatment completely. For example, a forty-five-year-old with elevated apoB has a lower ten-year risk than a seventy-five-year-old with low apoB. Medicine 2.0 would say to treat the seventy-five-year-old (because of their age), but not the forty-five-year-old. Medicine 3.0 says to disregard the ten-year risk and instead treat the causal agent in *both* cases—lowering the forty-five-year-old's apoB as much as possible.

Once you understand that apoB particles—LDL, VLDL, Lp(a)—are *causally* linked to ASCVD, the game completely changes. The only way to stop the disease is to remove the cause, and the best time to do that is now.

Still struggling with this idea? Consider the following example. We know that smoking is causally linked to lung cancer. Should we tell someone to stop smoking only after their ten-year risk of lung cancer reaches a certain threshold? That is, do we think it's okay for people to keep smoking until they are sixty-five and then quit? Or should we do everything we can to help young people, who have maybe just picked up the habit, quit altogether?

When viewed this way, the answer is unambiguous. The sooner you cut the head off the snake, the lower the risk that it will bite you.

Brief Overview of Lipid-Lowering Medications

While there are seven statins on the market, I tend to start with **rosuvastatin (Crestor)** and only pivot from that if there is some negative effect from the drug (e.g., a symptom or biomarker). My goal is aggressive: as rationalized by Peter Libby, I want to knock someone's apoB concentration down to 20 or 30 mg/dL, about where it would be for a child.

For people who can't tolerate statins, I like to use a newer drug, called **bempedoic acid (Nexletol)**, which manipulates a different pathway to accomplish much the same end: inhibiting cholesterol synthesis as a way to force the liver to increase LDLR and therefore LDL clearance. But where statins inhibit cholesterol synthesis throughout the body, and most notably in the muscles, bempedoic acid does so only in the liver. Therefore, it does not cause the side effects associated with statins, especially muscle soreness. The main issue with this drug is cost.

Another drug called **ezetimibe (Zetia)** blocks absorption of cholesterol in the GI tract.* That in turn depletes the amount of cholesterol in the liver, leading once again to increased LDLR expression and greater clearance of apoB particles, which is what we want. Ezetimibe pairs very well with statins because statins, which block cholesterol synthesis, tend to cause the body to reflexively increase cholesterol reabsorption in the gut—exactly the thing that ezetimibe so effectively prevents.

LDL receptors can be upregulated by a class of drugs that we mentioned earlier, called **PCSK9 inhibitors**, which attack a protein called PCSK9 that degrades LDL receptors. This increases the receptors' half-life, thus improving the liver's ability to clear apoB. As a monotherapy they have about the same apoB- or LDL-C-lowering potency as high-dose statins, but their most common use is in addition to statins; the combination of statins plus PCSK9 inhibitors is the most powerful pharmacological tool that we have against apoB. Alas, statins do not reduce Lp(a), but PCSK9 inhibitors do in most patients, typically to the tune of about 30 percent.

---

* Not the cholesterol you eat, which is not being absorbed anyway, but the cholesterol you make and recycle via your liver and biliary system.

Triglycerides also contribute to the apoB particle burden, because they are largely transported in VLDLs. Our dietary interventions are aimed at reducing triglycerides, but in cases where nutritional changes are insufficient, and in cases where genetics render dietary interventions useless, **fibrates** are the drug of choice.

Ethyl eicosapentaenoic acid (Vascepa), a drug derived from fish oil and consisting of four grams of pharmaceutical-grade eicosapentaenoic acid (EPA), also has FDA approval to reduce LDL in patients with elevated triglycerides.

Chapter 7 Summary:
• Most of the cholesterol our bodies have is made in house
– Most of the cholesterol we eat is excreted

CHAPTER 8

The Runaway Cell

New Ways to Address the Killer That Is Cancer

You may have to fight
a battle more than once to win it.

—Margaret Thatcher

Steve Rosenberg was still a young resident when he encountered the patient who would determine the course of his career—and, possibly, of cancer treatment in general. He was on a rotation at a VA hospital in Massachusetts in 1968 when a man in his sixties came in needing a relatively simple gallbladder operation. The man, whose name was James DeAngelo, already had a large scar across his abdomen, which he said was from a long-ago operation to remove a stomach tumor. He had also had metastatic tumors that had spread to his liver, he added, but the surgeons had not touched those.

Rosenberg was sure that his patient was confused. It would have been a miracle if he had survived even six months with metastatic stomach cancer.

But according to DeAngelo's hospital records, that was exactly what had happened. Twelve years earlier, he had walked into the same hospital complaining of malaise and low energy. At the time, his chart noted, he was drinking his way through three or four bottles of whiskey per week and smoking a pack or two of cigarettes every day. Surgeons had discovered a fist-sized tumor in his stomach and smaller metastatic tumors in his liver. They removed the stomach tumor, along with half of his stomach, but they left the liver tumors alone, deciding that it was too risky to try to remove them at the same time. And then they had sewn him up and sent him home to die, which he had obviously failed to do.

Rosenberg went ahead with the gallbladder operation, and while he was in there he decided to take a look around in DeAngelo's abdomen. He felt behind the liver, gingerly working his way under its soft purple lobes, expecting to feel lumps of remnant tumors—an unmistakable feeling, hard and round, almost alien-like—but he found absolutely no trace of any growths. "This man had had a virulent and untreatable cancer that should have killed him quickly," Rosenberg wrote in his 1992 book *The Transformed Cell*. "He had received no treatment whatsoever for his disease from us or from anyone else. And he had been cured."

How could this be? In all the medical literature, Rosenberg could find only four instances of complete and spontaneous remission of metastatic stomach cancer. He was mystified. But he eventually came up with a hypothesis: he believed that DeAngelo's own immune system had fought off the cancer and killed the remaining tumors in his liver, the way you or I might shake off a cold. His own body had cured his cancer. Somehow.

At the time, this notion was well out of the mainstream of cancer research. But Rosenberg suspected that he was onto something important. *The Transformed Cell* told the story of Rosenberg's quest to harness the immune system to fight cancer. Despite small successes peppered here and there, however, whatever phenomenon had erased James DeAngelo's tumors proved to be elusive; for the first ten years, not a single one of Rosenberg's patients had survived. Not one. But still he kept at it.

He was doing better as a cancer surgeon than a cancer researcher: he operated on President Ronald Reagan in 1985, removing cancerous polyps from his colon, and that had gone fine. But Rosenberg's goal was to eliminate the need for cancer surgeries, period. Finally, in the mid-1980s, he had a glimmer of success—just enough to keep him going.

As soon as I read *The Transformed Cell,* as a medical student, I knew that I wanted to be a surgical oncologist and that I had to work with Steve Rosenberg. Cancer had been on my mind since before I had even applied to medical school. During my postbac year, while taking med school prerequisite courses, I volunteered on the pediatric cancer ward of Kingston General Hospital in Ontario, spending time with kids who were undergoing cancer treatment. Thankfully, childhood leukemia is one area where Medicine 2.0 has made real progress. But not all the kids survived, and the bravery of these children, the pain that they and their parents endured, and the compassion of their medical teams moved me more deeply than any engineering or mathematical problem. It confirmed my decision to switch from engineering to medicine.

In my third year of medical school, I got the opportunity to spend four months in Rosenberg's lab, at the epicenter of American cancer research. By the time I arrived, it had been nearly three decades since Richard Nixon had declared a national War on Cancer in 1971. Initially, the hope was that cancer would be "cured" within five years, in time for the Bicentennial. Yet it remained stubbornly undefeated in 1976, and still by the time I finished medical school in 2001. And today, for all intents and purposes.

Despite well over $100 billion spent on research via the National Cancer Institute, plus many billions more from private industry and public charities—despite all the pink ribbons and yellow bracelets, and literally millions of published papers on the PubMed database—cancer is the second leading cause of death in the United States, right behind heart disease. Together, these two conditions account for almost one in every two American deaths. The difference is that we understand the genesis and progression of heart disease fairly well, and we have some effective tools with which to prevent and treat it. As a

Figure 6. Cancer Incidence by Age in the United States

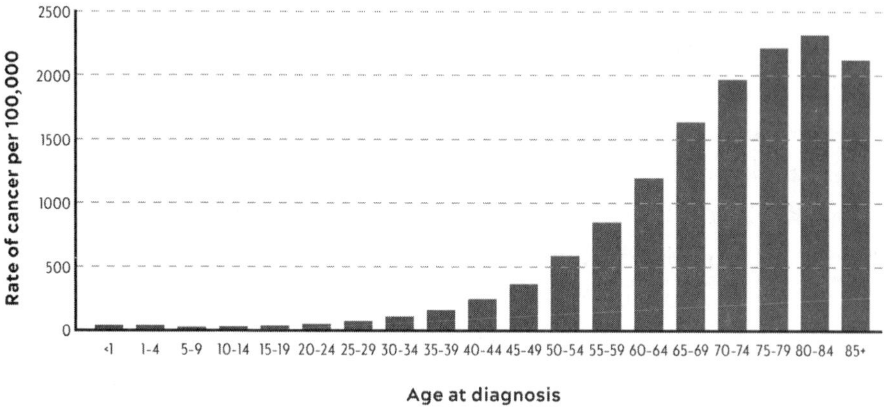

Source: National Cancer Institute (2021).

result, mortality rates from cardiovascular disease and cerebrovascular disease have dropped by two-thirds since the middle of the twentieth century. But cancer still kills Americans at almost exactly the same rate as it did fifty years ago.

We have made some progress against a few specific cancers, notably leukemia (especially childhood leukemia, as I noted earlier). For adults with leukemia, ten-year survival rates nearly doubled between 1975 and 2000, leaping from 23 percent to 44 percent. Survival rates for Hodgkin's and non-Hodgkin's lymphomas have increased as well, especially the former. Yet these represent relatively small victories in a "war" that has not gone particularly well.

Like heart disease, cancer is a disease of aging. That is, it becomes exponentially more prevalent with each decade of life, as figure 6 shows. But it can be deadly at almost any age, especially middle age. The median age of a cancer diagnosis is sixty-six, but in 2017 there were more cancer deaths among people between forty-five and sixty-four than from heart disease, liver disease, and stroke combined. This year, if recent trends continue, that same age group will also account for nearly 40 percent of the estimated 1.7 million new cases of cancer that are likely to be diagnosed in the United States, according to the

National Cancer Institute. By the time cancer is detected, however, it has probably already been progressing for years and possibly decades. As I write these words, I reflect sadly on three of my friends from high school who died of cancer in the past ten years, all younger than forty-five. I was able to say goodbye to only one of them before she died. Everyone reading this book probably has a few similar stories.

The problem we face is that once cancer is established, we lack highly effective treatments for it. Our toolbox is limited. Many (though not all) solid tumors can be removed surgically, a tactic that dates back to ancient Egypt. Combining surgery and radiation therapy is pretty effective against most local, solid-tumor cancers. But while we've gotten fairly good at this approach, we have essentially maxed out our ability to treat cancers this way. We are not getting any more juice from the squeeze. And surgery is of limited value when cancer has metastasized, or spread. Metastatic cancers can be slowed by chemotherapy, but they virtually always come back, often more resistant to treatment than ever. Our benchmark for success in a patient, or remission, is typically five-year survival, nothing more. We don't dare utter the word *cure*.

The second problem is that our ability to detect cancer at an early stage remains very weak. Far too often, we discover tumors only when they cause other symptoms, by which point they are often too locally advanced to be removed—or worse, the cancer has already spread to other parts of the body. I saw this happen many times during my training: we would remove a patient's tumor (or tumors), only to have them die a year later because that same cancer had taken hold elsewhere, like their liver or their lungs.

This experience informs our three-part strategy for dealing with cancer. Our first and most obvious wish is to avoid getting cancer at all, like the centenarians—in other words, prevention. But cancer prevention is tricky, because we do not yet fully understand what drives the initiation and progression of the disease with the same resolution that we have for atherosclerosis. Further, plain bad luck seems to play a major role in this largely stochastic process. But we do have some clues, which is what we'll talk about in the next two sections.

Next is the use of newer and smarter treatments targeting cancer's manifold weaknesses, including the insatiable metabolic hunger of fast-growing cancer cells and their vulnerability to new immune-based therapies, the outcome of decades of work by scientists like Steve Rosenberg. I feel that immunotherapy, in particular, has enormous promise.

Third, and perhaps most importantly, we need to try to detect cancer as early as possible so that our treatments can be deployed more effectively. I advocate early, aggressive, and broad screening for my patients—such as colonoscopy (or other colorectal cancer screening) at age forty, as opposed to the standard recommendation of forty-five or fifty—because the evidence is overwhelming that it's much easier to deal with most cancers in their early stages. I am also cautiously optimistic about pairing these tried-and-true staples of cancer screening with emerging methods, such as "liquid biopsies," which can detect trace amounts of cancer-cell DNA via a simple blood test.

Five decades into the war on cancer, it seems clear that no single "cure" is likely to be forthcoming. Rather, our best hope likely lies in figuring out better ways to attack cancer on all three of these fronts: prevention, more targeted and effective treatments, and comprehensive and accurate early detection.

What Is Cancer?

One major reason why cancer is so deadly—and so scary—is that we still know relatively little about how it begins and why it spreads.

Cancer cells are different from normal cells in two important ways. Contrary to popular belief, cancer cells don't grow faster than their noncancerous counterparts; they just don't *stop* growing when they are supposed to. For some reason, they stop listening to the body's signals that tell them when to grow and when to stop growing. This process is thought to begin when normal cells acquire certain genetic mutations. For example, a gene called *PTEN*, which normally stops cells from growing or dividing (and eventually becoming tumors), is often mutated or "lost" in people with cancer, including about

31 percent of men with prostate cancer and 70 percent of men with advanced prostate cancer. Such "tumor suppressor" genes are critically important to our understanding of the disease.

The second property that defines cancer cells is their ability to travel from one part of the body to a distant site where they should not be. This is called *metastasis,* and it is what enables a cancerous cell in the breast to spread to the lung. This spreading is what turns a cancer from a local, manageable problem to a fatal, systemic disease.

Beyond these two common properties, however, the similarities among different cancers largely end. One of the biggest obstacles to a "cure" is the fact that cancer is not one single, simple, straightforward disease, but a condition with mind-boggling complexity.

About two decades ago, the National Cancer Institute launched a huge and ambitious study called The Cancer Genome Atlas, whose goal was to sequence cancer tumor cells in hopes of finding the precise genetic changes that cause various types of cancer, such as breast, kidney, and liver cancer. Armed with this knowledge, scientists would be able to develop therapies targeted at these exact mutations. As one of the scientists who proposed the project said, "These are the starting blocks that we need to develop a cure."

But the early results of The Cancer Genome Atlas, published in a series of papers beginning in 2008, revealed more confusion than clarity. Rather than uncovering a definite pattern of genetic changes driving each type of cancer, the study found enormous complexity. Each tumor had more than one hundred different mutations, on average, and those mutations almost appeared to be random. A handful of genes emerged as drivers, including *TP53* (also known as p53, found in half of all cancers), *KRAS* (common in pancreatic cancer), *PIC3A* (common in breast cancer), and *BRAF* (common in melanoma), but very few if any of these well-known mutations were shared across all tumors. In fact, there didn't seem to be any individual genes that "caused" cancer at all; instead, it seemed to be random somatic mutations that *combined* to cause cancers. So not only is breast cancer genetically distinct from colon cancer (as the researchers expected), but no two breast cancer tumors are very much alike. If two women have breast cancer, at the same stage, their tumor genomes are likely to be very different from each other. Therefore, it would be difficult if not impossible to devise one treatment for both women based on the genetic profile of their tumors. Rather than revealing the shape of the forest, then, The Cancer Genome Atlas merely dragged us deeper into the maze of the trees.

Or so it seemed at the time. Ultimately, genome sequencing has proved to be a very powerful tool against cancer—just not in the way that was envisioned two decades ago.

Even when we treat a local cancer successfully, we can never be sure that it's entirely gone. We have no way of knowing whether cancer cells may have already spread and are lurking in other organs, waiting to establish a foothold there. It is this metastatic cancer that is responsible for most cancer deaths. If we want to reduce cancer mortality by a significant amount, we must do a better job of preventing, detecting, and treating metastatic cancers.

With a few exceptions, such as glioblastoma or other aggressive brain tumors, as well as certain lung and liver cancers, solid organ tumors typically kill you only when they spread to other organs. Breast cancer kills only when it becomes metastatic. Prostate cancer kills only when it becomes metastatic. You could live without either of those organs. So when you hear the sad story of someone dying from breast or prostate cancer, or even pancreatic or colon cancer, they died because the cancer spread to other, more critical organs such as the brain, the lungs, the liver, and bones. When cancer reaches those places, survival rates drop precipitously.

But what causes cancer to spread? We don't really know, and we are unlikely to find out anytime soon because only about 5 to 8 percent of US cancer research funding goes to the study of metastasis. Our ability to detect cancer metastasis is also very poor, although I do believe we are on the verge of some key breakthroughs in cancer screening, as we'll discuss later. Most of our energy has been focused on treating metastatic cancer, which is an extremely difficult problem. Once cancer has spread, the entire game changes: we need to treat it *systemically* rather than locally.

Right now, this usually means chemotherapy. Contrary to popular belief, killing cancer cells is actually pretty easy. I've got a dozen potential chemotherapy agents in my garage and under my kitchen sink. Their labels identify them as glass cleaner or drain openers, but they would easily kill cancer cells too. The problem, of course, is that these poisons will also slaughter every normal cell in between, likely killing the patient in the process. The game is won by killing cancers while sparing the normal cells. *Selective* killing is the key.

Traditional chemotherapy occupies a fuzzy region between poison and medicine; the mustard gas used as a weapon during World War I was a direct precursor to some of the earliest chemotherapy agents, some of which are still in use. These drugs attack the replicative cycle of cells, and because cancer cells are rapidly dividing, the chemo agents harm them more severely than normal cells. But many important noncancerous cells are also dividing frequently, such as those in the lining of the mouth and gut, the hair follicles, and the nails, which is why typical chemotherapy agents cause side effects like hair loss and gastrointestinal misery. Meanwhile, as cancer researcher Robert Gatenby points out, those cancer cells that do manage to survive chemotherapy often end up acquiring mutations that make them stronger, like cockroaches that develop resistance to insecticides.

The side effects of chemo might seem at the outset to be a fair trade for a "chance for a few more useful years," as the late author Christopher Hitchens noted in his cancer memoir *Mortality*. But as his treatment for metastatic esophageal cancer dragged on, he changed his mind. "I lay for days on end, trying in vain to postpone the moment when I would have to swallow. Every time I did swallow, a hellish tide of pain would flow up my throat, culminating in what felt like a mule kick in the small of my back. . . . And then I had an unprompted rogue thought: If I had been told about all this in advance, would I have opted for the treatment?"

Hitchens was experiencing the primary flaw of modern chemotherapy: It is systemic, but still not specific enough to target only cancerous cells and not normal healthy cells. Hence the horrible side effects he suffered. Ultimately, successful treatments will need to be both systemic *and* specific to a particular cancer type. They will be able to exploit some weakness that is unique to cancer cells, while largely sparing normal cells (and, obviously, the patient). But what might those weaknesses be?

Just because cancer is powerful does not mean it is invincible. In 2011, two leading cancer researchers named Douglas Hanahan and Robert Weinberg identified two key hallmarks of cancer that may lead—and in fact have led—to new treatments, as well as potential methods of reducing cancer risk. The first such hallmark is the fact that many cancer cells have an altered metabolism, consuming huge amounts of glucose. Second, cancer cells seem to have an uncanny ability to evade the immune system, which normally hunts down damaged and dangerous cells—such as cancerous cells—and targets them for destruction. This second problem is the one that Steve Rosenberg and others have been trying to solve for decades.

Metabolism and immune surveillance excite me because they are both systemic, a necessary condition for any new treatment to combat metastatic cancers. They both exploit features of cancer that are potentially more specific to tumors than simply runaway cell replication. But neither the metabolic nor the immune-based approaches to cancer are exactly new: dogged researchers have been laying the groundwork for progress in both of these areas for decades.

Cancer Metabolism

As you might have gathered by now, we tend to think of cancer as primarily a genetic disease, driven by mutations of unknown cause. Clearly, cancer cells are genetically distinct from normal human cells. But for the last century or so, a handful of researchers have been investigating another unique property of cancer cells, and that is their metabolism.

In the 1920s, a German physiologist named Otto Warburg discovered that cancer cells had a strangely gluttonous appetite for glucose, devouring it at up to forty times the rate of healthy tissues. But these cancer cells weren't "respiring" the way normal cells do, consuming oxygen and producing lots of

ATP, the energy currency of the cell, via the mitochondria. Rather, they appeared to be using a different pathway that cells normally use to produce energy under anaerobic conditions, meaning without sufficient oxygen, such as when we are sprinting. The strange thing was that these cancer cells were resorting to this inefficient metabolic pathway despite having plenty of oxygen available to them.

This struck Warburg as a very strange choice. In normal aerobic respiration, a cell can turn one molecule of glucose into as many as thirty-six units of ATP. But under anaerobic conditions, that same amount of glucose yields only two net units of ATP. This phenomenon was dubbed the Warburg effect, and even today, one way to locate potential tumors is by injecting the patient with radioactively labeled glucose and then doing a PET scan to see where most of the glucose is migrating. Areas with abnormally high glucose concentrations indicate the possible presence of a tumor.

Warburg was awarded the Nobel Prize in Physiology or Medicine in 1931 for his discovery of a crucial enzyme in the electron transport chain (a key mechanism for producing energy in the cell). By the time he died in 1970, the weird quirk of cancer metabolism that he had discovered had been all but forgotten. The discovery of the structure of DNA by James Watson, Francis Crick, Maurice Wilkins, and Rosalind Franklin in 1953 had caused a seismic paradigm shift, not just in cancer research but in biology in general.

As Watson recounted in a 2009 *New York Times* op-ed: "In the late 1940s, when I was working toward my doctorate, the top dogs of biology were its biochemists who were trying to discover how the intermediary molecules of metabolism were made and broken down. After my colleagues and I discovered the double helix of DNA, biology's top dogs then became its molecular biologists, whose primary role was finding out how the information encoded by DNA sequences was used to make the nucleic acid and protein components of cells."

Nearly forty years into the War on Cancer, however, Watson himself had become convinced that genetics did not hold the key to successful cancer treatment after all. "We may have to turn our main research focus away from decoding the genetic instructions behind cancer and toward understanding the chemical reactions within cancer cells," he wrote. It was time, he argued, to start looking at therapies that targeted cancer's metabolism as well as its genetics.

A handful of scientists had been pursuing the metabolic aspects of cancer all along. Lew Cantley, now at Harvard's Dana-Farber Cancer Center, has been investigating cancer metabolism since the 1980s, when the idea was unfashionable. One of the more puzzling questions he has tackled was *why* cancer cells needed to produce energy in this highly inefficient way. Because the inefficiency of the Warburg effect may be the point, as Cantley, Matthew Vander Heiden, and Craig Thompson argued in a 2009 paper. While it may not yield much in the way of energy, they found, the Warburg effect generates lots of by-products, such as lactate, a substance that is also produced during intense exercise. In fact, turning glucose into lactate creates so many extra molecules that the authors argued that the relatively small amount of energy it produces may actually be the "by-product."

There's a logic to this seeming madness: when a cell divides, it doesn't simply split into two smaller cells. The process requires not only the division of the nucleus, and all that stuff we learned in high school biology, but the actual physical materials required to construct a whole new cell. Those don't just appear out of nowhere. Normal aerobic cellular respiration produces only energy, in the form of ATP, plus water and carbon dioxide, which aren't much use as building materials (also, we exhale the latter two). The Warburg effect, also known as anaerobic glycolysis, turns the same amount of glucose into a little bit of energy and a whole lot of chemical building blocks—which are then used to build new cells rapidly. Thus, the Warburg effect is how cancer cells fuel their own proliferation. But it also represents a potential vulnerability in cancer's armor.*

This remains a controversial view in mainstream cancer circles, but it has

---

* This is not the only explanation for how the Warburg effect benefits a cancer cell. Another theory is that it helps protect the tumor from immune cells by making the tumor microenvironment less hospitable because of lower pH (i.e., more acidic) caused by the generation of lactic acid and reactive oxygen species. For an excellent review of these topics, see Liberti and Locasale (2016).

gotten harder and harder to ignore the link between cancer and metabolic dysfunction. In the 1990s and early 2000s, as rates of smoking and smoking-related cancers declined, a new threat emerged to take the place of tobacco smoke. Obesity and type 2 diabetes were snowballing into national and then global epidemics, and they seemed to be driving increased risk for many types of cancers, including esophageal, liver, and pancreatic cancer. The American Cancer Society reports that excess weight is a leading risk factor for both cancer cases and deaths, second only to smoking.

Globally, about 12 to 13 percent of all cancer cases are thought to be attributable to obesity. Obesity itself is strongly associated with thirteen different types of cancers, including pancreatic, esophageal, renal, ovarian, and breast cancers, as well as multiple myeloma (see figure 7). Type 2 diabetes also increases the risk of certain cancers, by as much as double in some cases (such as pancreatic and endometrial cancers). And extreme obesity (BMI $\geq$ 40) is associated with a 52 percent greater risk of death from all cancers in men, and 62 percent in women.

I suspect that the association between obesity, diabetes, and cancer is primarily driven by inflammation and growth factors such as insulin. Obesity, especially when accompanied by accumulation of visceral fat (and other fat outside of subcutaneous storage depots), helps promote inflammation, as dying fat cells secrete an array of inflammatory cytokines into the circulation (see figure 4 in chapter 6). This chronic inflammation helps create an environment that could induce cells to become cancerous. It also contributes to the development of insulin resistance, causing insulin levels to creep upwards—and, as we'll see shortly, insulin itself is a bad actor in cancer metabolism.

This insight comes courtesy of further work by Lew Cantley. He and his colleagues discovered a family of enzymes called PI3-kinases, or PI3K, that play a major role in fueling the Warburg effect by speeding up glucose uptake into the cell. In effect, PI3K helps to open a gate in the cell wall, allowing glucose to flood in to fuel its growth. Cancer cells possess specific mutations that turn up PI3K activity while shutting down the tumor-suppressing protein PTEN, which we talked about earlier in this chapter. When PI3K is activated

Figure 7. **Cancers Associated with Excess Weight and Obesity**

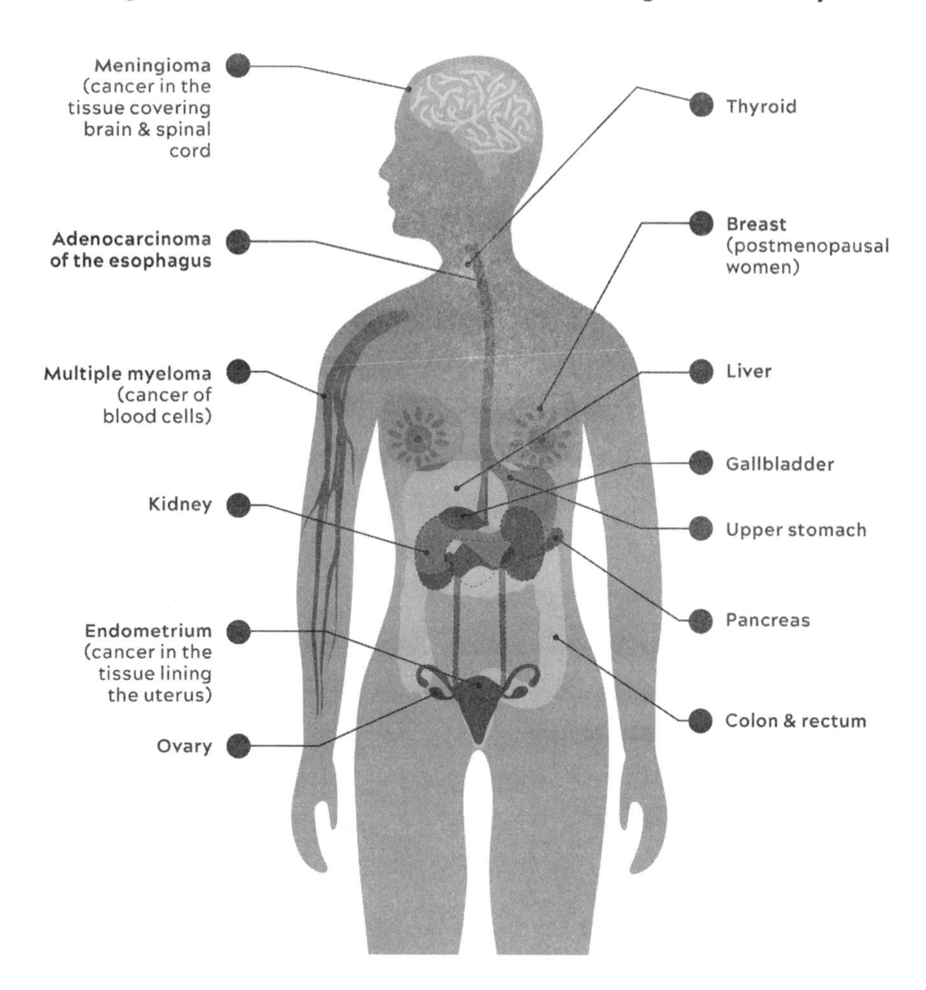

Source: NCI (2022a).

by insulin and IGF-1, the insulin-like growth factor, the cell is able to devour glucose at a great rate to fuel its growth. Thus, insulin acts as a kind of cancer enabler, accelerating its growth.

This in turn suggests that metabolic therapies, including dietary manipulations that lower insulin levels, could potentially help slow the growth of some cancers and reduce cancer risk. There is already some evidence that tinkering with metabolism can affect cancer rates. As we have seen, laboratory animals on calorically restricted (CR) diets tend to die from cancer at far lower rates than control animals on an *ad libitum* (all-they-can-eat) diet. Eating less appears to give them some degree of protection. The same may hold true in people: one study of caloric restriction in humans found that limiting caloric intake directly turns down the PI3K-associated pathway, albeit in muscle (which is not susceptible to cancer). This may be a function of lowered insulin rather than lower glucose levels.

While it's tricky to impossible to avoid or prevent the genetic mutations that help give rise to cancer, it is relatively easy to address the metabolic factors that feed it. I'm not suggesting that it's possible to "starve" cancer or that any particular diet will magically make cancer go away; cancer cells always seem to be able to obtain the energy supply they need. What I am saying is that we don't want to be anywhere on that spectrum of insulin resistance to type 2 diabetes, where our cancer risk is clearly elevated. To me, this is the low-hanging fruit of cancer prevention, right up there with quitting smoking. Getting our metabolic health in order is essential to our anticancer strategy. In the next section, we'll look at how metabolic interventions have also been used to potentiate other kinds of cancer therapy.

New Treatments

Lew Cantley's discovery of the PI3K pathway led to the development of a whole class of drugs that target cancer metabolism. Three of these drugs, known as PI3K inhibitors, have been FDA approved for certain relapsed leukemias and lymphomas, and a fourth was approved in late 2019 for breast cancer. But they haven't seemed to work as well as predicted, based on PI3K's prominent role in the growth pathways of cancer cells. Also, they had the annoying side effect of raising blood glucose, which in turn provoked a jump in insulin levels and IGF-1 as the cell tried to work around PI3K inhibition—the very thing we would want to avoid, under this theory.

This situation came up at a dinner in 2014, when I joined Cantley, who was then director of the Meyer Cancer Center at Weill Cornell Medical College in Manhattan, and Siddhartha Mukherjee, who is a practicing oncologist, research scientist, and Pulitzer Prize–winning author of *The Emperor of All Maladies,* a "biography" of cancer. I was a huge fan of Sid's work, so I was excited to sit down with these two giants in oncology.

Over dinner, I shared a story about a case where a PI3K-inhibiting drug treatment had been enhanced by a kind of metabolic therapy. I couldn't stop thinking about it, because the patient was the wife of a very close friend of mine. Sandra (not her real name) had been diagnosed with breast cancer six years earlier. It had already spread to her lymph nodes and her bones. Because of her poor prognosis, she qualified for a clinical trial of an experimental PI3K-inhibitor drug, in combination with standard therapies.

Sandra was a very motivated patient. From the day of her diagnosis, she had become obsessed with doing anything possible to stack the odds in her favor. She devoured everything she could read on the impact of nutrition on cancer, and she had concluded that a diet that reduced insulin and IGF-1 would aid in her treatment. So she worked out a regimen that consisted primarily of leafy vegetables, olive oil, avocados, nuts, and modest amounts of protein, mostly from fish, eggs, and poultry. The diet was just as notable for what it did not contain: added sugar and refined carbohydrates. All along, she underwent frequent blood tests to make sure her insulin and IGF-1 levels stayed low, which they did.

Over the next few years, every other woman who was enrolled at her trial site had died. Every single one. The patients had been on state-of-the-art chemotherapy plus the PI3K inhibitor, yet their metastatic breast cancer had still overtaken them. The trial had to be stopped because it was clear that the drugs were not working. Except for Sandra. Why was she still alive, while hundreds of other women with the same disease, at the same stage, were not? Was she merely lucky? Or could her very strict diet, which likely inhibited her insulin and IGF-1, have played a role in her fate?

I had a hunch that it may have. I believe that we have to pay attention to these outliers, these "miraculous" survivors. Even if they are only anecdotal, their stories may contain some useful insight into this lethal, mysterious disease. As Steve Rosenberg used to say, "These patients help us ask the right questions."

Yet in his 592-page magnum opus on cancer, published in 2010, Mukherjee had barely written a word about metabolism and metabolic therapies. It seemed premature to write about it, he later told me. Now, as I told this story over dinner, he seemed interested but skeptical. Cantley grabbed a napkin and started scribbling a graphic: the problem with PI3K inhibitors, he explained, is that by turning down the insulin-related PI3K pathway, they actually end up *raising* insulin and glucose levels. Because glucose is blocked from entering the cell, more of it stays in the bloodstream. The body then thinks it needs to produce more insulin to get rid of all that glucose, possibly negating some of the effects of the drug by activating PI3K. So what if we combined PI3K inhibitors with an insulin-minimizing or ketogenic diet?

From that crude drawing on a napkin, a study was born. Published in *Nature* in 2018, with Mukherjee and Cantley as senior authors, the study found that a combination of a ketogenic diet *and* PI3K inhibitors improved the responses to treatment of mice that had been implanted with human cancer tumors. The results are important because they show not only that a cancer cell's metabolism is a valid target for therapy but that a patient's metabolic state can affect the efficacy of a drug. In this case, the animals' ketogenic diet seemed to synergize with what was otherwise a somewhat disappointing treatment, and together they proved to be far more powerful than either one alone. It's like in boxing, where a combination often proves to be much more effective than any single punch. If your first punch misses, the second is already in motion, aimed directly at the spot where you anticipate your opponent will move. (Mukherjee and Cantley have since partnered on a start-up company to further explore this idea of combining drug treatment with nutritional interventions.)

Other types of dietary interventions have been found to help improve the effectiveness of chemotherapy, while limiting its collateral damage to healthy tissues. Work by Valter Longo of the University of Southern California and others has found that fasting, or a fasting-like diet, increases the ability of normal cells to resist chemotherapy, while rendering cancer cells more vulnerable to the treatment. It may seem counterintuitive to recommend fasting to cancer patients, but researchers have found that it caused no major adverse events in chemotherapy patients, and in some cases it may have improved the patients' quality of life. A randomized trial in 131 cancer patients undergoing chemotherapy found that those who were placed on a "fasting-mimicking diet" (basically, a very low-calorie diet designed to provide essential nutrients while reducing feelings of hunger) were more likely to respond to chemotherapy and to feel better physically and emotionally.

This flies in the face of traditional practice, which is to try to get patients on chemotherapy to eat as much as they can tolerate, typically in the form of high-calorie and even high-sugar diets. The American Cancer Society suggests using ice cream "as a topping on cake." But the results of these studies suggest that maybe it's not such a good idea to increase the level of insulin in someone who has cancer. More studies need to be done, but the working hypothesis is that because cancer cells are so metabolically greedy, they are therefore more vulnerable than normal cells to a reduction in nutrients—or more likely, a reduction in insulin, which activates the PI3K pathway essential to the Warburg effect.

This study and the Mukherjee-Cantley study we discussed earlier also point toward another important takeaway from this chapter, which is that there is rarely only *one* way to treat a cancer successfully. As Keith Flaherty, a medical oncologist and the director of developmental therapeutics at Massachusetts General Hospital, explained to me, the best strategy to target cancer is likely by targeting multiple vulnerabilities of the disease at one time, or in sequence. By stacking different therapies, such as combining a PI3K inhibitor with a ketogenic diet, we can attack cancer on multiple fronts, while also minimizing the likelihood of the cancer developing resistance (via mutations) to any single treatment. This is becoming a more common practice in conventional chemotherapy—but for it to be truly effective, we need more efficacious treatments to work with in the first place, ones that do a better job of singling out cancer cells for destruction while leaving healthy cells, and the patient, unharmed.

In the next section, we'll look at how the once-outlandish notion of immunotherapy has yielded multiple potentially game-changing cancer therapies that could fit this bill.

The Promise of Immunotherapy

Like metabolism, immunotherapy did not figure in *The Emperor of All Maladies*. It was barely on the radar when the book was published in 2010. But when Ken Burns made his documentary based on the book just five years later, both immunotherapy and Steve Rosenberg were featured very prominently, which speaks to the degree to which our thinking about cancer and especially immunotherapy has begun to change, just in the last decade.

The immune system is programmed to distinguish "nonself" from "self"—that is, to recognize invading pathogens and foreign bodies among our own healthy native cells, and then to kill or neutralize the harmful agents. An immunotherapy is any therapy that tries to boost or harness the patient's immune system to fight an infection or other condition (example: vaccines). The problem with trying to treat cancer this way is that while cancer cells are abnormal and dangerous, they are technically still our cells ("self"). They have cleverly evolved to hide from the immune system and specifically our T cells, the immune system's assassins that would ordinarily kill foreign cells. So for a cancer immunotherapy to succeed, we essentially need to teach the immune system to recognize and kill our own cells that have turned cancerous. It needs to be able to distinguish "bad self" (cancer) from "good self" (everything else).

Rosenberg wasn't the first person to try to harness the immune system against cancer. Back in the late nineteenth century, a Harvard-trained surgeon named William Coley had noticed that a patient with a serious tumor had been miraculously cured, apparently as a result of a major postsurgical infection. Coley began experimenting with bacterial inoculations that he hoped would trigger a similar immune response in other patients. But his medical colleagues were appalled by the notion of injecting patients with germs, and when others failed to reproduce Coley's results, his ideas were sidelined and condemned as quackery. Yet cases of spontaneous cancer remission—like the one Rosenberg had observed as a young resident—kept occurring, and nobody could really explain them. They offered a tantalizing glimpse of the healing power of the human body.

It wasn't an easy problem to solve. Rosenberg tried one approach after another, without success. A cancer "vaccine" went nowhere. He spent many years experimenting with interleukin-2 (IL-2), a cytokine that plays a major role in the immune response (it basically amplifies the activity of lymphocytes, white blood cells that fight infection). It had worked in animal models of metastatic cancer, but results in human subjects were more mixed: patients had to spend days, if not weeks, in the ICU, and they could very easily die from its massive side effects. Finally, in 1984, a late-stage melanoma patient named Linda Taylor was put into remission by high-dose IL-2 alone.

This was a huge turning point, because it showed that the immune system *could* fight off cancer. But the failures still outnumbered the successes, as high-dose IL-2 seemed to be effective only against melanoma and renal cell cancers, and only in 10 to 20 percent of patients with these two cancers.* It was a shotgun approach to a problem that required more precision. So Rosenberg turned his attention to the T cells directly. How could they be trained to spot and attack cancer cells?

It took many more years and several iterations, but Rosenberg and his team adapted a technique that had been developed in Israel that involved taking T cells from a patient's blood, then using genetic engineering to add antigen receptors that were specifically targeted to the patient's tumors. Now the T cells were programmed to attack the patient's cancer. Known as chimeric antigen receptor T cells (or CAR-T), these modified T cells could be multiplied in the lab and then infused back into the patient.

In 2010, Rosenberg and his team reported their first success with a CAR-T treatment, in a patient with advanced follicular lymphoma who had under-

---

* At the time it was not clear why this was the case, but today it's clear that this approach worked because these two cancers tend to have a large number of genetic mutations, meaning the immune system is more likely to recognize the cancerous cells as harmful and target them.

gone multiple rounds of conventional treatments, including chemotherapy, as well as a different kind of immunotherapy, all without success. Other groups pursued the technique as well, and finally, in 2017, the first two CAR-T-based treatments were approved by the FDA (making them the first cell and gene therapies ever approved by the FDA), one for adult lymphoma and another for acute lymphoblastic leukemia, the most common cancer in children. It had taken nearly fifty years, but Steve Rosenberg's once-outlandish theory had finally yielded a breakthrough.

As elegant as they are, however, CAR-T treatments have proven successful only against one specific type of cancer called B-cell lymphoma. All B-cells, normal and cancerous alike, express a protein called CD19, which is the target used by the CAR-T cell to zero in and kill them. Since we can live without B-cells, CAR-T works by obliterating *all* CD19-bearing cells. Unfortunately, we have not yet identified a similar marker for other cancers.

If we are going to bring down overall cancer death rates, we need a more broadly successful class of treatments. Luckily, the immunotherapy approach continued to evolve. Now, a bit more than a decade later, a handful of immunotherapy-based cancer drugs have been approved. In addition to CAR-T, there is a class of drugs called "checkpoint inhibitors," which take an opposite approach to the T cell–based therapies. Instead of activating T cells to go kill the cancer, the checkpoint inhibitors help make the cancer visible to the immune system.

To simplify a very long and fascinating story,* a researcher from Texas named James Allison, who has been working on immunotherapy for almost as long as Steve Rosenberg, figured out how cancer cells hide from the immune system by exploiting so-called checkpoints that are normally supposed to regulate our T cells and keep them from going overboard and attacking our nor-

* To learn more about the story of immunotherapy, read Charles Graeber's 2018 book, *Breakthrough*, which goes into greater detail about Jim Allison's work with checkpoint inhibitors.

mal cells, which would lead to autoimmune disease. Essentially, the checkpoints ask the T cells, one last time, "Are you *sure* you want to kill this cell?"

Allison found that if you blocked specific checkpoints, particularly one called CTLA-4, you effectively outed or unmasked the cancer cells, and the T cells would then destroy them. He tried the technique on cancer-prone mice, and in one early experiment he arrived at his lab one morning to find that all the mice who had received checkpoint-inhibiting therapy were still alive, while the ones without it were all dead. It's nice when your results are so clear that you don't even really need statistical analysis.

In 2018, Allison shared the Nobel Prize with a Japanese scientist named Tasuku Honjo, who had been working on a slightly different checkpoint called PD-1. The work of these two scientists has led to two approved checkpoint-inhibiting drugs, ipilimumab (Yervoy) and pembrolizumab (Keytruda), targeting CTLA-4 and PD-1, respectively.

All Nobel Prizes are impressive, but I'm pretty biased toward this one. Checkpoint inhibitors not only saved the life of a third Nobel laureate, former US president Jimmy Carter, who was treated with Keytruda for his metastatic melanoma in 2015, but they also came to the rescue of a very good friend of mine, a former colleague whom I'll call Michael. When he was only in his early forties, Michael was diagnosed with a very large colon tumor that required immediate surgery. I still remember the cover of the magazine that I flipped through anxiously while I was sitting in the waiting room during his procedure. Michael is the kindest soul I think I've ever known, and his brilliance and wit could make the worst day seem enjoyable during the years we worked together. I could not imagine losing him.

His family and friends were elated when his surgery proved successful and the pathology report found no sign of cancer in his nearby lymph nodes, despite the advanced size of the primary tumor. A few months later, our joy was turned back to despair as we learned that Michael's cancer had been the result of a genetic condition called Lynch syndrome. People with Lynch syndrome usually know they have it, because it is inherited in a dominant fashion. But Michael had been adopted, so he had no idea he was at risk. The mutations that define Lynch syndrome all but guarantee that its carriers develop early-onset colon cancer, as Michael had, but they are also at very high risk for other cancers. Sure enough, five years after he dodged that first bullet with colon cancer, Michael called to say that he now had pancreatic adenocarcinoma. This was even more distressing, because as we both well knew, this cancer is almost uniformly fatal.

Michael went to see the top pancreatic surgeon in his area, who confirmed the worst: surgery was not possible. The cancer was too advanced. Michael would have, at most, nine to twelve months to live. Making this all the more heart-wrenching was the fact that Michael and his wife had just welcomed their first children, twin girls, into the world that year. But *The New England Journal of Medicine* had recently reported that some patients with mismatch-repair deficiency (common in Lynch syndrome) had been successfully treated with Keytruda, the anti-PD-1 drug. It was a long shot, but it seemed at least plausible that Michael might benefit from this drug. Michael's doctors agreed to test him, and the tests confirmed that Michael was indeed a candidate for Keytruda. He was immediately enrolled in a clinical trial. While it is not guaranteed to work on all such patients, it worked on Michael, turning his immune system against his tumor and eventually eradicating all signs of pancreatic cancer in his body.

So now he is cancer-free for a second time, and beyond grateful to have survived a disease that should have killed him when his twin girls were still in diapers. Now he gets to see them grow up. The price he paid is that while his immune system was attacking his cancer, it went a bit overboard and also destroyed his pancreas in the process. As a result, he now has type 1 diabetes, since he can't produce insulin anymore. He lost his pancreas, but his life was saved. Seems like a fair trade, on balance.

Michael was one of the lucky ones. As of yet, the various immunotherapy treatments that have been approved still benefit only a fairly small percentage of patients. About a third of cancers can be treated with immunotherapy, and of those patients, just one-quarter will actually benefit (i.e., survive). That means that only 8 percent of potential cancer deaths could be prevented by immunotherapy, according to an analysis by oncologists Nathan Gay and Vinay Prasad. They work when they work, but only in a small number of patients and at great cost. Nevertheless, just two decades ago—when I was feeling frustration as a cancer surgeon in training—patients like former president Carter or my friend Michael, and countless others, would have all died.

But I (and others who know much more than I do) believe that we have had only a small taste of what can be accomplished via immunotherapy.

One idea that is currently being explored is the notion of combining immunotherapies with other treatments. One recent paper describes a clinical trial where a platinum-based chemotherapy was used in combination with a checkpoint inhibitor, resulting in improved overall survival in patients with lung cancer. These patients were not sensitive to the checkpoint inhibitor alone, but something about the chemo made the cancer more sensitive, or more "visible" if you will, to the immunotherapy. It's an extension of the idea of "stacking" therapies that we mentioned earlier.

To make immunotherapy more widely effective, we need to devise ways to help our immune cells detect and kill a broader array of cancers, not just a few specific types. Genetic analysis reveals that some 80 percent of epithelial cancers (that is, solid organ tumors) possess mutations that the immune system can recognize—thus making them potentially vulnerable to immune-based treatments.

One very promising technique is called adoptive cell therapy (or adoptive cell transfer, ACT). ACT is a class of immunotherapy whereby supplemental T cells are transferred into a patient, like adding reinforcements to an army, to bolster their ability to fight their own tumor. These T cells have been genetically programmed with antigens specifically targeted at the patient's individual tumor type. It is similar to CAR-T cell therapy, which we discussed earlier, but much broader in scope. As cancer grows, it quickly outruns the immune system's ability to detect and kill it; there simply aren't enough T cells to do the job, particularly when the cancer reaches the point of clinical detection. This is why spontaneous remission, as happened with James DeAngelo, is so rare. The idea behind ACT is basically to overwhelm the cancer with a huge number of targeted T cells, like supplementing an army with a brigade of trained assassins.

There are two ways to do ACT. First, we can take a sample of a patient's tumor and isolate those T cells that do recognize the tumor as a threat. These are called tumor-infiltrating lymphocytes (TILs), but there may only be a few million of them, not enough to mount a complete response against the tumor. By removing the TILs from the body and multiplying them by a factor of 1,000 or so, and then reinfusing them into the patient, we can expect to see a much better response. Alternatively, T cells can be harvested from the patient's blood and genetically modified to recognize his or her specific tumor. Each of these approaches has advantages and disadvantages,* but the interesting part is that ACT effectively means designing a new, customized anticancer drug for each individual patient.

This is obviously a costly proposition, and a very labor-intensive process, but it has a lot of promise. The proof of principle is here, but much more work is needed, not only to improve the efficacy of this approach but also to enable us to deliver this treatment more widely and more easily. And while the cost may seem prohibitive at first, I would point out that conventional chemotherapy is also very expensive—and its remissions are almost never permanent.

One striking feature of immune-based cancer treatment is that when it works, it really works. It is not uncommon for a patient with metastatic cancer to enter remission after chemotherapy. The problem is that it virtually never lasts. The cancer almost always comes back in some form. But when patients do respond to immunotherapy, and go into complete remission, they often *stay* in remission. Between 80 and 90 percent of so-called com-

---

* For example, TILs by definition have already demonstrated an affinity for the tumor; however, they may "age" more as they are multiplied (cells "age" every time they divide), losing some of their potency. Conversely, genetically modified T cells tend to be younger and easier to grow, but they don't necessarily have the same ability to kill tumors as TILs.

plete responders to immunotherapy remain disease-free fifteen years out. This is extraordinary—far better than the short-term, five-year time horizon at which we typically declare victory in conventional cancer treatment. One hesitates to use the word *cured*, but in patients who do respond to immunotherapy, it's safe to assume that the cancer is pretty much gone.

The important message here is that there is hope. For the first time in my lifetime, we are making progress in the War on Cancer, if we can even still call it that. There are now treatments that can, and do, save the lives of thousands of people who would have inevitably died just a decade ago. Twenty years ago, someone with metastatic melanoma could expect to live about six more months, on average. Now that number is twenty-four months, with about 20 percent of such patients being completely cured. This represents measurable progress—almost entirely thanks to immunotherapy. Improved early detection of cancer, which we will discuss in the final section of this chapter, will likely make our immunotherapy treatments still more effective.

Immunotherapy traveled a rocky road, and it could have been abandoned completely at many points along the way. In the end, it survived because its chimerical successes turned out to be not so chimerical, but mostly thanks to the determination and persistence of visionary scientists such as James Allison, Tasuku Honjo, Steve Rosenberg, and others, who kept going even when their work seemed pointless and possibly crazy.

Early Detection

The final and perhaps most important tool in our anticancer arsenal is early, aggressive screening. This remains a controversial topic, but the evidence is overwhelming that catching cancer early is almost always net beneficial.

Unfortunately, the same problem that I encountered in residency applies today: too many cancers are detected too late, after they've grown and spread via metastasis. Very few treatments work against these advanced cancers; in most cases, outside of the few cancers that respond to immunotherapies, the best we can hope for is to delay death slightly. The ten-year survival rate for patients with metastatic cancer is virtually the same now as it was fifty years ago: zero. We need to do more than hope for novel therapies.

When cancers are detected early, in stage I, survival rates skyrocket. This is partly because of simple math: these earlier-stage cancers comprise fewer total cancerous cells, with fewer mutations, and thus are more vulnerable to treatment with the drugs that we do have, including some immunotherapies. I would go so far as to argue that early detection is our *best* hope for radically reducing cancer mortality.

This claim makes intuitive sense, but it is supported by even a cursory look at data comparing success rates at treating specific cancers in the metastatic versus adjuvant (i.e., postsurgical) setting. Let's look at colon cancer first. A patient with metastatic colon cancer, which means the cancer has spread past their colon and adjacent lymph nodes to another part of the body, such as the liver, will typically be treated with a combination of three drugs known as the FOLFOX regimen. This treatment yields a median survival time of about 31.5 months,* meaning about half of patients live longer than this, and half do not. Regardless, virtually none of these patients will be alive in ten years. If a patient undergoes successful surgery for stage III colon cancer, which means all the cancer was removed and there was no visible spread to distant organs, then the follow-up care is treatment with the exact same FOLFOX treatment regimen. But in this scenario, fully 78.5 percent of these patients will survive for another *six* years—more than twice as long as median survival for the metastatic patients—and 67 percent of them will still be alive ten years after surgery. That's an impressive difference.

So what explains it? The difference has to do with the overall burden of cancer cells in each patient. In advanced, metastatic cancer there are tens if not hundreds of billions of cancer cells in need of treatment. In less advanced cancers, while there are undoubtedly still millions, if not billions, of cancer cells that have escaped the surgeon's scalpel, the far lower population means they will also have fewer mutations and thus less resistance to the treatment.

---

* For 95 percent of metastatic colon cancer patients.

It's a similar story for patients with breast cancer. Patients with HER2-positive metastatic breast cancer[*] can expect a median survival time of just under five years, with standard treatment consisting of three chemotherapy drugs. But if our patient has a smaller (< 3 cm), localized, HER2+ tumor that is removed surgically, plus adjuvant treatment with just two of these chemo drugs, she will have a 93 percent chance of living for at least another seven years without disease. The lower a patient's overall tumor burden, the more effective our drugs tend to be—and the greater the patient's odds of survival. Again, just as with immunotherapy treatment, it's a numbers game: the fewer cancerous cells we have, the greater our likelihood of success.

The problem is that we're still not very good at detecting cancer in these early stages—yet. Out of dozens of different types of cancers, we have agreed-upon, reliable screening methods for only *five:* lung (for smokers), breast, prostate, colorectal, and cervical. Even so, mainstream guidelines have been waving people away from some types of early screening, such as mammography in women and blood testing for PSA, prostate-specific antigen, in men. In part this has to do with cost, and in part this has to do with the risk of false positives that may lead to unnecessary or even dangerous treatment (entailing further costs). Both are valid issues, but let's set aside the cost issue and focus on the false-positive problem.

Medicine 2.0 says that because there are significant false positives with certain tests, we shouldn't do these tests on most people, period. But if we put on our Medicine 3.0 glasses, we see it differently: these tests are potentially useful, and they're just about all we have. So how can we make them more useful and accurate?

With all diagnostic tests, there is a trade-off between *sensitivity,* or the ability of the test to detect an existing condition (i.e., its true positive rate, expressed as a percentage), and *specificity,* which is the ability to determine that someone does *not* have that condition (i.e., true negative rate). Together,

---

* This means the expression of human epidermal growth factor receptor 2, which is a protein receptor on the surface of breast cancer cells that promotes growth. This is overexpressed in roughly 30 percent of breast cancers.

these represent the test's overall accuracy. In addition, however, we have to consider the prevalence of the disease in our target population. How likely is it that the person we are testing actually has this condition? Mammography has a sensitivity in the mideighties and a specificity in the low nineties. But if we are examining a relatively low-risk population, perhaps 1 percent of whom actually have breast cancer, then even a test with decent sensitivity is going to generate a fairly large number of false positives. In fact, in this low-risk group, the "positive predictive value" of mammography is only about 10 percent—meaning that if you do test positive, there is only about a one-in-ten chance that you actually have breast cancer. In other populations, with greater overall prevalence (and risk), the test performs much better.

The situation with mammography illustrates why we need to be very strategic about *who* we are testing and what their risk profile might be, and to understand what our test can and can't tell us. No single diagnostic test, for anything, is 100 percent accurate. So it is foolish to rely on just one test, not only for breast cancer but in many other areas as well. We need to think in terms of stacking test modalities—incorporating ultrasound and MRI in addition to mammography, for example, when looking for breast cancer. With multiple tests, our resolution improves and fewer unnecessary procedures will be performed.

In short, the problem is not the tests themselves but how we use them. Prostate cancer screening provides an even better example. It's no longer as simple as "Your PSA number is X or higher, and therefore we must biopsy your prostate, a painful procedure with many unpleasant possible side effects." Now we know to look at other parameters, such as PSA velocity (the speed at which PSA has been changing over time), PSA density (PSA value normalized to the volume of the prostate gland), and free PSA (comparing the amount of PSA that is bound versus unbound to carrier proteins in the blood). When those factors are taken into account, PSA becomes a much better indicator of prostate cancer risk.

Then there are other tests, such as the 4K blood test, which looks for specific proteins that might give us a better idea of how aggressive and potentially dangerous the patient's prostate cancer might be. The key question we want to answer is, Will our patient die *with* prostate cancer, as many men do, or will he die *from* it? We'd rather not disrupt his life, and potentially do him harm, in the course of finding out. Combining these blood tests I've just described with the techniques of multiparametric MRI imaging means that the likelihood of performing an unnecessary biopsy or surgery is now very low.

There is a similar but quieter controversy around screening for colorectal cancer (CRC), which has long been a rite of passage for those in middle age.* The purpose of the colonoscopy is to look not only for full-fledged tumors but also for polyps, which are growths that form in the lining of the colon. Most polyps remain small and harmless and never become cancerous, but some have the potential to become malignant and invade the wall of the colon. Not all polyps become cancer, but all colon cancers came from polyps. This is what makes a colonoscopy such a powerful tool. The endoscopist is able not only to spot potentially cancerous growths before they become dangerous but also to intervene on the spot, using instruments on the colonoscope to remove polyps for later examination. It combines screening and surgery into one procedure. It's an amazing tool.

Traditional guidelines have recommended colorectal cancer screenings for average-risk people ages fifty to seventy-five. These preventive screenings are fully covered under the Affordable Care Act, and if no polyps are found and the patient is of average risk, the procedure needs to be repeated only every ten years, according to consensus guidelines. But there is ample evidence out there that age fifty may be too old for a first screening, even in patients with average risk factors (that is, no family history of colon cancer and no personal history of inflammatory bowel disease). About 70 percent of

---

* There are several different colon cancer screening methods that break down into two categories: stool-based tests and direct visualization tests. Stool-based tests are essentially a screening test for a screening test. A positive stool-based test prompts a direct visualization test of the colon: either a flexible sigmoidoscopy, which allows the endoscopist to view the lower part of the colon (including the sigmoid and descending colon), or a traditional colonoscopy, which examines the entire colon. In my opinion, none of the other tests compares to a colonoscopy.

people who are diagnosed with CRC before the age of fifty have no family history or hereditary conditions linked to the disease. In 2020, some 3,640 Americans died from colorectal cancer before they turned fifty—and given the slow-moving nature of the disease, it's likely that many of those who died later than that already had the disease on their fiftieth birthday. This is why the American Cancer Society updated its guidelines in 2018, lowering the age to forty-five for people at average risk.

In my practice, we go further, typically encouraging average-risk individuals to get a colonoscopy by age forty—and even sooner if anything in their history suggests they may be at higher risk. We then repeat the procedure as often as every two to three years, depending on the findings from the previous colonoscopy. If a sessile (flat) polyp is found, for example, we're inclined to do it sooner than if the endoscopist finds nothing at all. Two or three years might seem like a very short window of time to repeat such an involved procedure, but colon cancer has been documented to appear within the span of as little as six months to two years after a normal colonoscopy. Better safe than sorry.*

Why do I generally recommend a colonoscopy before the guidelines do? Mostly because, of all the major cancers, colorectal cancer is one of the easiest to detect, with the greatest payoff in terms of risk reduction. It remains one of the top five deadliest cancers in the United States, behind lung (#1) and breast/prostate (#2 for women/men), and just ahead of pancreas (#4) and liver (#5) cancers. Of these five, though, CRC is the one we have the best shot at catching early. As it grows in a relatively accessible location, the colon, we can see it without any need for imaging techniques or a surgical biopsy. Because it is so easily observed, we understand its progression from normal tissue to polyp to tumor. Finding it early makes a huge difference, since we

* A study published in 2022 found that the risk of CRC was only reduced by 18 percent (relative) and 0.22 percent (absolute) in people advised to get one colonoscopy in a ten-year period versus those not advised to. However, only 42 percent of those advised to get a colonoscopy actually did so, and they only underwent one during the period of the study. I would argue this was not a test of the efficacy of frequent colonoscopy for prevention of CRC and was, instead, a test of the effectiveness of telling people to get (infrequent) colonoscopy.

can effectively eliminate polyps or growths on the spot. If only we could do that with arterial plaques.

My bottom line is that it is far better to screen early than risk doing it too late. Think asymmetric risk: It's possible that *not* screening early and frequently enough is the most dangerous option.*

Other cancers that are relatively easy to spot on visual examination include skin cancer and melanomas. The pap smear for cervical cancer is another well-established, minimally invasive test that I recommend my patients do yearly. When we're talking about cancers that develop *inside* the body, in our internal organs, things get trickier. We can't see them directly, so we must rely on imaging technologies such as low-dose CT scans for lung cancer. These scans are currently recommended in smokers and former smokers, but (as always) I think they should be used more widely, because about 15 percent of lung cancers are diagnosed in people who have *never* smoked. Lung cancer is the #1 cause of cancer deaths overall, but lung cancer in never-smokers ranks seventh, all by itself.

MRI has a distinct advantage over CT in that it does not produce any ionizing radiation but still provides good resolution. One newer technique that can enhance the ability of a screening MRI to differentiate between a cancer and noncancer is something called diffusion-weighted imaging with background subtraction, or DWI for short. The idea behind DWI is to look at water movement in and around tissue, at different points in time very close to each other (between ten and fifty microseconds, typically). If the water is held

---

* For those seeking more detailed guidance, this is what I wrote (Attia 2020a) in a blog post on CRC screening a few years ago: "Before you get your first colonoscopy, there are [a] few things you can do that may improve your risk-to-benefit ratio. You should ask what your endoscopist's adenoma detection rate (ADR) is. The ADR is the proportion of individuals undergoing a colonoscopy who have one or more adenomas (or colon polyps) detected. The benchmarks for ADR are greater than 30% in men and greater than 20% in women. You should also ask your endoscopist how many perforations he or she has caused, specifically, as well as any other serious complications, like major intestinal bleeding episodes (in a routine screening setting). Another question you should ask is what is your endoscopist's withdrawal time, defined as the amount of time spent viewing as the colonoscope is withdrawn during a colonoscopy. A longer withdrawal time suggests a more thorough inspection. A 6-minute withdrawal time is currently the standard of care."

or trapped, then it could indicate the presence of a tightly packed cluster of cells, a possible tumor. So the higher the density of cells, the brighter the signal on the DWI phase of MRI, making DWI functionally a radiographic "lump detector." Right now, DWI works best in the brain because it suffers least from movement artifacts.

I remain optimistic that this technique can be improved over time, with optimization of software and standardization of technique. Despite all this, even something as advanced as the best DWI MRI is not without problems, if used in isolation. While the sensitivity of this test is very high (meaning it's very good at finding cancer if cancer is there, hence very few false negatives), the specificity is relatively low (which means it's not as good at telling you when you don't have cancer, hence a lot of false positives). This is the inevitable trade-off, the yin and yang if you will, between sensitivity and specificity. The more you increase one, the more you decrease the other.*

I tell patients, if you're going to have a whole-body screening MRI, there is a good chance we'll be chasing down an insignificant thyroid (or other) nodule in exchange for getting such a good look at your other organs. As a result of this, about a quarter of my patients, understandably, elect not to undergo such screening. Which brings me to the next tool in the cancer screening tool kit, a tool that can complement the high sensitivity / low specificity problem of imaging tests.

I am cautiously optimistic about the emergence of so-called "liquid biopsies" that seek to detect the presence of cancers via a blood test.† These are used in two settings: to detect recurrences of cancer in patients following treatment and to screen for cancers in otherwise healthy patients, a fast-moving and exciting field called multicancer early detection.

Max Diehn, one of my med school classmates and now an oncology professor at Stanford, has been on the forefront of this research since 2012.

---

* The specificity of MRI is particularly reduced by glandular tissue. MRI is so good at detecting glandular cancer that it significantly overdoes it. The thyroid gland might be the worst offender.

† These are called liquid biopsies to distinguish them from traditional solid-tissue biopsies.

Max and his colleagues set out to ask a seemingly simple question initially. After a patient with lung cancer has had a resection of their tumor, is there any way a blood test can be used to screen the patient for signs of tumor recurrence?

Historically, this has been done via imaging tests, such as CT scans, that enable us to "see" a tumor. Radiation exposure notwithstanding, the main issue is that these tests don't have very high resolution. It's very difficult for these imaging technologies to discern a cancer smaller than about one centimeter in diameter. Even if you assume that this one-centimeter nodule is the only collection of cancer cells in the patient's body (not a great assumption; the mouse in the trap is rarely the only one in the house), you're still talking about more than a billion cancer cells by the time you reach the threshold of traditional detection. If we could catch these recurrent cancers sooner, we might have a better shot at keeping patients in remission—for the same reasons that it's easier to treat adjuvant versus metastatic cancer, as we discussed a few pages ago.

Max and his colleagues came up with a wholly different method. Because cancer cells are growing constantly, they tend to shed cellular matter, including bits of tumor DNA, into the circulation. What if there were a blood test that could detect this so-called cell-free DNA? We would already know the genetic signature of the tumor from the surgery—that is, how the lung cancer cells differ from normal lung cells. Thus, it should be possible to screen for this cell-free DNA in a patient's plasma and thereby determine the presence of cancer.

Make no mistake about it, this is still akin to looking for a needle in a haystack. In an early-stage cancer, which are the cancers we would most want to find via liquid biopsies, we might be talking about 0.01 to 0.001 percent of cell-free DNA coming from the cancer (or about one part in ten thousand to one hundred thousand). Only with the help of next-generation, high-throughput DNA screening technology is this possible. These tests are becoming more widely used in the postsurgical setting, but the technology is still relatively young. The key is that you must know what you're looking for—the patterns of mutations that distinguish cancer from normal cells.

Some researchers are beginning to develop ways to use blood tests to screen for cancer generally, in otherwise healthy people. This is an order of magnitude more difficult, like looking for a needle in ten haystacks; worse, in this case we don't even know what the needle should look like. We don't know anything about the patient's tumor mutation patterns, because we're not yet certain that they *have* cancer. Thus we must look for other potential markers. One company leading the charge with this type of assay is called Grail, a subsidiary of the genetic-sequencing company Illumina. The Grail test, known as Galleri, looks at methylation patterns of the cell-free DNA, which are basically chemical changes to the DNA molecules that suggest the presence of cancer. Using very-high-throughput screening and a massive AI engine, the Galleri test can glean two crucial pieces of information from this sample of blood: Is cancer present? And if so, where is it? From what part of the body did it most likely originate?

With any diagnostic test, a decision has to be made with respect to how to calibrate, or tune, it. Is it going to be geared toward higher sensitivity or higher specificity? Galleri has been validated against a database called the Circulating Cell-free Genome Atlas (CCGA), which is based on blood samples from more than fifteen thousand patients both with and without cancer. In this study, the Galleri test proved to have a very high specificity, about 99.5 percent, meaning only 0.5 percent of tests yielded a false positive. If the test says you have cancer, somewhere in your body, then it is likely that you do. The trade-off is that the resulting sensitivity can be low, depending on the stage. (That is, even if the test says you don't have cancer, you are not necessarily in the clear.)

The thing to keep in mind here, however, is that this test still has much higher resolution than radiographic tests such as MRI or mammogram. Those imaging-based tests require "seeing" the tumor, which can happen only when the tumor reaches a certain size. With Galleri, the test is looking at cell-free DNA, which can come from any size tumor—even ones that remain invisible to imaging tests.

One early observation of the CCGA study was that detectability was associated not only with the stage of tumor, which would be expected (the more advanced the tumor, the greater the likelihood of finding cell-free DNA in the blood), but also with the subtype of tumor. For example, the detection rate for stage I/II hormone receptor–*positive* breast cancer is about 25 percent, while the detection rate for stage I/II hormone receptor–*negative* breast cancer is about 75 percent. What does this difference tell us? We know that breast cancer is not a uniform disease and that the hormone receptor–negative tumors are more lethal than hormone receptor–positive tumors. Thus, the test proves more accurate in detecting the more lethal subtype of breast cancer.

Liquid biopsies could be viewed as having two functions: first, to determine cancer's presence or absence, a binary question; and second and perhaps more important, to gain insight into the specific cancer's biology. How dangerous is this particular cancer likely to be? The cancers that shed more cell-free DNA, it seems, also tend to be more aggressive and deadly—and are thus the cancers that we want to detect, and treat, as soon as possible. This technology is still in its infancy, but I'm hopeful that pairing different diagnostic tests ranging from radiographic (e.g., MRI) to direct visualization (e.g., colonoscopy) to biological/genetic (e.g., liquid biopsy) will allow us to correctly identify the cancers that need treatment the soonest, with the fewest possible false positives.

The implications of this, I think, are seismic: if liquid biopsies deliver on their promise, we could completely flip the time line of cancer so that we are routinely intervening early, when we have a chance of controlling or even eliminating the cancer—rather than the way we typically do it now, coming in at a late stage, when the odds are already stacked against the patient, and hoping for a miracle.

Of all the Horsemen, cancer is probably the hardest to prevent. It is probably also the one where bad luck in various forms plays the greatest role, such as in the form of accumulated somatic mutations. The only modifiable risks that really stand out in the data are smoking, insulin resistance, and obesity (all to be avoided)—and maybe pollution (air, water, etc.), but the data here are less clear.

We do have some treatment options for cancer, unlike with Alzheimer's disease (as we'll see in the next chapter), and immunotherapy in particular has great promise. Yet our treatment and prevention strategies remain far less effective than the tools that we have to address cardiovascular disease and the spectrum of metabolic dysfunction from insulin resistance to type 2 diabetes.

Until we learn how to prevent or "cure" cancer entirely, something I do not see happening in our lifetime, short of some miraculous breakthroughs, we need to focus far more energy on early detection of cancer, to enable better targeting of specific treatments at specific cancers while they are at their most vulnerable stages. If the first rule of cancer is "Don't get cancer," the second rule is "Catch it as soon as possible."

This is why I'm such an advocate for early screening. It's a simple truth that treating smaller tumors with fewer mutations is far easier than if we wait for the cancer to advance and potentially acquire mutations that help it evade our treatments. The only way to catch it early is with aggressive screening.

This does come at significant cost, which is why Medicine 2.0 tends to be more conservative about screening. There is a financial cost, of course, but there is also an emotional cost, particularly of tests that may generate false positives. And there are other, incidental risks, such as the slight risk from a colonoscopy or the more significant risk from an unnecessary biopsy. These three costs must be weighed against the cost of missing a cancer, or not spotting it early, when it is still susceptible to treatment.

Nobody said this was going to be easy. We still have a very long way to go. But there is finally hope, on multiple fronts—far more than when I was training to become a cancer surgeon. More than fifty years into the War on Cancer, we can finally see a path to a world where a cancer diagnosis typically means an early detection of a treatable problem rather than a late discovery of a grim one. Thanks to better screening and more effective treatments such as immunotherapy, cancer could someday become a manageable disease, perhaps no longer even qualifying as a Horseman.

CHAPTER 9

Chasing Memory

Understanding Alzheimer's Disease and Other Neurodegenerative Diseases

The greatest obstacle to discovery is not ignorance—it is the illusion of knowledge.

—Daniel J. Boorstin

Most people tend to go to the doctor when they are sick or think they might be. Nearly all my patients first come to see me when they are relatively healthy, or think they are. That was the case with Stephanie, a forty-year-old woman who walked into my office for her initial visit in early 2018 with no real complaint. She was simply "interested" in longevity.

Her family history was not remarkable. Three of her four grandparents had died from complications of atherosclerosis in their late seventies or early eighties, and the fourth had died from cancer. Pretty much par for the course for the Greatest Generation. The only red flag was the fact that her mother, otherwise healthy at seventy, was beginning to suffer some memory loss, which Stephanie attributed to "old age."

We set up another appointment for a week later to review her initial blood work. I rely as much as one can on biomarkers, so we run a comprehensive array of tests, but there are a few things that I immediately scan for when I get a new patient's results back. Among them is their level of Lp(a), the high-risk lipoprotein that we talked about in chapter 7, along with their apoB concentration. A third thing that I always check is their *APOE* genotype, the gene related to Alzheimer's disease risk that we mentioned in chapter 4.

Stephanie's labs revealed that she had the *APOE e4* allele, which is associated with a greater risk of Alzheimer's disease—and not just one copy, but two (*e4/e4*), which meant her risk of developing Alzheimer's disease was up to twelve times greater than that of someone with two copies of the common *e3* allele. The *e2* version of *APOE* appears to protect carriers against Alzheimer's disease: 10 percent reduced risk for someone with *e2/e3,* and about 20 percent for *e2/e2*. Stephanie was unlucky.

She was only the fourth patient I'd ever encountered with this quite uncommon genotype, shared by only about 2 to 3 percent of the population, and she had no idea she was at risk—although in hindsight, her mother's forgetfulness may have been a symptom of early Alzheimer's disease. Now I faced a double challenge: how to break the news to her, directly but gently; and even more tricky, how to explain what it meant and what it didn't mean.

In circumstances like this I generally think it's best to get right to the point, so after we sat down, I said something like, "Stephanie, there is something we found in your blood test that may be of concern to me—not because anything is wrong now, but because of the risk it poses in twenty to thirty years or so. You have a combination of genes that increases your risk of developing Alzheimer's disease. But it's also important that you understand that what we're about to discuss is only a marker for risk, not a fait accompli, and I am convinced that we can mitigate this risk going forward."

Stephanie was devastated. She was dealing with a lot of stress to begin with: a divorce, a difficult work situation, and now this. It's tough to explain the nuances of genes and risk to somebody when their eyes are wide with fear and they are hearing only *I'm doomed.* It took several discussions over the course of many weeks before she began to understand the rest of the message, which was that she was not, in fact, doomed.

Alzheimer's disease is perhaps the most difficult, most intractable of the Horsemen diseases. We have a much more limited understanding of how and why it begins, and how to slow or prevent it, than we do with atherosclerosis. Unlike with cancer, we currently have no way to treat it once symptoms begin. And unlike type 2 diabetes and metabolic dysfunction, it does not appear to be readily reversible (although the jury is still out on that). This is why, almost without exception, my patients fear dementia more than any other consequence of aging, including death. They would rather die from cancer or heart disease than lose their minds, their very selves.

Alzheimer's disease is the most common, but there are other neurodegenerative diseases that concern us. The most prevalent of these are Lewy body dementia and Parkinson's disease, which are actually different forms of a related disorder known (confusingly) as "dementia with Lewy bodies." The primary difference between them is that Lewy body dementia is primarily a dementing disorder, meaning it affects cognition, while Parkinson's disease is considered primarily (but not entirely) a movement disorder, although it does also result in cognitive decline. In the United States, about 6 million people are diagnosed with Alzheimer's disease, while about 1.4 million have Lewy body dementia, and 1 million have been diagnosed with Parkinson's, which is the fastest-growing neurodegenerative disease. Beyond that, there are a variety of less common but also serious neurodegenerative conditions such as amyotrophic lateral sclerosis (ALS, or Lou Gehrig's disease) and Huntington's disease.

All these result from some form of neurodegeneration, and as yet, there is no cure for any of them—despite the billions and billions of dollars that have been spent chasing these complex conditions. Maybe there will be a breakthrough in the near future, but for now our best and only strategy is to try to prevent them. The only shred of good news here is that while these disorders have traditionally been considered as completely separate and distinct diseases, evolving evidence suggests that there is a larger continuum between them than has previously been recognized, which means that some of our prevention strategies could apply to more than one of them as well.

Many doctors shy away from *APOE* gene testing. The conventional wisdom holds that someone with the high-risk *e4* allele is all but guaranteed to develop Alzheimer's disease, and there is nothing we can do for them. So why burden the patient with this terrible knowledge?

Because there are two types of bad news: that about things we can change, and about things that we can't. Assuming that a patient's *e4* status falls into the latter category is, in my opinion, a mistake. While it is true that over one-half of people with Alzheimer's disease have at least one copy of *e4*, merely possessing this risk gene is not the same as being diagnosed with dementia due to Alzheimer's disease. There are *e4/e4*-carrying centenarians without any signs of dementia, likely because they have other genes that protect them from *e4;* for example, a certain variant of the gene *Klotho (KL),* called *kl-vs,* seems to protect carriers of *e4* from developing dementia. And plenty of "normal" *e3/e3* carriers will still go on to develop Alzheimer's. Having the *e4* gene variant merely signals increased risk. It's not a done deal.

The other point I tried to make to Stephanie was that time was on her side. The disease rarely progresses to a clinical stage before about age sixty-five, even in patients with two copies of *e4*. That gives us about twenty-five years to try to prevent or delay her from developing this horrible illness with the tools currently at our disposal. In the interim, hopefully, researchers might come up with more effective treatments. It was a classic asymmetric situation, where doing nothing was actually the riskiest course of action.

Understanding Alzheimer's

Although Alzheimer's disease was first named in the early 1900s, the phenomenon of "senility" has been remarked on since ancient times. Plato believed that because advancing age seemingly "gives rise to all manners of forgetfulness as well as stupidity," older men were unsuited for leadership positions requiring acumen or judgment. William Shakespeare gave us an unforgettable portrayal of an old man struggling with his failing mind in *King Lear.*

The notion that this might be a disease was first suggested by Dr. Alois Alzheimer, a psychiatrist who worked as medical director at the state asylum in Frankfurt, Germany. In 1906, while performing an autopsy on a patient named Auguste Deter, a woman in her midfifties who had suffered from memory loss, hallucinations, aggression, and confusion in her final years, he noticed that something was clearly wrong with her brain. Her neurons were entangled and spiderweb-like, coated with a strange white dental substance. He was so struck by their odd appearance that he made drawings of them.

Another colleague later dubbed this condition "Alzheimer's disease," but after Alzheimer himself died in 1915 (of complications from a cold, at fifty-one), the disease he had identified was more or less forgotten for fifty years, relegated to obscurity along with other less common neurological conditions such as Huntington's and Parkinson's diseases as well as Lewy body dementia. Patients with the kinds of symptoms that we now associate with these conditions, including mood changes, depression, memory loss, irritability, and irrationality, were routinely institutionalized, as Auguste Deter had been. Plain old "senility," meanwhile, was considered to be an inevitable part of aging, as it had been since Plato's day.

It wasn't until the late 1960s that scientists began to accept that "senile dementia" was a disease state and not just a normal consequence of aging. Three British psychiatrists, Garry Blessed, Bernard Tomlinson, and Martin Roth, examined the brains of seventy patients who had died with dementia and found that many of them exhibited the same kinds of plaques and tangles that Alois Alzheimer had observed. Further studies revealed that a patient's degree of cognitive impairment seemed to correlate with the extent of plaques found in his or her brain. These patients, they concluded, also had Alzheimer's disease. A bit more than a decade later, in the early 1980s, other researchers identified the substance in the plaques as a peptide called amyloid-beta. Because it is often found at the scene of the crime, amyloid-beta was immediately suspected to be a primary cause of Alzheimer's disease.

Amyloid-beta is a by-product that is created when a normally occurring substance called amyloid precursor protein, or APP, a membrane protein that is found in neuronal synapses, is cleaved into three pieces. Normally, APP is split into two pieces, and everything is fine. But when APP is cut in thirds, one of the resulting fragments then becomes "misfolded," meaning it loses its normal structure (and thus its function) and becomes chemically stickier, prone to aggregating in clumps. This is amyloid-beta, and it is clearly bad stuff. Laboratory mice that have been genetically engineered to accumulate amyloid-beta (they don't naturally) have difficulty performing cognitive tasks that are normally easy, such as finding food in a simple maze. At the same time, amyloid also triggers the aggregation of another protein called tau, which in turn leads to neuronal inflammation and, ultimately, brain shrinkage. Tau was likely responsible for the neuronal "tangles" that Alois Alzheimer observed in Auguste Deter.

Scientists have identified a handful of genetic mutations that promote very rapid amyloid-beta accumulation, all but ensuring that someone will develop the disease, often at a fairly young age. These mutations, the most common of which are called *APP, PSEN1,* and *PSEN2*, typically affect the APP cleavage. In families carrying these genes, very-early-onset Alzheimer's disease is rampant, with family members often developing symptoms in their thirties and forties. Luckily, these mutations are very rare, but they occur in 10 percent of early-onset Alzheimer's cases (or about 1 percent of total cases). People with Down syndrome also tend to accumulate large amounts of amyloid plaques over time, because of genes related to APP cleavage that reside on chromosome 21.

It was not a huge leap to conclude, on the basis of available evidence, that Alzheimer's disease is caused directly by this accumulation of amyloid-beta in the brain. The "amyloid hypothesis," as it's called, has been the dominant theory of Alzheimer's disease since the 1980s, and it has driven the research priorities of the National Institutes of Health and the pharmaceutical industry alike. If you could eliminate the amyloid, the thinking has been, then you could halt or even reverse the progression of the disease. But it hasn't worked out that way. Several dozen drugs have been developed that target amyloid-beta in one way or another. But even when they succeed in clearing amyloid or slowing its production, these drugs have yet to show benefit in improving patients' cognitive function or slowing the progression of the disease. Every single one of them failed.

One hypothesis that emerged, as these drugs were failing one after another, was that patients were being given the drugs too late, when the disease had already taken hold. It is well known that Alzheimer's develops slowly, over decades. What if we gave the drugs earlier? This promising hypothesis has been tested in large and well-publicized clinical trials involving people with an inherited mutation that basically predestines them to early-onset Alzheimer's, but those, too, failed in the end. A broader clinical trial was launched in 2022 by Roche and Genentech, testing early administration of an anti-amyloid compound in genetically normal people with verified amyloid accumulation in their brains but no clear symptoms of dementia; results are expected in 2026. Some researchers think that the disease process might be reversible at the point where amyloid is present but not tau, which appears later. That theory is being tested in yet another ongoing study.

Meanwhile, in June of 2021, the FDA gave approval to an amyloid-targeting drug called aducanumab (Aduhelm). The drug's maker, Biogen, had submitted data for approval twice before and had been turned down. This was its third try. The agency's expert advisory panel recommended against approving the drug this time as well, saying the evidence of a benefit was weak or conflicted, but the agency went ahead and approved it anyway. It has met a tepid reception in the marketplace, with Medicare and some insurers refusing to pay its $28,000 annual cost unless it is being used in a clinical trial at a university.

This cascade of drug failures has caused frustration and confusion in the Alzheimer's field, because amyloid has long been considered a signature of the disease. As Dr. Ronald Petersen, director of the Mayo Clinic Alzheimer's Disease Research Center, put it to *The New York Times* in 2020: "Amyloid and tau define the disease. . . . To not attack amyloid doesn't make sense."

But some scientists have begun to openly question the notion that amyloid causes *all* cases of Alzheimer's disease, citing these drug failures above all. Their doubts seemed to be validated in July of 2022, when *Science* published an article calling into question a widely cited 2006 study that had given new impetus to the amyloid theory, at a time when it had already seemed to be weakening. The 2006 study had pinpointed a particular subtype of amyloid that it claimed directly caused neurodegeneration. That in turn inspired numerous investigations into that subtype. But according to the *Science* article, key images in that study had been falsified.

There was already plenty of other evidence calling into question the causal relationship that has long been assumed between amyloid and neurodegeneration. Autopsy studies have found that more than 25 percent of cognitively normal people nevertheless had large deposits of amyloid in their brains when they died—some of them with the same degree of plaque buildup as patients who died with severe dementia. But for some reason, these people displayed no cognitive symptoms. This was not actually a new observation: Blessed, Tomlinson, and Roth noted back in 1968 that other researchers had observed "plaque formation and other changes [that] were sometimes just as intense in normal subjects as in cases of senile dementia."

Some experts maintain that these patients actually did have the disease but that its symptoms had been slower to emerge, or that they had somehow masked or compensated for the damage to their brains. But more recent studies have found that the reverse can also be true: some patients with all the symptoms of Alzheimer's disease, including significant cognitive decline, have little to no amyloid in their brains, according to amyloid PET scans and/or cerebrospinal fluid (CSF) biomarker testing, two common diagnostic techniques. Researchers from the Memory and Aging Center at University of California San Francisco found via PET scans that close to one in three patients with mild to moderate dementia had no evidence of amyloid in their brains. Still other studies have found only a weak correlation between the degree of amyloid burden and the severity of disease. It appears, then, that the presence of amyloid-beta plaques may be neither necessary for the development of Alzheimer's disease nor sufficient to cause it.

This raises another possibility: that the condition that Alois Alzheimer observed in 1906 was not the same condition as the Alzheimer's disease that afflicts millions of people around the world. One major clue has to do with the age of onset. Typically, what we call Alzheimer's disease (or late-onset Alzheimer's disease) does not present in significant numbers until age sixty-five. But Dr. Alzheimer's own Patient Zero, Auguste Deter, showed severe symptoms by the time she was fifty, a trajectory more in line with early-onset Alzheimer's disease than with the dementia that slowly begins to afflict people in their late sixties, seventies, and eighties. A 2013 analysis of preserved tissue from Auguste Deter's brain found that she did in fact carry the *PSEN1* mutation, one of the early-onset dementia genes. (It affects the cleavage of the amyloid precursor protein, producing loads of amyloid.) She did have Alzheimer's disease, but a form of it that you get *only* because you have one of these highly deterministic genes. Our mistake might have been to assume that the other 99 percent of Alzheimer's disease cases progress the way hers did.

This is not all that uncommon in medicine, where the index case for a particular disease turns out to be the exception rather than the rule; extrapolating from this one case can lead to problems and misunderstanding down the road. At the same time, if Auguste Deter's illness had appeared when she was seventy-five instead of fifty, then perhaps it might not have seemed remarkable at all.

Just as Alzheimer's disease is defined (rightly or wrongly) by accumulations of amyloid and tau, Lewy body dementia and Parkinson's disease are associated with the accumulation of a neurotoxic protein called alpha-synuclein, which builds up in aggregates known as Lewy bodies (first observed by a colleague of Alois Alzheimer's named Friedrich Lewy). The *APOE e4* variant not only increases someone's risk for Alzheimer's but also significantly raises their risk of Lewy body dementia as well as Parkinson's disease with dementia, further supporting the notion that these conditions are related on some level.

All this places high-risk patients like Stephanie in a terrible predicament: they are at increased risk of developing a disorder or disorders whose causes we still do not fully understand, and for which we lack effective treatments. That means we need to focus on what until fairly recently was considered a taboo topic in neurodegenerative disease: prevention.

Can Neurodegenerative Disease Be Prevented?

Stephanie was terrified. I had treated other patients who had two copies of the *e4* gene, but none of them had responded with this much fear and anxiety. It took us four long discussions over a period of two months just to get her past the initial shock of the news. Then it was time to have a talk about what to do.

She had no obvious signs of cognitive impairment or memory loss. Yet. Just as a side note, some of my patients freak out because they lose their car keys or their phones from time to time. As I keep reminding them, that does not mean they have Alzheimer's disease. It generally only means that they are busy and distracted (ironically, often by the very same cell phones that they keep misplacing). Stephanie was different. Her risk was real, and now she knew it.

With my highest-risk patients, like Stephanie, I would at that time typically collaborate with Dr. Richard Isaacson, who opened the first Alzheimer's disease prevention clinic in the United States in 2013. Richard remembers that when he first interviewed with the dean of Weill Cornell Medical College and described his proposed venture, she seemed taken aback by his then-radical idea; the disease was not considered to be preventable. Also, at barely thirty, he did not look the part. "She was expecting Oliver Sacks," he told me. "Instead, she was sitting across the table from Doogie Howser."

Isaacson had gone off to college at age seventeen, to a joint BA-MD program at the University of Missouri in Kansas City and had his medical degree in hand by the time he was twenty-three. He was motivated to learn as much as possible about Alzheimer's disease by the fear that it ran in his family.

When he was growing up in Commack, Long Island, Richard had watched his favorite relative, his great-uncle Bob, succumb to Alzheimer's disease. When Richard was three, Uncle Bob had saved him from drowning in a swimming pool at a family party. But about a decade later, his beloved uncle had begun to change. He repeated his stories. His sense of humor faded. He developed a vacant stare and seemed to take longer to process what he was hearing around him. He was eventually diagnosed with Alzheimer's disease—but he was already gone. It was as if he had vanished.

Richard couldn't stand to watch his favorite uncle be reduced to nothing. He was also worried because he knew that Alzheimer's disease risk has a strong genetic component. Was he also at risk? Were his parents? Could the disease somehow be prevented?

As he went into medical practice at the University of Miami, he began collecting every single study and recommendation he could find on possible ways to reduce risk of Alzheimer's disease, gathering them into a photocopied sheaf that he would hand out to patients. He published the sheaf in 2010 as a book titled *Alzheimer's Treatment, Alzheimer's Prevention: A Patient and Family Guide.*

He soon found out that in Alzheimer's disease circles, the very word *prevention* was somewhat off limits. "The Alzheimer's Association said, you can't really say this," he remembers. That same year, a panel of experts assembled by the National Institutes of Health declared, "Currently, firm conclusions cannot be drawn about the association of any modifiable risk factor with cognitive decline or Alzheimer's disease."

That led Isaacson to realize that the only way to show that a preventive approach could work, in a way that would be accepted by the broader medical community, was in a large academic setting. Cornell was willing to take a chance, and his clinic opened in 2013. It was the first of its kind in the country, but now there are half a dozen similar centers, including one in Puerto Rico. (Isaacson now works for a private company in New York; his colleague, Dr. Kellyann Niotis, has joined my practice, focusing on patients at risk for neurodegenerative diseases.)

Meanwhile, the idea of preventing Alzheimer's disease began to gain scientific support. A two-year randomized controlled trial in Finland, published in 2015, found that interventions around nutrition, physical activity, and cognitive training helped maintain cognitive function and prevent cognitive decline among a group of more than 1,200 at-risk older adults. Two other large European trials have found that multidomain lifestyle-based interventions have improved cognitive performance among at-risk adults. So there were signs of hope.

To a typical doctor, Stephanie's case would have seemed pointless: She had no symptoms and she was still relatively young, in her early forties, a good two decades before any clinical dementia was likely to develop. Medicine 2.0 would say there was nothing for us to treat yet. In Medicine 3.0, this makes her an ideal patient, and her case an urgent one. If there were ever a disease that called for a Medicine 3.0 approach—where prevention is not only important but our *only option*—Alzheimer's disease and related neurodegenerative diseases are it.

Frankly, it might seem odd for a practice as small as ours to employ a full-time preventive neurologist such as Kellyann Niotis. Why do we do this? Because we think we can really move the needle by starting early and being very rigorous in how we quantify and then try to address each patient's risk. Some patients, like Stephanie, are at obvious higher risk; but in a broader sense, all of us are at some risk of Alzheimer's disease and other neurodegenerative disease.

Viewed through the lens of prevention, the fact that Stephanie had the *APOE e4/e4* genotype was actually good news, in a way. Yes, she was at far higher risk than someone with *e3/e3*, but at least we knew the genes we were up against and the likely trajectory of the disease, if she developed it. It is more worrying when a patient presents with an overwhelming family history of dementia, or the early signs of cognitive decline, but does *not* carry any of the known Alzheimer's risk genes, such as *APOE e4* and a few others. That means that there could be some other risk genes in play, and we don't know what they might be. Stephanie was at least facing a known risk. This was a start.

She had two additional risk factors that were out of her control: being Caucasian and being female. While those of African descent are at an overall increased risk of developing Alzheimer's disease, for unclear reasons, *APOE e4* seems to present *less* risk to them than to people of Caucasian, Asian, and Hispanic descent. Regardless of *APOE* genotype, however, Alzheimer's disease is almost twice as common in women than in men. It is tempting to attribute this to the fact that more women live to age eighty-five and above, where incidence of the disease pushes 40 percent. But that alone does not explain the differential. Some scientists believe there may be something about menopause, and the abrupt decline in hormonal signaling, that sharply increases the risk of neurodegeneration in older women. In particular, it appears that a rapid drop in estradiol in women with an *e4* allele is a driver of risk; that, in turn, suggests a possible role for perimenopausal hormone replacement therapy in these women.

Menopause is not the only issue here. Other reproductive history factors, such as the number of children the woman has had, age of first menstruation, and exposure to oral contraceptives, may also have a significant impact on Alzheimer's risk and later life cognition. And new research suggests that women are more prone to accumulate tau, the neurotoxic protein we mentioned earlier. The end result is that women have a greater age-adjusted risk of Alzheimer's, as well as faster rates of disease progression overall, regardless of age and educational level.

While female Alzheimer's patients outnumber men by two to one, the reverse holds true for Lewy body dementia and Parkinson's, both of which are twice as prevalent in men. Yet Parkinson's also appears to progress more rapidly in women than in men, for reasons that are not clear.

Parkinson's is also tricky genetically: While we have identified several gene variants that increase risk for PD, such as *LRRK2* and *SNCA*, about 15 percent of patients diagnosed have some family history of the disease—and are therefore presumed to have a genetic component—but do not have any known risk genes or SNPs.

Like the other Horsemen, dementia has an extremely long prologue. Its beginnings are so subtle that very often the disease isn't recognized until someone is well into its early stages. This is when their symptoms go beyond occasional lapses and forgetfulness to noticeable memory problems such as forgetting common words and frequently losing important objects (forgetting passwords becomes a problem too). Friends and loved ones notice changes, and performance on cognitive tests begins to slip.

Medicine 2.0 has begun to recognize this early clinical stage of Alzheimer's, which is known as mild cognitive impairment (MCI). But MCI is not the first stage on the long road to dementia: a large 2011 analysis of data from the UK's Whitehall II cohort study found that subtler signs of cognitive changes often become apparent well before patients meet the criteria for MCI. This is called stage I preclinical Alzheimer's disease, and in the United States alone over forty-six million people are estimated to be in this stage, where the disease is slowly laying the pathological scaffolding in and around neurons but major symptoms are still largely absent. While it's not clear how many of these patients will go on to develop Alzheimer's, what is clear is that just as most of an iceberg lies beneath the surface of the ocean, dementia can progress unnoticed for years before any symptoms appear.

The same is true of other neurodegenerative diseases, although they each have different early warning signs. Parkinson's may show up as subtle changes in movement patterns, a frozen facial expression, stooped posture or shuffling gait, a mild tremor, or even changes in a person's handwriting (which may become small and cramped). Someone in the early stages of Lewy body dementia may exhibit similar physical symptoms, but with slight cognitive changes as well; both may exhibit alterations in mood, such as depression or anxiety. Something seems "off," but it's hard for a layperson to pinpoint.

This is why an important first step with any patient who may have cognitive issues is to subject them to a grueling battery of tests. One reason I like to have a preventive neurologist on staff is that these tests are so complicated and difficult to administer that I feel they are best left to specialists. They are also critically important to a correct diagnosis—assessing whether the patient is already on the road to Alzheimer's disease or to another form of neurodegenerative dementia, and how far along they might be. These are clinically validated, highly complex tests that cover every domain of cognition and memory, including executive function, attention, processing speed, verbal fluency and memory (recalling a list of words), logical memory (recalling a phrase in the middle of a paragraph), associative memory (linking a name to a face), spatial memory (location of items in a room), and semantic memory (how many animals you can name in a minute, for example). My patients almost always come back complaining about the difficulty of the tests. I just smile and nod.

The intricacies and nuances of the tests give us important clues about what might be happening inside the brains of patients who are still very early in the process of cognitive change that goes along with age. Most importantly, they enable us to distinguish between normal brain aging and changes that may lead to dementia. One important section of the cognitive testing evaluates the patient's sense of smell. Can they correctly identify scents such as coffee, for example? Olfactory neurons are among the first to be affected by Alzheimer's disease.

Specialists such as Richard and Kellyann also become attuned to other, less quantifiable changes in people on the road to Alzheimer's disease, including changes in gait, facial expressions during conversations, even visual tracking. These changes could be subtle and not recognizable to the average person, but someone more skilled can spot them.

The trickiest part of the testing is interpreting the results to distinguish among different types of neurodegenerative disease and dementia. Kellyann dissects the test results to try and trace the likely location of the pathology in the brain, and the specific neurotransmitters that are involved; these determine the pathological features of the disease. Frontal and vascular dementias primarily affect the frontal lobe, a region of the brain responsible for executive functioning such as attention, organization, processing speed, and problem solving. So these forms of dementia rob an individual of such higher-order cognitive features. Alzheimer's disease, on the other hand, predominantly affects the temporal lobes, so the most distinct symptoms relate to memory, language, and auditory processing (forming and comprehending speech)—although researchers are beginning to identify different possible subtypes of Alzheimer's disease, based on which brain regions are most affected. Parkinson's is a bit different in that it manifests primarily as a movement disorder, resulting from (in part) a deficiency in producing dopamine, a key neurotransmitter. While Alzheimer's can be confirmed by testing for amyloid in the cerebrospinal fluid, these other forms of neurodegeneration are largely clinical diagnoses, based on testing and interpretation. Thus, they can be more subjective, but with all these conditions it is critical to identify them as soon as possible, to allow more time for preventive strategies to work.

One reason why Alzheimer's and related dementias can be so tricky to diagnose is that our highly complex brains are adept at compensating for damage, in a way that conceals these early stages of neurodegeneration. When we have a thought or a perception, it's not just one neural network that is responsible for that insight, or that decision, but many individual networks working simultaneously on the same problem, according to Francisco Gonzalez-Lima, a behavioral neuroscientist at the University of Texas in Austin. These parallel networks can reach different conclusions, so when we use the expression "I am of two minds about something," that is not scientifically inaccurate. The brain then picks the most common response. There is redundancy built into the system.

The more of these networks and subnetworks that we have built up over our lifetime, via education or experience, or by developing complex skills such as speaking a foreign language or playing a musical instrument, the more resistant to cognitive decline we will tend to be. The brain can continue functioning more or less normally, even as some of these networks begin to fail. This is called "cognitive reserve," and it has been shown to help some patients to resist the symptoms of Alzheimer's disease. It seems to take a longer time for the disease to affect their ability to function. "People that have Alzheimer's disease and are very cognitively engaged, and have a good backup pathway, they're not going to decline as quickly," Richard says.

There is a parallel concept known as "movement reserve" that becomes relevant with Parkinson's disease. People with better movement patterns, and a longer history of moving their bodies, such as trained or frequent athletes, tend to resist or slow the progression of the disease as compared to sedentary people. This is also why movement and exercise, not merely aerobic exercise but also more complex activities like boxing workouts, are a primary treatment/prevention strategy for Parkinson's. Exercise is the only intervention shown to delay the progression of Parkinson's.

But it's difficult to disentangle cognitive reserve from other factors, such as socioeconomic status and education, which are in turn linked to better metabolic health and other factors (also known as "healthy user bias"). Thus, the evidence on whether cognitive reserve can be "trained" or used as a preventive strategy, such as by learning to play a musical instrument or other forms of "brain training," is highly conflicted and not conclusive—although neither of these can hurt, so why not?

The evidence suggests that tasks or activities that present more varied challenges, requiring more nimble thinking and processing, are more productive at building and maintaining cognitive reserve. Simply doing a crossword puzzle every day, on the other hand, seems only to make people better at doing crossword puzzles. The same goes for movement reserve: dancing appears to be more effective than walking at delaying symptoms of Parkinson's disease, possibly because it involves more complex movement.

This was one thing that Stephanie, a high-performing, well-educated professional, had in her favor. Her cognitive reserve was very robust, and her baseline scores were strong. This meant that we likely had plenty of time to devise a prevention strategy for her, perhaps decades—but given her increased genetic risk, we could not afford to delay. We needed to come up with a plan. What would that plan look like? How can this seemingly unstoppable disease be prevented?

We'll start by taking a closer look at changes that might be happening *inside* the brain of someone on the road to Alzheimer's. How are these changes contributing to the progression of the disease, and can we do anything to stop them or limit the damage?

Once we begin looking at Alzheimer's disease outside the prism of the amyloid theory, we start to see certain other defining characteristics of dementia that might offer opportunities for prevention—weaknesses in our opponent's armor.

Alternatives to Amyloid

For decades, almost in tandem with observations of plaques and tangles, researchers have also noted problems with cerebral blood flow, or "perfusion," in patients with dementia. On autopsy, Alzheimer's brains often display marked calcification* of the blood vessels and capillaries that feed them. This is not a new observation: In their seminal 1968 paper that defined Alzheimer's disease as a common age-related condition, Blessed, Tomlinson, and Roth had also noted severe vascular damage in the brains of their deceased study subjects. The phenomenon had been noted in passing for decades, as far back as 1927. But it was generally considered to be a consequence of neurodegeneration and not a potential cause.

In the early 1990s, a Case Western Reserve neurologist named Jack de la Torre was flying to Paris for a conference and thinking about the origins of Alzheimer's disease. The amyloid hypothesis was still fairly new, but it didn't sit well with de la Torre because of what he had observed in his own lab. On the flight, he had a eureka moment. "The evidence from dozens of rat experiments seemed to be screaming at me," he later wrote. In those experiments, he had restricted the amount of blood flowing to the rats' brains, and over time they had developed symptoms remarkably similar to those of Alzheimer's disease in humans: memory loss and severe atrophy of the cortex and hippocampus. Restoring blood flow could halt or reverse the damage to some extent, but it seemed to be more severe and more lasting in older animals than younger ones. The key insight was that robust blood flow seemed to be critical to maintaining brain health.

The brain is a greedy organ. It makes up just 2 percent of our body weight, yet it accounts for about 20 percent of our total energy expenditure. Its eighty-six billion neurons *each* have between one thousand and ten thousand synapses connecting them to other neurons or target cells, creating our thoughts, our personalities, our memories, and the reasoning behind both our good

---

* You may recall from chapter 7 that calcification is part of the repair process for blood vessels damaged by the forces of atherosclerosis.

and bad decisions. There are computers that are bigger and faster, but no machine yet made by man can match the brain's ability to intuit and learn, much less feel or create. No computer possesses anything approaching the multidimensionality of the human self. Where a computer is powered by electricity, the beautiful machine that is the human brain depends on a steady supply of glucose and oxygen, delivered via a huge and delicate network of blood vessels. Even slight disruptions to this vascular network can result in a crippling or even fatal stroke.

On top of this, brain cells metabolize glucose in a different way from the rest of the body; they do not depend on insulin, instead absorbing circulating glucose directly, via transporters that essentially open a gate in the cell membrane. This enables the brain to take top priority to fuel itself when blood glucose levels are low. If we lack new sources of glucose, the brain's preferred fuel, the liver converts our fat into ketone bodies, as an alternative energy source that can sustain us for a very long time, depending on the extent of our fat stores. (Unlike muscle or liver, the brain itself does not store energy.) When our fat runs out, we will begin to consume our own muscle tissue, then our other organs, and even bone, all in order to keep the brain running at all costs. The brain is the last thing to shut off.

As his plane crossed the Atlantic, de la Torre scribbled down his ideas on the only available writing surface, which happened to be an air-sickness bag. The flight attendants looked grim as he asked for another bag, and another. His "barf bag theory," as he jokingly called it, was that Alzheimer's disease is primarily a vascular disorder of the brain. The dementia symptoms that we see result from a gradual reduction in blood flow, which eventually creates what he calls a "neuronal energy crisis," which in turn triggers a cascade of unfortunate events that harms the neurons and ultimately causes neurodegeneration. The amyloid plaques and tangles come later, as a consequence rather than a cause. "We believed, and still do, that amyloid-beta is an important pathological *product* of neurodegeneration," de la Torre wrote recently, ". . . [but] it is not the cause of Alzheimer's disease."

There was already evidence to support his theory. Alzheimer's is more likely to be diagnosed in patients who have suffered a stroke, which typically results from a sudden blockage of blood flow in specific regions of the brain. In these cases, symptoms emerge abruptly, as if a switch has been flipped. Additionally, it has been established that people with a history of cardiovascular disease are at a higher risk of developing Alzheimer's disease. Evidence also demonstrates a linear relationship between cognitive decline and increased intimal media thickness in the carotid artery, a major blood vessel that feeds the brain. Cerebral blood flow already declines naturally during the aging process, and this arterial thickening, a measure of arterial aging, could cause a further reduction in cerebral blood supply. Vascular disease is not the only culprit here either. In all, some two dozen known risk factors for Alzheimer's disease also happen to reduce blood flow, including high blood pressure, smoking, head injury, and depression, among others. The circumstantial evidence is strong.

Improved neuroimaging techniques have confirmed not only that cerebral perfusion is decreased in brains affected by Alzheimer's disease but also that a drop in blood flow seems to predict *when* a person will transition from preclinical Alzheimer's disease to MCI, and on to full-fledged dementia. Although vascular dementia is currently considered distinct from dementia due to Alzheimer's, making up roughly 15 to 20 percent of dementia diagnoses in North America and Europe, and up to 30 percent in Asia and developing countries, its symptoms and pathology overlap so significantly that de la Torre considers them different manifestations of the same basic condition.

Another compelling and perhaps parallel theory of Alzheimer's disease says that it stems from abnormal glucose metabolism in the brain. Scientists and physicians have long noted a connection between Alzheimer's disease and metabolic dysfunction. Having type 2 diabetes doubles or triples your risk of developing Alzheimer's disease, about the same as having one copy of the *APOE e4* gene. On a purely mechanistic level, chronically elevated blood glucose, as seen in type 2 diabetes and prediabetes/insulin resistance, can directly damage the vasculature of the brain. But insulin resistance alone is enough to elevate one's risk

Insulin seems to play a key role in memory function. Insulin receptors are highly concentrated in the hippocampus, the memory center of the brain.

Several studies have found that spraying insulin right into subjects' noses—administering it as directly as possible into their brains—quickly improves cognitive performance and memory, even in people who have already been diagnosed with Alzheimer's disease. One study found that intranasal insulin helped preserve brain volume in Alzheimer's patients. Clearly, it is helpful to get glucose into neurons; insulin resistance blocks this. As the authors wrote, "Several lines of evidence converge to suggest that central insulin resistance plays a causal role in the development and progression of Alzheimer's disease."

The signal event here (again) appears to be a drop in energy delivery to the brain, similar to what is seen in the onset of vascular dementia. Brain imaging studies reveal lower brain glucose metabolism, decades before the onset of other symptoms of vascular dementia. Intriguingly, this reduction appears to be especially dramatic in brain regions that are also affected in Alzheimer's disease, including the parietal lobe, which is important for processing and integrating sensory information; and the hippocampus of the temporal lobe, which is critical to memory. Just like reduced blood flow, reduced glucose metabolism essentially starves these neurons of energy, provoking a cascade of responses that include inflammation, increased oxidative stress, mitochondrial dysfunction—and ultimately neurodegeneration itself.

The Role of *APOE e4*

It is still not completely clear how or why, but *e4* seems to accelerate other risk factors and driver mechanisms for Alzheimer's—particularly metabolic factors such as reduced brain glucose metabolism, which we've just discussed. Simply put, it appears to make everything worse, including the Alzheimer's gender gap: a woman with one copy of *e4* is four times more likely to develop the disease than a man with the same genotype.

The protein for which it codes, APOE (apolipoprotein E), plays an important role in both cholesterol transport and glucose metabolism. It serves as the main cholesterol carrier in the brain, moving cholesterol across the blood-brain barrier to supply the neurons with the large amounts of it they require. Hussain Yassine, a neuroscientist at the University of Southern California who studies the role of APOE in Alzheimer's disease, compares its role to that of an orchestra conductor. For some reason, he says, people with the *e4* allele appear to have defects in both cholesterol transport and glucose metabolism, to a degree not seen in those with *e2* or *e3*. Even though the higher risk APOE e4 protein differs from the harmless e3 one by just one amino acid, it appears to be less efficient at moving cholesterol into and especially out of the brain. There is also some evidence that the APOE e4 protein may also cause early breakdown of the blood-brain barrier itself, making the brain more susceptible to injury and eventual degeneration.

Curiously, *APOE e4* was not always a bad actor. For millions of years, *all* our post-primate ancestors were *e4/e4*. It was the original human allele. The *e3* mutation showed up about 225,000 years ago, while *e2* is a relative latecomer, arriving only in the last 10,000 years. Data from present-day populations with a high prevalence of *e4* suggest that it may have been helpful for survival in environments with high levels of infectious disease: children carrying *APOE e4* in Brazilian favelas are more resistant to diarrhea and have stronger cognitive development, for example. In environments where infectious disease was a leading cause of death, *APOE e4* carriers may have been the lucky ones, in terms of longevity.

This survival benefit may have been due to the role of APOE e4 in promoting inflammation, which can be beneficial in some situations (e.g., fighting infection) but harmful in others (e.g., modern life). As we saw in chapter 7, inflammation promotes atherosclerotic damage to our blood vessels, setting the stage for Alzheimer's disease and dementia. People with Alzheimer's disease often have high levels of inflammatory cytokines such as TNF-alpha and IL-6 in their brains, and studies have also found higher levels of neuroinflammation in *e4* carriers. None of these, obviously, are good for our long-term brain health; as noted earlier, *e4* just seems to make every risk factor for Alzheimer's disease worse.

The *e4* variant also seems to be maladaptive in other ways, such as in dealing with our modern diets. Not only are *e4* carriers more likely to develop metabolic syndrome in the first place, but the APOE e4 protein may be partially responsible for this, by disrupting the brain's ability to regulate insulin levels and maintain glucose homeostasis in the body. This phenomenon becomes apparent when these patients are on continuous glucose monitoring, or CGM (which we'll discuss in more detail in chapter 15). Even young patients with *e4* show dramatic blood glucose spikes after eating carbohydrate-rich foods, although the clinical significance of this is unclear.

Thus, *e4* itself could help drive the very same metabolic dysfunction that also increases risk of dementia. At the same time, it appears to intensify the damage done *to* the brain by metabolic dysfunction. Researchers have found that in high-glucose environments, the aberrant form of the APOE protein encoded by *APOE e4* works to block insulin receptors in the brain, forming sticky clumps or aggregates that prevent neurons from taking in energy.

But not everyone with the *APOE e4* genotype is affected by it in the same way. Its effects on disease risk and the course of disease are highly variable. Factors like biological sex, ethnicity, and lifestyle clearly play a role, but it is now believed that Alzheimer's risk and the effect of *APOE* are also powerfully dependent on *other* Alzheimer's-risk-related genes that a person might carry, such as *Klotho,* the protective gene we mentioned earlier. This could explain, for example, why some people with *e4* may never go on to develop Alzheimer's disease, while others do so quickly.

All this suggests that metabolic and vascular causes of dementia may be somewhat overlapping, just as patients with insulin resistance are also prone to vascular disease. And it tells us that with high-risk patients like Stephanie, we need to pay special attention to their metabolic health.

The Preventive Plan

In spite of everything, I remain cautiously optimistic for patients like Stephanie, even with her highly elevated genetic risk. The very concept of Alzheimer's prevention is still relatively new; we have only begun to scratch the surface of what might be accomplished here. As we better understand the disease, our treatments and interventions can become more sophisticated and hopefully effective.

I actually think we know more about preventing Alzheimer's than we do about preventing cancer. Our primary tool for preventing cancer is to not smoke and to keep our metabolic health on track, but that's a very broad-brush approach that only takes us so far. We still need to screen aggressively and hope we somehow manage to find any cancers that do develop before it's too late. With Alzheimer's disease, we have a much larger preventive tool kit at our disposal, and much better diagnostic methods as well. It's relatively easy to spot cognitive decline in its early stages, if we're looking carefully. And we're learning more about genetic factors as well, including those that at least partially offset high-risk genes like *APOE e4*.

Because metabolism plays such an outsize role with at-risk *e4* patients like Stephanie, our first step is to address any metabolic issues they may have. Our goal is to improve glucose metabolism, inflammation, and oxidative stress. One possible recommendation for someone like her would be to switch to a Mediterranean-style diet, relying on more monounsaturated fats and fewer refined carbohydrates, in addition to regular consumption of fatty fish. There is some evidence that supplementation with the omega-3 fatty acid DHA, found in fish oil, may help maintain brain health, especially in *e4/e4* carriers. Higher doses of DHA may be required because of *e4*-induced metabolic changes and dysfunction of the blood-brain barrier.

This is also one area where a ketogenic diet may offer a real functional advantage: when someone is in ketosis, their brain relies on a mix of ketones and glucose for fuel. Studies in Alzheimer's patients find that while their brains become less able to utilize glucose, their ability to metabolize ketones does not decline. So it may make sense to try to diversify the brain's fuel source from only glucose to both glucose and ketones. A systematic review of randomized controlled trials found that ketogenic therapies improved general cognition and memory in subjects with mild cognitive impairment and early-stage Alzheimer's disease. Think of it as a flex-fuel strategy.

In Stephanie's case, she cut out not only added sugar and highly refined carbohydrates but also alcohol. The precise role of alcohol in relation to Alzheimer's disease remains somewhat controversial: some evidence suggests that alcohol may be slightly protective against Alzheimer's, while other evidence shows that heavier drinking is itself a risk factor for the disease, and *e4* carriers may be more susceptible to alcohol's deleterious effects. I'm inclined to err on the side of caution, and so is Stephanie.

The single most powerful item in our preventive tool kit is exercise, which has a two-pronged impact on Alzheimer's disease risk: it helps maintain glucose homeostasis, and it improves the health of our vasculature. So along with changing Stephanie's diet, we put her back on a regular exercise program, focusing on steady endurance exercise to improve her mitochondrial efficiency. This had a side benefit in that it helped manage her off-the-charts high cortisol levels, due to stress; stress and anxiety-related risk seem more significant in females. As we'll see in chapter 11, endurance exercise produces factors that directly target regions of the brain responsible for cognition and memory. It also helps lower inflammation and oxidative stress.

Strength training is likely just as important. A study looking at nearly half a million patients in the United Kingdom found that grip strength, an excellent proxy for overall strength, was strongly and inversely associated with the incidence of dementia (see figure 8). People in the lowest quartile of grip strength (i.e., the weakest) had a 72 percent higher incidence of dementia, compared to those in the top quartile. The authors found that this association held up even after adjusting for the usual confounders such as age, sex, socioeconomic status, diseases such as diabetes and cancer, smoking, and lifestyle factors such as sleep patterns, walking pace, and time spent watching TV. And there appeared to be no upper limit or "plateau" to this relationship; the greater someone's grip strength, the lower their risk of dementia.

It's tempting to dismiss findings like these for the same reasons we should be skeptical of epidemiology. But unlike epidemiology in nutrition (much more on that in chapter 14), the epidemiology linking strength and cardiorespiratory fitness to lower risk for neurodegeneration is so uniform in its direction and magnitude that my own skepticism of the power of exercise, circa 2012, has slowly melted away. I now tell patients that exercise is, full stop and

Figure 8. **Association of Handgrip Strength with Dementia Incidence**

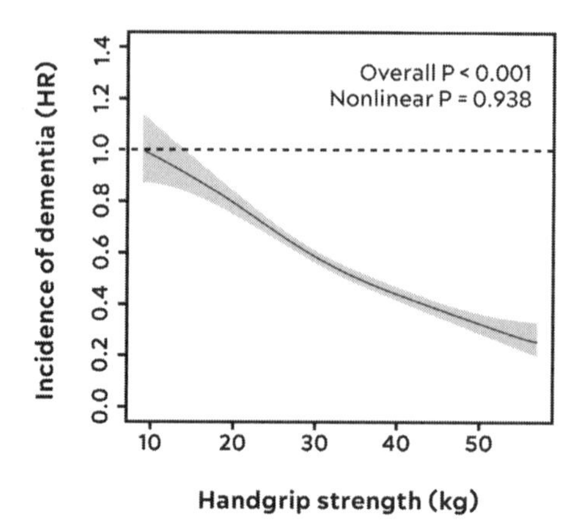

This graph shows how the incidence of dementia declines with increasing handgrip strength. Note that data are presented as hazard ratios in comparison with the weakest group; e.g., 0.4 = 40 percent. Thus someone with 40 kg grip strength has about 40 percent as much risk of dementia as someone with 10 kg.

Source: Esteban-Cornejo et al. (2022).

hands down, the best tool we have in the neurodegeneration prevention tool kit. (We'll explore the ins and outs of this in great detail in chapters 11 and 12.)

Sleep is also a very powerful tool against Alzheimer's disease, as we'll see in chapter 16. Sleep is when our brain heals itself; while we are in deep sleep our brains are essentially "cleaning house," sweeping away intracellular waste that can build up between our neurons. Sleep disruptions and poor sleep are potential drivers of increased risk of dementia. If poor sleep is accompanied by high stress and elevated cortisol levels, as in Stephanie's case, that acts almost as a multiplier of risk, as it contributes to insulin resistance and damaging the hippocampus at the same time. Furthermore, hypercortisolemia (excess cortisol due to stress) impairs the release of melatonin, the hormone that normally signals to our brains that it is time to go to sleep (and that may also help prevent neuronal loss and cognitive impairment). Addressing Stephanie's difficulties with sleep was therefore urgent. Her divorce and her work situation were making it almost impossible for her to get more than four hours of uninterrupted sleep on any given night.

Another somewhat surprising risk factor that has emerged is hearing loss. Studies have found that hearing loss is clearly associated with Alzheimer's disease, but it's not a direct symptom. Rather, it seems hearing loss may be causally linked to cognitive decline, because folks with hearing loss tend to pull back and withdraw from interactions with others. When the brain is deprived of inputs—in this case auditory inputs—it withers. Patients with hearing loss miss out on socializing, intellectual stimulation, and feeling connected; prescribing them hearing aids may help relieve some symptoms. This is just a hypothesis for the moment, but it is being tested right now in a clinical trial called ACHIEVE (Aging and Cognitive Health Evaluation in Elders) that is currently ongoing.

While depression is also associated with Alzheimer's disease, it appears to be more of a symptom than a risk factor or driver of the disease. Nevertheless, treating depression in patients with MCI or early Alzheimer's disease does appear to help reduce some other symptoms of cognitive decline.

Another surprising intervention that may help reduce systemic inflammation, and possibly Alzheimer's disease risk, is brushing *and* flossing one's teeth. (You heard me: *Floss.*) There is a growing body of research linking oral health, particularly the state of one's gum tissue, with overall health. Researchers have found that one pathogen in particular, a microbe called *P. gingivalis* that commonly causes gum disease, is responsible for large increases in levels of inflammatory markers such as IL-6. Even stranger, *P. gingivalis* has also shown up inside the brains of patients with Alzheimer's disease, although scientists are not certain that this bacterium is directly causing dementia, notes Dr. Patricia Corby, a professor of dental health at New York University. Nevertheless, the association is too strong to be ignored. (Also, better oral health correlates strongly with better overall health, particularly in terms of cardiovascular disease risk, so I pay much more attention to flossing and gum health than I used to.)

One other somewhat recent addition to my thinking on dementia (and

ASCVD while we're at it) prevention is the use of dry saunas. Until about 2019 I was very skeptical of the data linking sauna use to brain and heart health. However, the more time I spend buried in this literature, the more I become convinced by the magnitude of the benefit, the uniformity of the studies, and the mechanisms providing plausibility. I'm not quite as confident that regular sauna use will reduce your risk of Alzheimer's disease as I am that exercise will do so, but I am much more confident than I was at the outset of my journey. The best interpretation I can draw from the literature suggests that at least four sessions per week, of at least twenty minutes per session, at 179 degrees Fahrenheit (82 degrees Celsius) or hotter seems to be the sweet spot to reduce the risk of Alzheimer's by about 65 percent (and the risk of ASCVD by 50 percent).

Other potential interventions that have shown some promise in studies include lowering homocysteine with B vitamins, while optimizing omega-3 fatty acids. Higher vitamin D levels have been correlated with better memory in *e4/e4* patients but it's difficult to know from the current literature if this means supplementing with vitamin D will reduce risk of AD. And as mentioned earlier, hormone replacement therapy for women during the transition from perimenopause to menopause seems promising, especially for women with at least one copy of *e4*.

The scariest aspect of Alzheimer's disease boils down to this: Medicine 2.0 cannot help us. At all. The point at which Medicine 2.0 steps in, the point of diagnosis, is also likely near the point of no return for most Alzheimer's patients, beyond which little or nothing can be done. Once dementia is diagnosed, it is extremely difficult to slow and maybe impossible to reverse (though we're not certain of that). So we are forced to leave the familiar territory of the medicine that we know, with its promise of certainty, and embrace the Medicine 3.0 concepts of prevention and risk reduction.

As it stands now, Alzheimer's disease is the last of the Horsemen that we must bypass on our way to becoming centenarians; it's the last obstacle we face. Typically, it is diagnosed later in life—and centenarians develop it *much* later in life, if at all. The longer we can go without developing dementia, the better our odds of living longer, and living in better health. (Remember, cognition is one of the three key vectors of healthspan.) But until science comes up with more effective treatments, prevention is our only option. Therefore, we need to adopt a very early and comprehensive approach to preventing Alzheimer's and other forms of neurodegenerative disease.

Broadly, our strategy should be based on the following principles:

1. WHAT'S GOOD FOR THE HEART IS GOOD FOR THE BRAIN. That is, vascular health (meaning low apoB, low inflammation, and low oxidative stress) is crucial to brain health.
2. WHAT'S GOOD FOR THE LIVER (AND PANCREAS) IS GOOD FOR THE BRAIN. Metabolic health is crucial to brain health.
3. TIME IS KEY. We need to think about prevention early, and the more the deck is stacked against you genetically, the harder you need to work and the sooner you need to start. As with cardiovascular disease, we need to play a very long game.
4. OUR MOST POWERFUL TOOL FOR PREVENTING COGNITIVE DECLINE IS EXERCISE. We've talked a lot about diet and metabolism, but exercise appears to act in multiple ways (vascular, metabolic) to preserve brain health; we'll get into more detail in Part III, but exercise—lots of it—is a foundation of our Alzheimer's-prevention program.

I have great hope that in the future we will learn much more about how to prevent and treat all forms of dementia. But it's going to take hard work and creative thinking from scientists researching the disease, a significant investment in new theories and approaches, much more attention to strategies of prevention, and courage on the part of patients such as Stephanie who must face down this most feared and least understood of all the Horsemen.

PART III

CHAPTER 10

Thinking Tactically

Building a Framework of Principles That Work for You

Absorb what is useful, discard what is useless,
and add what is specifically your own.

—Bruce Lee

In the mid-nineteenth century, a French physician named Stanislas Tanchou observed that cancer was becoming ever more prevalent in the fast-growing cities of Europe. The Industrial Revolution was charging ahead at full speed, changing society in unimaginable ways. He saw a connection between the two: "Cancer, like insanity, seems to increase with the progress of civilization."

He was prescient. Eventually cancer, as well as heart disease, type 2 diabetes, and dementia (along with a few others), became collectively known as "diseases of civilization," because they seemed to have spread in lockstep with the industrialization and urbanization of Europe and the United States.

This doesn't mean that civilization is somehow "bad" and that we all need to return to a hunter-gatherer lifestyle. I would much rather live in our modern world, where I worry about losing my iPhone or missing a plane flight, than endure the rampant disease, random violence, and lawlessness that our ancestors suffered through for millennia (and that people in some parts of our world still experience). But even as modern life has helped extend our lifespans and improve living standards, it has also created conditions that conspire to *limit* our longevity in certain ways.

The conundrum we face is that our environment has changed dramatically over the last century or two, in almost every imaginable way—our food supply and eating habits, our activity levels, and the structure of our social networks—while our genes have scarcely changed at all. We saw a classic example of this in chapter 6, with the changing role that fructose has played in our diet. Long ago, when we consumed fructose mainly in the form of fruit and honey, it enabled us to store energy as fat to survive cold winters and periods of scarcity. Fructose was our friend. Now fructose is vastly overabundant in our diet, too much of it in liquid form, which disrupts our metabolism and our overall energy balance. We can easily take in far more fructose calories than our bodies can safely handle.

This new environment we have created is potentially toxic with respect to what we eat (chronically, not acutely),* how we move (or don't move), how we sleep (or don't sleep), and its overall effect on our emotional health (just spend a few hours on social media). It's as foreign to our evolved genome as an airport would have been to, say, Hippocrates. That, coupled with our newfound ability to survive epidemics, injuries, and illnesses that formerly killed us, has added up to almost a defiance of natural selection. Our genes no longer match our environment. Thus, we must be cunning in our tactics if we are to adapt and thrive in this new and hazardous world.

This is why we have navigated through the preceding two hundred pages about our objective and our strategy. To figure out what to do, we need to

* Acutely, our food supply is safer than ever thanks to refrigeration and advances in food processing, and regulations that prevent toxic substances from being used in food. Chronically, not so much (see chapter 15).

know our adversary inside and out, the way Ali knew Foreman. By now, we should understand our strategy fairly well. Hopefully, I have at least given you some understanding of the biological mechanisms that help predispose us to certain diseases, and how those diseases progress.

Now it's time to explore our tactics, the means and methods by which we will try to navigate this strange and sometimes perilous new environment. How are we going to outlive our old expectations and live our best Bonus Decades? What concrete actions can we take to reduce our risk of disease and death and improve the quality of our lives as we age?

In Medicine 3.0, we have five tactical domains that we can address in order to alter someone's health. The first is *exercise,* which I consider to be by far the most potent domain in terms of its impact on both lifespan and healthspan. Of course, exercise is not just one thing, so I break it down into its components of aerobic efficiency, maximum aerobic output (VO_2 max), strength, and stability, all of which we'll discuss in more detail. Next is diet or nutrition—or as I prefer to call it, *nutritional biochemistry.* The third domain is *sleep,* which has gone underappreciated by Medicine 2.0 until relatively recently. The fourth domain encompasses a set of tools and techniques to manage and improve *emotional health.* Our fifth and final domain consists of the various drugs, supplements, and hormones that doctors learn about in medical school and beyond. I lump these into one bucket called *exogenous molecules,* meaning molecules we ingest that come from outside the body.

In this section I will not be talking much about exogenous molecules, beyond those that I have already mentioned specifically (e.g., lipid-lowering drugs, rapamycin, and metformin, the diabetes drug that is being tested for possible longevity effects). Instead, I want to focus on the other four domains, none of which were really covered, or even mentioned, in medical school or residency. We learned next to nothing about exercise, nutrition, sleep, or emotional health. That may be changing, slowly, but if some doctors understand these things today, and are actually able to help you, it's likely because they have sought out that information on their own.

At first glance, some of our tactics might seem a bit obvious. Exercise. Nutrition. Sleep. Emotional health. Of course, we want to optimize all of these. But the devil (or, to me, the delight) is in the details. *In what way(s)* should we be exercising? *How* are we going to improve our diet? *How* can we sleep longer and better?

In each of these cases, while the broad-brush goals are clear, the specifics and the nuances are not. Our options are almost infinite. This requires us to really drill down and figure out how to come up with an effective tactical game plan—and to be able to change course as needed. We have to dig deeper to get beyond the obvious.

What constitutes an effective tactic?

One way I like to explain this is through the example of car accidents, which also happen to be a minor obsession of mine. They kill far too many people across all age groups—one person every twelve minutes, according to the National Highway Traffic Safety Administration—yet I believe that a fair number of these deaths could be prevented, with the proper tactics.

What can we do to reduce our risk of dying behind the wheel? Is it even possible to avoid car accidents, when they seem so random?

The obvious tactics we already know about: wear a seat belt, don't text and drive (seemingly difficult for many people), and don't drink and drive, since alcohol is a factor in up to a third of fatalities. Automotive fatality statistics also reveal that almost 30 percent of deaths involve excessive speed. These are helpful reminders, but not really surprising or insightful.

Recognizing the danger points is the first step in developing good tactics. I had almost automatically assumed that freeways would prove to be the deadliest place to drive because of the high speeds involved. But decades' worth of auto accident data reveal that, in fact, a very high proportion of fatalities occur at intersections. The most common way to be killed, as a driver, is by another car that hits yours from the left, on the driver's side, having run a red light or traveling at high speed. It's typically a T-bone or broadside crash, and often the driver who dies is not the one at fault.

The good news is that at intersections we have choices. We have agency. We can decide whether and when to drive into the crossroads. This gives us an opportunity to develop specific tactics to try to avoid getting hit in an intersection. We are most concerned about cars coming from our left, toward our driver's side door, so we should pay special attention to that side. At busy intersections, it makes sense to look left, then right, then left again, in case we missed something the first time. A high school friend who is now a long-haul truck driver agrees: before entering *any* intersection, even if he has the right of way (i.e., a green light), he *always* looks left first, then right, specifically to avoid this type of crash. And keep in mind, he's in a huge truck.

Now we have a *specific, actionable* tactic that we can employ every time we drive. Even if it can't guarantee that we are 100 percent safe, it reduces our risk in a small but demonstrable way. Better yet, our tactic has leverage: a relatively minor effort yields a potentially significant risk reduction.

We approach our tactics the same way, zooming in from the vague and general to the specific and targeted. We use data and intuition to figure out where to focus our efforts, and feedback to determine what is and isn't working. And seemingly small tweaks can yield a significant advantage if compounded over time.

My car accident analogy may seem like a bit of a tangent, but it's really not that dissimilar from the situation we face in our quest for longevity. The automobile is ubiquitous in our society, an environmental hazard that we need to learn to live with. Similarly, in order to stay healthy as we grow older, we must learn to navigate a world that is filled with ever more hazards and risks to our health. In this third and final section of the book, we will explore various methods by which we can mitigate or eliminate those risks, and improve and increase our healthspan—and how to apply them to each unique patient.

Our two most complex tactical domains are nutrition and exercise, and I find that most people need to make changes in both—rarely just one or the other. When I evaluate new patients, I'm always asking three key questions:

a. Are they overnourished or undernourished? That is, are they taking in too many or too few calories?
b. Are they undermuscled or adequately muscled?
c. Are they metabolically healthy or not?

Not surprisingly, there is a high degree of overlap between the overnourished camp and those with poor metabolic health, but I've taken care of many thin patients with metabolic problems as well. Almost always, though, poor metabolic health goes along with being undermuscled, which speaks to the interplay between nutrition and exercise.

We will talk about all these different situations in much more detail, but briefly, this is why it's important to coordinate between *all* the different tactical interventions we employ. For example, with a patient who is overnourished, we want to find a way to reduce their caloric intake (there are three ways to do this, as you'll see in chapter 15). But if they are also undermuscled, which is common, we want to be careful to make sure they are still getting enough protein, since the goal is not weight loss but fat loss coupled with muscle gain. It can get complicated.

None of our tactical domains is fully separate from the others. In chapter 16, for example, we will see how sleep has a tremendous effect on our insulin sensitivity and our exercise performance (and our emotional well-being, as well). That said, with most patients I devote a great deal of attention to their fitness and their nutrition, which are closely linked. We rely heavily on data in our decision-making and developing our tactics, including static biomarkers such as triglycerides and liver function tests, as well as dynamic biomarkers such as oral glucose tolerance tests, along with anthropometric measures such as data on body composition, visceral adipose tissue, bone density, and lean mass.

Much of what you are about to read mirrors the discussions I have with my patients every single day. We talk about their objectives, and the science underpinning our strategy. When it comes to specific tactics, I give them direction to help them create their own playbook. I almost never write out a prescription for them to follow blindly. My goal is to empower them to take action to fix their fitness, nutrition, sleep, and emotional help. (Note that for most of these things, I don't actually even *need* a prescription pad.) But the *action* part is their responsibility; not much of this stuff is easy. It requires them to change their habits and do the work.

What follows is not a step-by-step plan to be followed blindly. There is no blanket solution for every person. Providing very granular exercise, dietary, or lifestyle advice requires individual feedback and iteration, something I can't safely or accurately accomplish in a book. Rather, I hope you will learn a framework for managing your movement, nutrition, sleep, and emotional health that will take you much further than any broad prescription for how many grams of this or that macronutrient every single person on earth must eat. I believe this represents the best we can do right now, on the basis of our current understanding of the relevant science and my own clinical experience (which is where the "art" comes in). I'm constantly tinkering, experimenting, switching things up in my own regimen and in that of my patients. And my patients themselves are constantly changing.

We are not bound by any specific ideology or school of thought, or labels of any kind. We are not "keto" or "low-fat," and we do not emphasize aerobic training at the expense of strength, or vice versa. We range widely and pick and choose and test tactics that will hopefully work for us. We are open to changing our minds. For example, I used to recommend long periods of water-only fasting for some of my patients—and practiced it myself. But I no longer do so, because I've become convinced that the drawbacks (mostly having to do with muscle loss and undernourishment) outweigh its metabolic benefits in all but my most overnourished patients. We adapt our tactics on the basis of our changing needs and our changing understanding of the best science out there.

Our only goal is to live longer and live better—to *outlive*. To do that, we must rewrite the narrative of decline that so many others before us have endured and figure out a plan to make each decade better than the one before.

CHAPTER 11

Exercise

The Most Powerful Longevity Drug

I never won a fight in the ring;
I always won in preparation.

—MUHAMMAD ALI

Several years ago, my friend John Griffin pinged me with a question about how he should be exercising: Should he be doing more cardio or more weights? What did I think?

"I'm really confused by all the contradictory stuff I'm seeing out there," he wrote.

Behind his seemingly simple question, I heard a plea for help. John is a smart guy with an incisive mind, and yet even he was frustrated by all the conflicting advice from "experts" touting this or that workout as the sure path to perfect health. He couldn't figure out what he needed to be doing in the gym or why.

This was before I had gotten back into the full-time practice of medicine.

At the time, I was immersed in the world of nutrition research, which if anything is even *more* confounding than exercise science, rife with contradictory findings and passionately held dogmas backed by flimsy data. Are eggs bad or good? What about coffee? It was driving me nuts too.

I started typing out a reply and kept on writing. By the time I hit SEND, I had written close to two thousand words, way more than he asked for. The poor guy just wanted a quick answer, not a memo. I didn't stop there either. I later expanded that email into a ten-thousand-word manifesto on longevity, which eventually grew into the book you are holding in your hands.*

Clearly, something about John's question triggered me. It's not that I was a passionate devotee of strength training over endurance, or vice versa; I'd done plenty of both. I was reacting to the binary nature of his question. In case you haven't figured it out by now, I'm not fond of the way we reduce these complex, nuanced, vitally important questions down to simple either-ors. Cardio or weights? Low-carb or plant-based? Olive oil or beef tallow?

I don't know. Must we really take sides?

The problem, and we will see this again in the nutrition chapters, is that we have this need to turn everything into a kind of religious war over which is the One True Church. Some experts insist that strength training is superior to cardio, while an equal number assert the opposite. The debate is as endless as it is pointless, sacrificing science on the altar of advocacy. The problem is that we are looking at these hugely important domains of life—exercise, but also nutrition—through a far too narrow lens. It's not about which side of the gym you prefer. It's so much more essential than that.

More than any other tactical domain we discuss in this book, exercise has the greatest power to determine how you will live out the rest of your life. There are reams of data supporting the notion that even a fairly minimal amount of exercise can lengthen your life by several years. It delays the onset of chronic diseases, pretty much across the board, but it is also amazingly effective at extending and improving healthspan. Not only does it reverse physical decline, which I suppose is somewhat obvious, but it can slow or reverse

---

* So if you like this book, please thank John Griffin. If you don't, blame me.

cognitive decline as well. (It also has benefits in terms of emotional health, although those are harder to quantify.)

So if you adopt only one new set of habits based on reading this book, it *must* be in the realm of exercise. If you currently exercise, you will likely want to rethink and modify your program. And if exercise is not a part of your life at the moment, you are not alone—77 percent of the US population is like you. Now is the time to change that. Right now. Even a little bit of daily activity is much better than nothing. Going from zero weekly exercise to just ninety minutes per week can reduce your risk of dying from all causes by 14 percent. It's very hard to find a drug that can do that.

Thus, my answer to questions like the one my friend John Griffin asked me is yes and yes. Yes, you should be doing more cardio. And yes, you should be lifting more weights.

At the other end of the spectrum, if you're someone like me who has been exercising since kindergarten, I promise you these chapters will offer you insights about how to better structure your program—not to achieve a faster marathon time or bragging rights at your gym, but to live a longer and better life, and most important, a life in which you can continue enjoying physical activity well into your later years.

It's obviously not a revelation that exercise is good for you; so is chicken soup if you have a sore throat. But not many people realize how profound its effects really are. Study after study has found that regular exercisers live as much as a *decade* longer than sedentary people. Not only do habitual runners and cyclists tend to live longer, but they stay in better health, with less morbidity from causes related to metabolic dysfunction. For those who are not habitual exercisers (yet), you're in luck: The benefits of exercise begin with any amount of activity north of zero—even brisk walking—and go up from there. Just as almost any diet represents a vast improvement over eating only fast food, almost any exercise is better than remaining sedentary.

Although my medical school classmates and I learned almost zilch about exercise, let alone how to "prescribe" it to patients, Medicine 2.0 does at least recognize its value. Unfortunately, the advice rarely goes beyond generic recommendations to move more and sit less. The US government's physical activity guidelines suggest that "active adults" engage in at least 30 minutes of "moderate-intensity aerobic activity," five times per week (or 150 minutes in total). This is to be supplemented with two days of strength training, targeting "all major muscle groups."

Imagine if doctors were this vague about cancer treatment:

DOCTOR: Ms. Smith, I'm sorry to have to tell you this, but you have colon cancer.

MS. SMITH: That's terrible news, Doctor. What should I do?

DOCTOR: You need chemotherapy treatment.

MS. SMITH: What kind of chemotherapy? What dose? How often? For how long? What about the side effects?

DOCTOR: ¯_(ツ)_/¯

We need more specific guidance to help us achieve our goals, and to do so in a way that is efficient but also safe. But first, I want to spend some time exploring *why* exercise is so important, because I find the data around it to be so persuasive. When I share these data with my patients, they are rarely surprised by the fact that high aerobic fitness and strength are associated with longer lifespan and healthspan—but they are always amazed by the *magnitude* of the benefit. The data on exercise tell us, with great clarity, that the more we do, the better off we will be.

Let's start with cardiorespiratory or aerobic fitness. This means how efficiently your body can deliver oxygen to your muscles, and how efficiently your muscles can extract that oxygen, enabling you to run (or walk) or cycle or swim long distances. It also comes into play in daily life, manifesting as physical stamina. The more aerobically fit you are, the more energy you will have for whatever you enjoy doing—even if your favorite activity is shopping.

It turns out that peak aerobic cardiorespiratory fitness, measured in terms of VO_2 max, is perhaps the single most powerful marker for longevity. VO_2 max represents the maximum rate at which a person can utilize oxygen. This is measured, naturally, while a person is exercising at essentially their upper limit of effort. (If you've ever had this test done, you will know just how unpleasant it is.) The more oxygen your body is able to use, the higher your VO_2 max.

Our human body has an amazing ability to respond to the demands placed on it. Let's say I'm just sitting on the couch, watching a movie. At rest, someone my size might require about 300 ml of oxygen per minute in order to generate enough ATP, the chemical "fuel" that powers our cells, to perform all the physiological functions necessary to stay alive and watch the movie. This is a pretty low level of energy demand, but if I go outside and jog around my neighborhood, the energy demands ramp up. My breathing quickens, and my heart rate accelerates to help me extract and utilize ever more oxygen from the air I breathe, in order to keep my muscles working. At this level of intensity, someone my size might require 2,500 to 3,000 ml of oxygen per minute, an eight- to tenfold increase from when I was sitting on the couch. Now, if I start running up a hill as fast as I can, my body's oxygen demand will increase from there: 4,000 ml, 4,500 ml, even 5,000 ml or more depending on the pace and my fitness level. The fitter I am, the more oxygen I can consume to make ATP, and the faster I can run up that hill.

Eventually, I will reach the point at which I just can't produce any more energy via oxygen-dependent pathways, and I'll be forced to switch over to less efficient, less sustainable ways of producing power, such as those used in sprinting. The amount of oxygen that I am using at this level of effort represents my VO_2 max. (And not long after that, I will "fail," meaning I am no longer able to continue running up the hill at that pace.) VO_2 max is typically expressed in terms of the volume of oxygen a person can use, per kilogram of body weight, per minute. An average forty-five-year-old man will have a VO_2 max around 40 ml/kg/min, while an elite endurance athlete will likely score in the high 60s and above. An unfit person in their thirties or forties, on the other hand, might score only in the high 20s on a VO_2 max test, according to Mike Joyner, an exercise physiologist and researcher at the Mayo Clinic. They simply won't be able to run up that hill at all.* The higher someone's VO_2 max, the more oxygen they can consume to make ATP, and the faster they can ride or run—in short, the more they can do.

This number is not just relevant to athletes; it turns out to be highly correlated with longevity. A 2018 study in *JAMA* that followed more than 120,000 people found that higher VO_2 max (measured via a treadmill test) was associated with lower mortality across the board. The fittest people had the lowest mortality rates—by a surprising margin. Consider this: A person who smokes has a 40 percent greater risk of all-cause mortality (that is, risk of dying at any moment) than someone who does not smoke, representing a hazard ratio or (HR) of 1.40. This study found that someone of below-average VO_2 max for their age and sex (that is, between the 25th and 50th percentiles) is at *double* the risk of all-cause mortality compared to someone in the top quartile (75th to 97.6th percentiles). Thus, poor cardiorespiratory fitness carries a greater relative risk of death than smoking.

That's only the beginning. Someone in the bottom quartile of VO_2 max for their age group (i.e., the least fit 25 percent) is nearly four times likelier to die than someone in the top quartile—and five times likelier to die than a person with elite-level (top 2.3 percent) VO_2 max. That's stunning. These benefits are not limited to the very fittest people either; even just climbing from the bottom 25 percent into the 25th to 50th percentile (e.g., least fit to below average) means you have cut your risk of death nearly in half, according to this study.

These results were confirmed by a much larger and more recent study, published in 2022 in the *Journal of the American College of Cardiology*, looking at data from 750,000 US veterans ages thirty to ninety-five (see figure 9). This was a completely different population that encompassed both sexes and all races, yet the researchers found a nearly identical result: someone in the

---

* Most of the top riders in the Tour de France will have a VO_2 max in the high 70s or low 80s. The *highest* VO_2 max ever recorded was an absolutely mind-bending 97.5 ml/kg/min.

Figure 9. Mortality Risk For Non-Elite Fitness and Select Comorbidities

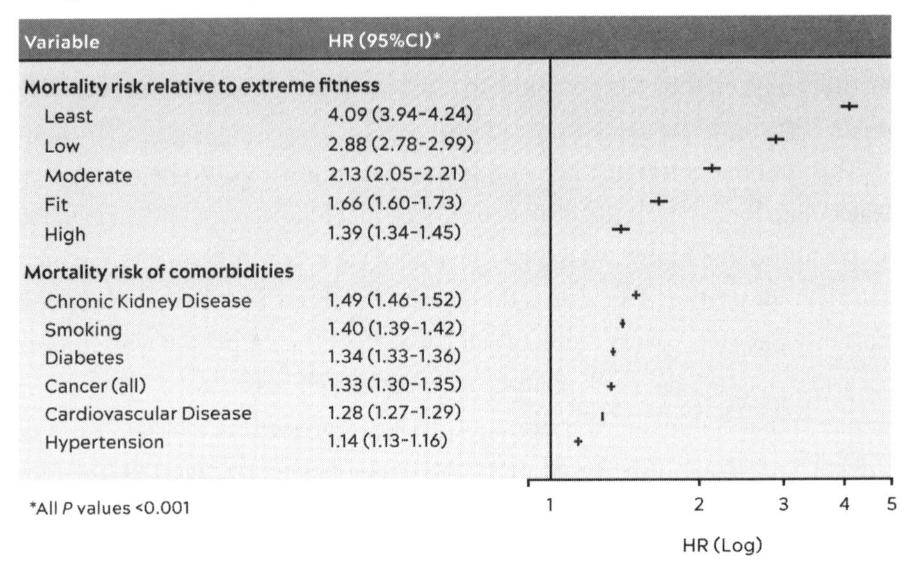

Variable	HR (95%CI)*
Mortality risk relative to extreme fitness	
Least	4.09 (3.94-4.24)
Low	2.88 (2.78-2.99)
Moderate	2.13 (2.05-2.21)
Fit	1.66 (1.60-1.73)
High	1.39 (1.34-1.45)
Mortality risk of comorbidities	
Chronic Kidney Disease	1.49 (1.46-1.52)
Smoking	1.40 (1.39-1.42)
Diabetes	1.34 (1.33-1.36)
Cancer (all)	1.33 (1.30-1.35)
Cardiovascular Disease	1.28 (1.27-1.29)
Hypertension	1.14 (1.13-1.16)

*All *P* values <0.001

This table expresses all-cause mortality risk for different fitness levels compared to individuals in the top 2% of VO_2 max for their age and sex ("extreme fitness") [TOP] and for various comorbidities—that is, people with versus without each illness. [BOTTOM] Fitness groups are divided by percentile: Least (<20th percentile); Low (21st to 40th percentile); Moderate (41st to 60th percentile); Fit (61st to 80th percentile); High (81st to 97th percentile).

Source: Kokkinos et al. (2022).

least fit 20 percent has a 4.09 times greater risk of dying than a person in the top 2 percent of their age and sex category. Even someone of moderate fitness (40th to 60th percentile) is still at more than double the risk of all-cause mortality than the fittest group, this study found. "Being unfit carried a greater risk than any of the cardiac risk factors examined," the authors concluded.

Of course, there are almost certainly confounders here, just as with all observational study, including that of nutrition. But at least five factors* increase my confidence in at least the partial causality of this relationship.

---

* These factors represent five of the nine criteria defined in the 1930s by Austin Bradford Hill, one of the godfathers of scientific methodology, as a tool for evaluating epidemiological and laboratory findings. We will meet Bradford Hill again in the chapters on nutrition.

First, the *magnitude* of the effect size is very large. Second, the data are consistent and *reproducible* across many studies of disparate populations. Third, there is a *dose-dependent response* (the fitter you are, the longer you live). Fourth, there is great *biologic plausibility* to this effect, via the known mechanisms of action of exercise on lifespan and healthspan. And fifth, virtually all *experimental data* on exercise in humans suggest that it supports improved health.

As the authors of the *JAMA* study concluded, "Cardiorespiratory fitness is inversely associated with long-term mortality *with no observed upper limit of benefit* [emphasis mine]. Extremely high aerobic fitness was associated with the greatest survival."

I can't tell you, from these data, that simply having a high VO_2 max will offset your high blood pressure or your smoking habit, as much as these hazard ratios suggest it might. Without a randomized controlled trial, we can't know for sure, but I kind of doubt it. But I can say with a very high degree of certainty that having a higher VO_2 max is better for your overall health and longevity than having a lower VO_2 max. Period.

Even better news, for our purposes, is that VO_2 max can be increased via training. We can move the needle a lot on this measure of fitness, as we'll see.

The strong association between cardiorespiratory fitness and longevity has long been known. It might surprise you, as it did me, to learn that muscle may be almost as powerfully correlated with living longer. A ten-year observational study of roughly 4,500 subjects ages fifty and older found that those with low muscle mass were at 40 to 50 percent greater risk of mortality than controls, over the study period. Further analysis revealed that it's not the mere muscle *mass* that matters but the *strength* of those muscles, their ability to generate force. It's not enough to build up big pecs or biceps in the gym—those muscles also have to be strong. They have to be capable of creating force. Subjects with low muscle strength were at double the risk of death, while those with low muscle mass and/or low muscle strength, plus metabolic syndrome, had a 3 to 3.33 times greater risk of all-cause mortality.

Strength may even trump cardiorespiratory fitness, at least one study suggests. Researchers following a group of approximately 1,500 men over forty with hypertension, for an average of about eighteen years, found that even if a man was in the bottom half of cardiorespiratory fitness, his risk of all-cause mortality was still almost 48 percent lower if he was in the top third of the group in terms of strength versus the bottom third.*

It's pretty much the same story we saw with VO_2 max: The fitter you are, the lower your risk of death. Again, there is no other intervention, drug or otherwise, that can rival this magnitude of benefit. Exercise is so effective against diseases of aging—the Horsemen—that it has often been compared to medicine.

John Ioannidis, a Stanford scientist with a penchant for asking provocative questions, decided to test this metaphor literally, running a side-by-side comparison of exercise studies versus drug studies. He found that in numerous randomized clinical trials, exercise-based interventions performed as well as *or better than* multiple classes of pharmaceutical drugs at reducing mortality from coronary heart disease,† prediabetes or diabetes, and stroke.

Even better: You don't need a doctor to prescribe exercise for you.

Much of this effect, I think, likely has to do with improved mechanics: exercise strengthens the heart and helps maintain the circulatory system. As we'll see later in this chapter, it also improves the health of the mitochondria, the crucial little organelles that produce energy in our cells (among other things). That, in turn, improves our ability to metabolize both glucose and fat. Having more muscle mass and stronger muscles helps support and protect the body—and also maintains metabolic health, because those muscles consume energy efficiently. The list goes on and on, but simply put, exercise helps the human "machine" perform far better for longer.

At a deeper biochemical level, exercise really does act like a drug. To be

* Cardiorespiratory fitness was measured on a treadmill using a modified Balke Protocol, and strength was measured by one-rep max in bench press and leg extension.

† The exception, in Ioannidis's analysis, was heart failure, which responded more favorably to treatment with diuretic drugs than to exercise-based interventions.

more precise, it prompts the body to produce its own, endogenous drug-like chemicals. When we are exercising, our muscles generate molecules known as cytokines that send signals to other parts of our bodies, helping to strengthen our immune system and stimulate the growth of new muscle and stronger bones. Endurance exercise such as running or cycling helps generate another potent molecule called brain-derived neurotrophic factor, or BDNF, that improves the health and function of the hippocampus, a part of the brain that plays an essential role in memory. Exercise helps keep the brain vasculature healthy, and it may also help preserve brain volume. This is why I view exercise as a particularly important part of the tool kit for patients at risk of developing Alzheimer's disease—such as Stephanie, the woman with the high-risk Alzheimer's genes whom we met in chapter 9.

The data demonstrating the effectiveness of exercise on lifespan are as close to irrefutable as one can find in all human biology. Yet if anything, I think exercise is even *more* effective at preserving healthspan than extending lifespan. There is less hard evidence here, but I believe that this is where exercise really works its magic when applied correctly. I tell my patients that even if exercise *shortened* your life by a year (which it clearly does not), it would still be worthwhile purely for the healthspan benefits, especially in middle age and beyond.

One of the prime hallmarks of aging is that our physical capacity erodes. Our cardiorespiratory fitness declines for various reasons that begin with lower cardiac output, primarily due to reduced maximum heart rate. We lose strength and muscle mass with each passing decade, our bones grow fragile and our joints stiffen, and our balance falters, a fact that many men and women discover the hard way, by falling off a ladder or while stepping off a curb.

To paraphrase Hemingway, this process happens in two ways: gradually, and then suddenly. The reality of the situation is that old age can be really tough on our bodies. Longitudinal and cross-sectional studies find that fat-free mass (meaning mostly muscle mass) and activity levels remain relatively consistent as people age from their twenties and thirties into middle age. But both physical activity levels *and* muscle mass decline steeply after about age sixty-five, and then even more steeply after about seventy-five. It's as if people just fall off a cliff sometime in their mid-seventies.

By age eighty, the average person will have lost eight kilograms of muscle, or about eighteen pounds, from their peak. But people who maintain higher activity levels lose much less muscle, more like three to four kilograms on average. While it's not clear which direction the causation flows here, I suspect it's probably both ways: people are less active because they are weaker, and they are weaker because they are less active.

Continued muscle loss and inactivity literally puts our lives at risk. Seniors with the least muscle mass (also known as lean mass) are at the greatest risk of dying from all causes. One Chilean study looked at about one thousand men and four hundred women, with an average age of seventy-four at enrollment. The researchers divided the subjects into quartiles, based on their appendicular lean mass index (technically, the muscle mass of their extremities, arms and legs, normalized to height), and followed them over time. After twelve years, approximately 50 percent of those in the lowest quartile were dead, compared to only 20 percent of those in the highest quartile for lean mass. While we can't establish causality here, the strength and reproducibility of findings like this suggest this is more than just a correlation. Muscle helps us survive old age.

This is another area where lifespan and healthspan overlap to a great extent. That is, I suspect that having more muscle mass delays death precisely *because* it also preserves healthspan. This is why I place so much emphasis on maintaining our musculoskeletal structure—which I call the "exoskeleton," à la *Terminator,* for lack of a better term.

Your exoskeleton (muscle) is what keeps your actual skeleton (bones) upright and intact. Having more muscle mass on your exoskeleton appears to protect you from all kinds of trouble, even adverse outcomes following surgery—but most important, it is highly correlated with a lower risk of falling, a leading but oft-ignored cause of death and disability in the elderly. As figure 10 reveals, falls are by *far* the leading cause of accidental deaths in those ages sixty-five and older—and this is without even counting the people who die three or six or twelve months after their nonfatal but still serious fall pushed them into a long and painful decline. Eight hundred thousand older people are hospitalized for falls each year, according to the CDC.

I believe this association likely works both ways: someone with more muscle mass is less likely to fall and injure themselves, while those who are less likely to fall for other reasons (better balance, more body awareness) will also have an easier time maintaining muscle mass. Conversely, muscle atrophy and sarcopenia (age-related muscle loss) increase our risk of falling and possibly requiring surgery—while at the same time worsening our odds of surviving said surgery without complications. Just as with VO_2 max, it is important to maintain muscle mass at all costs.

Exercise in all its forms is our most powerful tool for fighting this misery and reducing our risk of death across the board. It slows the decline, not just physically but across all three vectors of healthspan, including cognitive and emotional health. A recent study of older British adults found that those with sarcopenia at baseline were nearly *six times* likelier to report having a low quality of life a decade later than those who had maintained more muscle mass.

Figure 10. **Accidental Deaths in the United States**

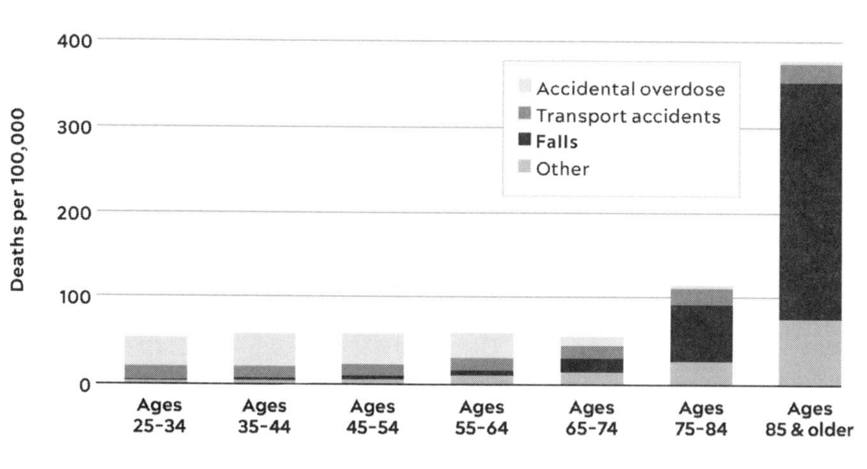

Source: CDC (2021).

This is what we want to avoid. We can avoid it with the help of this powerful "drug" called exercise that miraculously extends lifespan and improves healthspan. The difference is that it requires much more work and knowledge than simply taking a pill. But the more effort you're willing to put in now, the more benefit you'll reap in the future.

This is why I place such an emphasis on weight training—and doing it now, no matter your age. It is never too late to start; my mom did not begin lifting weights until she was sixty-seven, and it has changed her life. There are dozens of studies showing that strength training programs can significantly improve the mobility and physical function of subjects who are obese, or recovering from cancer treatment, even those who are already elderly and frail. Therefore, I will find a way to lift heavy weights in some way, shape, or form four times per week, no matter what else I am doing or where I might be traveling.

But exercise gets little more than lip service from Medicine 2.0. When was the last time your doctor tested your grip strength or asked you a detailed question about your strength training? Does your doctor know your VO_2 max? Or have they offered training suggestions for how to improve it? I'm guessing that has happened precisely never—because none of what we're covering here or in the nutrition chapters is even considered to rise to the level of "healthcare," at least in our system. Some insurance companies offer discounts or incentives for members to go to the gym, but the kind of focused attention that I think we all need (including me) is well beyond the purview of most physicians.

It's only *after* we get injured or become so weak that we are in danger of losing our independence, that we are deemed eligible for physical therapy and rehabilitation. So think of what follows as "prehab"—physical therapy *before* you need it.

When I talk with my patients about exercise, I often bring them back to the story of Sophie, my friend Becky's mother, whom we met in chapter 3. She had actually been relatively active, even in retirement. She played golf once or twice a week and worked most days in her garden. It wasn't a structured "exercise" program; she was just doing the things she loved. But then she injured her shoulder, and then her knee, requiring surgery for both. Even after surgery, she was never able to recover fully. Her activity level dropped off almost to zero. As Becky related to me, her mother mostly sat around the house, feeling depressed. Her cognitive decline fairly quickly ensued.

This made me incredibly sad as I sat there in the pew at her funeral. Yet her story was all too familiar. We have all seen older friends and relatives go through a similar ordeal, slowly (or not slowly) weakening until they can no longer find enjoyment in the things they once loved to do. What could have been done, I wondered, to change Sophie's fate?

Had she simply needed to "exercise more"? Go to the gym and use the elliptical? Would that have saved her somehow?

It wasn't clear that the answer was that simple. I had done plenty of exercise in my own life, but by the time of Sophie's funeral I was nursing a handful of my own pet injuries that I'd accumulated over the years. Fit as I was, it wasn't clear that I was on a better path than Sophie had been.

For most of my life, I've been obsessive with respect to fitness, always focusing on one particular sport—and inevitably taking it to an extreme. After I had burned myself out on boxing, then running, and finally long-distance open-water swimming, I turned to cycling. I went all in: My primary aim in life was to win the local cycling time-trial series, a twenty-kilometer individual race against the clock that almost nobody else cared about. I spent hours analyzing power-meter data and calculating my coefficient of aerodynamic drag, looking to shave precious seconds from my times using dorky models that I built in Excel.

The truth was that I had become pretty useless at everything *except* pedaling my road bike as fast as possible for twenty kilometers. I possessed a high VO_2 max and I could produce a lot of power through the pedals, but I was not truly strong or flexible, and I did not have great balance or stability. I was a one-dimensional athlete, and if I had kept it up, I might have ended up with my spine fused into my time-trial tuck, still able to ride my bike but unable to do anything else useful, especially with my upper body. I eventually quit competitive cycling because it became clear that at a certain point this obsessive approach becomes unsustainable in virtually any activity. How many old marathon runners do you know who are still racing? Probably not many.

After that, I fell into this kind of directionless limbo where I bounced around between different fitness activities. I tried to get back into running. I went to Pilates classes with my wife. I did Barry's Bootcamp for a time. You name it, I tried it. Then I joined a boutique fitness chain that specializes in high-intensity interval training, or HIIT. I enjoyed the fast-paced workouts they offered, like treadmill sprints and thirty-second Burpee blocks, and it took a fraction of the time that cycling did, so I was happy about that. But I didn't have a goal other than "exercising."

All this changed as I sat in the pew of the church at Sophie's funeral. Officially, she had died from pneumonia, but what had really killed her, I realized, was the slow gravitational pull of aging on her body. That had not begun in the last year or even the last decade of her life. It had been working against her, pulling her down, since before I had met her—for decades. And it was killing the rest of us, too: her daughter, Becky, along with my patients, myself, and everyone reading this book is likely headed for the same steep decline.

That thought saddened me to my core. But then it hit me: the only way that we would be able to fight this was to adopt the philosophy of a decathlete—and apply it to aging.

Of all Olympic athletes, the decathletes are most revered. The male and female winners of the gold medal are declared the "World's Greatest Athletes." Yet they are not the best at any of the ten individual events in which they compete; they likely would not even medal. But they are still considered the greatest because they are remarkably good at so many different events. They are true generalists—yet they train like specialists.

We need to adopt a similar approach to aging, I decided: each of us needs to be training for the Centenarian Decathlon.

The Centenarian Decathlon

What in the world is the Centenarian Decathlon?

I'm not talking about an actual competition among hundred-year-olds, although similar events do already exist: the National Senior Games, held every other year, brings together remarkable older athletes, some of them in their nineties and beyond. The record for the hundred-meter dash for women ages one hundred and up is about forty-one seconds.

The Centenarian Decathlon is a framework I use to organize my patients' physical aspirations for the later decades of their lives, especially their Marginal Decade. I know, it's a somewhat morbid topic, thinking about our own physical decline. But not thinking about it won't make it any less inevitable.

Think of the Centenarian Decathlon as the ten most important physical tasks you will want to be able to do for the rest of your life. Some of the items on the list resemble actual athletic events, while some are closer to activities of daily living, and still others might reflect your own personal interests. I find it useful because it helps us visualize, with great precision, exactly what kind of fitness we need to build and maintain as we get older. It creates a template for our training.

I start by presenting my patients with a long list of physical tasks that might include some of the following:

1. Hike 1.5 miles on a hilly trail.
2. Get up off the floor under your own power, using a maximum of one arm for support.
3. Pick up a young child from the floor.
4. Carry two five-pound bags of groceries for five blocks.
5. Lift a twenty-pound suitcase into the overhead compartment of a plane.
6. Balance on one leg for thirty seconds, eyes open. (Bonus points: eyes closed, fifteen seconds.)
7. Have sex.

8. Climb four flights of stairs in three minutes.
9. Open a jar.
10. Do thirty consecutive jump-rope skips.

The full list is much longer, with more than fifty different items, but you get the idea. Once they've read it I ask them to please select which of these tasks they want to be able to perform in their ninth, or better yet tenth, decade. Which ones do they choose?

All of them, typically. They want to be able to hike a mile and a half, or carry their own groceries, or pick up a great-grandchild, or get up if they fall down. Or play eighteen holes of golf, or open a jar, or fly somewhere on a plane. Of course they do.

That's great, I say. You'll make that kid's day when you pick her up like that. But now let's do a little math. Let's say the kid weighs twenty-five or thirty pounds. That's basically the same as doing a squat while holding a thirty-pound dumbbell in front of you (i.e., a goblet squat). Can you do that now, at age forty? Most likely. But now let's look into the future. Over the next thirty or forty years, your muscle strength will decline by about 8 to 17 percent per decade—accelerating as time goes on. So if you want to pick up that thirty-pound grandkid or great-grandkid when you're eighty, you're going to have to be able to lift about fifty to fifty-five pounds now. Without hurting yourself. Can you do that?

I press the issue. You also want to be able to hike on a hilly trail? To do that comfortably requires a VO_2 max of roughly 30 ml/kg/min. Let's take a look at the results of your latest VO_2 max test—and guess what, you only scored a 30. You're average for your age, but I'm afraid that's not good enough, because your VO_2 max is also going to decline. So we're going to have to go ahead and cross that hike off your list. You can pull it off now, but you likely won't be able to do it when you're older.

On it goes. To lift that twenty-pound suitcase overhead when you are older means doing so with forty or fifty pounds now. To be able to climb four flights of stairs in your eighties means you should be able to pretty much sprint up those same stairs today. In every case, you need to be doing *much more now,* to armor yourself against the natural and precipitous decline in strength and aerobic capacity that you will undergo as you age.

Eventually, my patients get it. Together, we come up with a list of ten or fifteen events in their personal Centenarian Decathlon, representing their goals for their later decades. This then determines how they should be training.

The beauty of the Centenarian Decathlon is that it is broad yet unique to each individual. Nor is it limited to ten events; for most people it ends up being more, depending on their goals. My version of the Decathlon is tailored to my own particular interests, such as swimming and archery. It's also fairly aggressive, I admit, reflecting the importance of a high level of fitness in my life. So I would probably add in some of the following events:

11. Swim half a mile in twenty minutes.
12. Walk with a thirty-pound dumbbell in each hand for one minute.
13. Draw back and fire a fifty-pound compound bow.
14. Do five pull-ups.
15. Climb ninety steps in two minutes (VO_2 max = 32).
16. Dead-hang for one minute.
17. Drive a race car within 5 to 8 percent of the pace I can do so today.
18. Hike with a twenty-pound backpack for an hour.
19. Carry my own luggage.
20. Walk up a steep hill.

In the end, most people's Centenarian Decathlons will probably overlap to a degree. Someone who enjoys stand-up paddleboarding, for example, would perhaps choose "events" focused around building core and cross-body strength. But she will likely be training the same muscle groups as I am doing for archery, and maintaining a similar degree of stamina and balance.

The Centenarian Decathlon is ambitious, no question. A ninety-year-old who is even able to board a plane under her own power, let alone hoist a carry-on bag, is doing extremely well. But there is a method to the madness. These individual tasks are not out of reach. There are octogenarians, nonagenarians, and even centenarians right now who are running marathons, racing bicycles, lifting weights, flying airplanes, jumping *out* of airplanes, skiing the Rocky Mountains, competing in actual decathlons, and doing all sorts of other amazing things. So all these events are within the realm of possibility.

One purpose of the Centenarian Decathlon, in fact, is to help us redefine what is possible in our later years and wipe away the default assumption that most people will be weak and incapable at that point in their lives. We need to abolish that decrepit stereotype and create a new narrative—perhaps modeled after the old-school fitness guru Jack LaLanne, who kept doing his usual rigorous daily workout right up until his death at age ninety-six. Unlike most very long-lived individuals, he didn't just get there by accident or luck. He built and maintained a high level of fitness throughout his life, beginning in the 1930s, when very few people exercised regularly and "fitness centers" did not yet exist. As he got older, he set out very deliberately to defy the stereotype of aging as a period of misery and decline. He did the work, and he succeeded, giving us a glimpse of what an older person is truly capable of achieving.

If we are to follow in LaLanne's footsteps, we must stop pointlessly "exercising," just because we think we are supposed to, banging away on the elliptical trainer at lunch hour. I promise, you can do better. I suggest you join me and start *training*, with a very specific purpose, which is to be kick-ass one-hundred-year-olds. When my patients say they are more interested in being kick-ass fifty-year-olds than Centenarian Decathletes, I reply that there is no better way to make that happen than to set a trajectory toward being vibrant at one hundred (or ninety, or eighty) just as an archer who trains at 100 yards will be more accurate at 50. By fixing our aim on the Centenarian Decathlon, we can make every decade between now and then better as well.

With the Centenarian Decathlon as my goal, I now work out with the focus that I once directed exclusively toward cycling, swimming, or boxing. It's not about being great at any one pursuit, but about being pretty good at just about everything. As Centenarian Decathletes, we are no longer training for a specific event, but to become a different sort of athlete altogether: an athlete of life.

CHAPTER 12

Training 101

How to Prepare for the Centenarian Decathlon

It is impossible to produce superior performance unless you do something different from the majority.

—Sir John Templeton

Most treatments of exercise are either very specific (e.g., how to train for your first marathon) or overly vague (e.g., "Just keep moving!"). Or they emphasize "cardio" over "weights," or vice versa. In this chapter, we are seeking to optimize our exercise regimen around the principle of longevity. What combination of modalities will help us delay the onset of chronic disease and death, while simultaneously maintaining healthspan for as long as possible?

This question turns out to be more complicated than how to lower your risk of cardiovascular disease, because there are more variables, and more choices within each variable. It is not a one-dimensional problem but more of a three-dimensional one. The three dimensions in which we want to optimize our fitness are aerobic endurance and efficiency (aka cardio), strength, and stability. All three of these are key to maintaining your health and strength as you age. (And as we've seen, they also extend lifespan.) But both cardio and strength are far more nuanced than most people realize—and stability may be the least understood component of all.

When we say "cardio," we are talking about not one thing, but a physiologic continuum, ranging from an easy walk to an all-out sprint. The various levels of intensity all count as cardio but are fueled by multiple different energy systems. For our purposes, we are interested in two particular regions of this continuum: long, steady endurance work, such as jogging or cycling or swimming, where we are training in what physiologists call zone 2, and maximal aerobic efforts, where VO_2 max comes into play.

The strength side of the equation seems simpler, at first: if you use your muscles to counter some resistance, in the form of weights or other forces (e.g., gravity, or elastic bands), they will adapt and grow stronger. That's how muscle works, and it's really quite wonderful. There are a few specific movements that I consider to be foundational, but here our most important goal is not only to build strength and muscle mass. It's equally important that we avoid injury in the process.

This is where stability comes in. We will talk about it in much more detail in the next chapter, but I consider stability to be just as important as aerobic fitness and strength. It's a bit hard to define, but I think of stability as the solid foundation that enables us to do everything else that we do, without getting injured. Stability makes us bulletproof. Sophie was relatively fit for her age, but she likely lacked stability, making her vulnerable to injury. Many people are in the same boat without even realizing it—even me, in my twenties. It almost doesn't matter how fit you are; you can still be at risk. This is why our approach to exercise must increase not only our conventional measures of fitness, such as our VO_2 max and our muscular strength, but above all our resistance to injury.

In the sections to follow, we will be building a framework around each of these, to help you craft your training program for your own Centenarian Decathlon.

Aerobic Efficiency: Zone 2

Notice that one word has been missing in our discussion of exercise thus far: *calories*. Most people think that one of the primary benefits of exercise, if not the primary benefit, is that it "burns calories." And it does, but we are more interested in a finer distinction—not calories, but *fuels*. How we utilize different fuels, glucose and fatty acids, is critical not only to our fitness but also to our metabolic and overall health. Aerobic exercise, done in a very specific way, improves our ability to utilize glucose and especially fat as fuel.

The key here are the mitochondria, those tiny little intracellular organelles that produce much of our energy. These cellular "engines" can burn both glucose and fat, and thus they are fundamental to our metabolic health. Healthy mitochondria are also important to maintaining the health of our brain, and to controlling potential bad actors like oxidative stress and inflammation. I am convinced that it is impossible to be healthy without also having healthy mitochondria, which is why I place a great deal of emphasis on long, steady endurance training in zone 2.

Zone 2 is one of five levels of intensity used by coaches and trainers in endurance sports to structure their athletes' training programs. It can get confusing, because some coaches define training zones in terms of heart rate, while others focus on different levels of power output; adding to the confusion, some models have five zones, but others have six or seven. Typically, zone 1 is a walk in the park and zone 5 (or 6, or 7) is an all-out sprint. Zone 2 is more or less the same in all training models: going at a speed slow enough that one can still maintain a conversation but fast enough that the conversation might be a little strained. It translates to aerobic activity at a pace somewhere between easy and moderate.

I had done plenty of zone 2 workouts in my cycling days; this type of training is foundational for any endurance sport. But I had never fully grasped the importance of zone 2 training to our overall health until I happened to meet a very bright exercise scientist named Iñigo San Millán in 2018. I had flown to the United Arab Emirates for a meeting, and shortly after landing, at 11 p.m. on a cool December evening, I was introduced to San Millán, an assistant professor at the University of Colorado School of Medicine who had recently been hired as director of performance for the UAE Team Emirates professional cycling team. He was there to do preseason testing of some of the UAE team riders, and when he found out I was a former cyclist, he put me on a stationary bike right then and there, in the middle of the night, to do a VO_2 max test. My kind of guy.

A native of Spain and a former professional cyclist himself, San Millán has worked with all kinds of athletes and coaches in many sports, including hundreds of top professional cyclists. He is also the personal coach of 2020 and 2021 Tour de France champion (and 2022 runner-up) Tadej Pogačar. Despite his impressive sports résumé, San Millán's true passion is studying the relationship between exercise, mitochondrial health, and diseases such as cancer and type 2 diabetes. As he explained, he hopes to use his insights into the fittest people on the planet, professional cyclists and other elite endurance athletes, in order to help the very least fit people—the one-third to one-half of the population with metabolic diseases or derangements.

In San Millán's view, healthy mitochondria are key to both athletic performance *and* metabolic health. Our mitochondria can convert both glucose and fatty acids to energy—but while glucose can be metabolized in multiple different ways, fatty acids can be converted to energy *only* in the mitochondria. Typically, someone working at a lower relative intensity will be burning more fat, while at higher intensities they would rely more on glucose. The healthier and more efficient your mitochondria, the greater your ability to utilize fat, which is by far the body's most efficient and abundant fuel source. This ability to use both fuels, fat and glucose, is called "metabolic flexibility," and it is what we want: in chapters 6 and 7, we saw how the relentless accumulation and spillover of fat drives conditions such as diabetes and cardiovascular disease. Healthy mitochondria (fostered by zone 2 training) help us keep this fat accumulation in check.

A few years ago, San Millán and his colleague George Brooks published a fascinating study that helps illustrate this point. They compared three groups of subjects: professional cyclists, moderately active healthy males, and sedentary men who met the criteria for the metabolic syndrome, meaning essentially that they were insulin resistant. They had each group ride a stationary bicycle at a given level of intensity relative to their fitness (about 80 percent of their maximum heart rate), while the scientists analyzed the amount of oxygen they consumed and the CO_2 they exhaled in order to determine how efficiently they produced power—and what primary fuels they were using. The differences they found were striking. The professional cyclists could zoom along, producing a huge amount of power while still burning primarily fat. But the subjects with metabolic syndrome relied almost entirely on glucose for their fuel source, even from the first pedal stroke. They had virtually zero ability to tap into their fat stores, meaning they were metabolically *inflexible:* able to use only glucose but not fat.

Obviously, these two groups—professional athletes and sedentary, unhealthy people—were as dissimilar as could be. San Millán's insight was that the sedentary subjects needed to be training in a manner similar to the Tour de France–bound cyclists he worked with. A professional cyclist might spend thirty to thirty-five hours a week training on his or her bike, and 80 percent of that time in zone 2. For an athlete, this builds a foundation for all their other, more intense training. (The catch is that a professional rider's zone 2 output feels like zone 5 for most people.)

As fundamental as zone 2 training is for professional cyclists, however, San Millán believes that it's even *more* important for nonathletes, for two reasons. First, it builds a base of endurance for anything else you do in life, whether that is riding your bike in a one-hundred-mile century ride or playing with your kids or grandkids. The other reason is that he believes it plays a crucial role in preventing chronic disease by improving the health and efficiency of your mitochondria, which is why training aerobic endurance and efficiency (i.e., zone 2 work) is the first element of my Centenarian Decathlon training program.

When we are exercising in zone 2, most of the work is being done by our type 1, or "slow-twitch," muscle fibers. These are extremely dense with mitochondria and thus well-suited for slow-paced, efficient endurance work. We can go for a long time without feeling fatigued. If we pick up the pace, we begin to recruit more type 2 ("fast-twitch") muscle fibers, which are less efficient but more forceful. They also generate more lactate in the process, because of the way they create ATP. Lactate itself is not bad; trained athletes are able to recycle it as a type of fuel. The problem is that lactate becomes lactic acid when paired with hydrogen ions, which is what causes that acute burning you feel in your muscles* during a hard effort.

In technical terms, San Millán describes zone 2 as the maximum level of effort that we can maintain without accumulating lactate. We still produce it, but we're able to match production with clearance. The more efficient our mitochondrial "engine," the more rapidly we can clear lactate, and the greater effort we can sustain while remaining in zone 2. If we are "feeling the burn" in this type of workout, then we are likely going too hard, creating more lactate than we can eliminate.

Because I am a numbers guy and I love biomarkers and feedback, I often test my own lactate while I am working out this way, using a small handheld lactate monitor, to make sure my pacing is correct. The goal is to keep lactate levels constant, ideally between 1.7 and 2.0 millimoles. This is the zone 2 threshold for most people. If I'm working too hard, lactate levels will rise, so I'll slow down. (It's sometimes tempting to go too hard in zone 2, because the workout feels relatively "easy" on good days.) I make a point of this because lactate is literally what defines zone 2. It's all about keeping lactate levels steady in this range, and the effort sustainable.

If you don't happen to have a portable lactate meter on hand, like most people, there are other ways to estimate your zone 2 range that are reasonably accurate. If you know your maximum heart rate—not estimated, but your actual maximum, the highest number you've ever seen on a heart rate monitor—your zone 2 will correspond to between approximately 70 and 85 percent of that peak number, depending on your fitness levels. That's a big range, so when starting people out, I prefer they rely on their rate of perceived

---

* This is because the hydrogen ion does not allow the actin and myosin filaments in your muscles to relax, causing pain and stiffness in the muscle.

exertion, or RPE, also known as the "talk test." How hard are you working? How easy is it to speak? If you're at the top of zone 2, you should be able to talk but not particularly interested in holding a conversation. If you can't speak in complete sentences at all, you're likely into zone 3, which means you're going too hard, but if you can comfortably converse, you're likely in zone 1, which is too easy.

Zone 2 output is highly variable, depending on one's fitness. In San Millán and Brooks's study, the professional cyclists produced about three hundred watts of power in zone 2, while the sedentary, metabolically unhealthy subjects could generate only about one hundred watts at the same relative level of intensity. That's a huge difference. If we express this output in terms of watts per kilogram of body weight, the difference becomes even more stark: The seventy-kilogram cyclists put out more than four watts per kilogram of body weight, while the one-hundred-plus kilogram sedentary subjects could only manage about one watt per kilogram.

This pronounced difference comes back to the fact that the unhealthy subjects' mitochondria—their engine(s)—were much less efficient than those of the athletes, so they very quickly switched over from aerobic respiration, burning fat and glucose in the mitochondria with oxygen, to the much less efficient glycolysis, an energy-producing pathway that consumes only glucose and produces loads of lactate (similar to the way cancer cells produce energy, via the Warburg effect). Once we start producing energy this way, lactate accumulates and our effort quickly becomes unsustainable. There are other (fortunately rare) genetic diseases that target the mitochondria and produce far more severe sequelae, but in terms of mass-acquired chronic conditions, type 2 diabetes does a real number on the mitochondria, and San Millán's data very elegantly demonstrate the disability that it creates.

Even when we are at rest, our lactate levels tell us much about our metabolic health. People with obesity or other metabolic problems will tend to have much higher resting lactate levels, a clear sign that their mitochondria are not functioning optimally, because they are already working too hard just to maintain baseline energy levels. This means that they are relying almost totally on glucose (or glycogen) for all their energy needs—and that they are totally unable to access their fat stores. It seems unjust, but the people who most need to burn their fat, the people with the most of it, are unable to unlock virtually *any* of that fat to use as energy, while the lean, well-trained professional athletes are able to do so easily because they possess greater metabolic flexibility (and healthier mitochondria).[*]

Mitochondrial health becomes especially important as we grow older, because one of the most significant hallmarks of aging is a decline in the number and quality of our mitochondria. But the decline is not necessarily a one-way street. Mitochondria are incredibly plastic, and when we do aerobic exercise, it stimulates the creation of many new and more efficient mitochondria through a process called mitochondrial biogenesis, while eliminating ones that have become dysfunctional via a recycling process called mitophagy (which is like autophagy, touched on in chapter 5, but for mitochondria). A person who exercises frequently in zone 2 is improving their mitochondria with every run, swim, or bike ride. But if you don't use them, you lose them.

This is another reason why zone 2 is such a powerful mediator of metabolic health and glucose homeostasis. Muscle is the largest glycogen storage sink in the body, and as we create more mitochondria, we greatly increase our capacity for disposing of that stored fuel, rather than having it end up as fat or remaining in our plasma. Chronic blood glucose elevations damage organs from our heart to our brain to our kidneys and nearly everything in between—even contributing to erectile dysfunction in men. Studies have found that while we are exercising, our overall glucose uptake increases as much as one-hundred-fold compared to when we are at rest. What's interesting is that this glucose uptake occurs via multiple pathways. There is the

---

* Patients on chemotherapy treatment for cancer often seem to be similarly compromised in terms of mitochondrial efficiency. There is also speculation, with some evidence, that this afflicts patients with so-called "long COVID."

usual, insulin-signaled way that we're familiar with, but exercise also activates other pathways, including one called non-insulin-mediated glucose uptake, or NIMGU, where glucose is transported directly across the cell membrane without insulin being involved at all.

This in turn explains why exercise, especially in zone 2, can be so effective in managing both type 1 and type 2 diabetes: It enables the body to essentially bypass insulin resistance in the muscles to draw down blood glucose levels. I have one patient with type 1 diabetes, meaning he produces zero insulin, who keeps his glucose in check almost entirely by walking briskly for six to ten miles every day, and sometimes more. As he walks, his muscle cells are vacuuming glucose out of his bloodstream via NIMGU. He still needs to inject himself with insulin, but only a tiny fraction of the amount that he would otherwise require.

One other plus of zone 2 is that it is very easy to do, even for someone who has been sedentary. For some people, a brisk walk might get them into zone 2; for those in better condition, zone 2 means walking uphill. There are many different ways to do it: you can ride a stationary bicycle at the gym, or walk or jog or run around the track at the local high school, or swim some laps in the pool. The key is to find an activity that fits into your lifestyle, that you enjoy doing, and that enables you to work at a steady pace that meets the zone 2 test: You're able to talk in full sentences, but just barely.

How much zone 2 training you need depends on who you are. Someone who is just being introduced to this type of training will derive enormous benefit from even two 30-minute sessions per week to start with. Based on multiple discussions with San Millán and other exercise physiologists, it seems that about three hours per week of zone 2, or four 45-minute sessions, is the minimum required for most people to derive a benefit and make improvements, once you get over the initial hump of trying it for the first time. (People who are training for major endurance events, such as running a marathon, obviously need to do more than this.) I am so persuaded of the benefits of zone 2 that it has become a cornerstone of my training plan. Four times a week, I will spend about an hour riding my stationary bike at my zone 2 threshold.

One way to track your progression in zone 2 is to measure your output in watts at this level of intensity. (Many stationary bikes can measure your wattage as you ride.) You take your average wattage output for a zone 2 session and divide it by your weight to get your watts per kilogram, which is the number we care about. So if you weigh 60 kilos (about 132 pounds) and can generate 125 watts in zone 2, that works out to a bit more than 2 watts/kg, which is about what one would expect from a reasonably fit person. These are rough benchmarks, but someone who is very fit will be able to produce 3 watts/kg, while professional cyclists put out 4 watts/kg and up. It's not the number that matters, but how much you are improving over time. (If you're a runner or a walker, the same principle applies: As you improve, your zone 2 pace will get faster.)

Zone 2 can be a bit boring on its own, so I typically use the time to listen to podcasts or audiobooks, or just think about issues that I'm working on—a side benefit of zone 2 is that it also helps with cognition, by increasing cerebral blood flow and by stimulating the production of BDNF, brain-derived neurotrophic factor, which we touched on earlier. This is another reason why zone 2 is such an important part of our Alzheimer's disease prevention program.

I think of zone 2 as akin to building a foundation for a house. Most people will never see it, but it is nevertheless important work that helps support virtually everything else we do, in our exercise regimen and in our lives.

Maximum Aerobic Output: VO_2 Max

If zone 2 represents a steady state, where you are kind of cruising along at a sustainable pace, VO_2 max efforts are almost the opposite. This is a much higher level of intensity—a hard, minutes-long effort, but still well short of an all-out sprint. At VO_2 max, we are using a combination of aerobic and anaerobic pathways to produce energy, but we are at our maximum rate of oxygen consumption. Oxygen consumption is the key.

Besides improving mitochondrial health and glucose uptake and metabolic flexibility, and all those other good things, zone 2 training also increases your VO_2 max somewhat. But if you really want to raise your VO_2 max, you need to train this zone more specifically. Typically, for patients who are new to exercising, we introduce VO_2 max training after about five or six months of steady zone 2 work.

One reason why I emphasize this so much is that this measure of peak aerobic capacity is powerfully correlated with longevity, as we saw in chapter 11. I have all my patients undergo VO_2 max testing and then train to improve their score. Even if you are not competing in high-level endurance sports, your VO_2 max is an important number that you can and should know.

Testing is widely available, even from some of the larger fitness chains. The bad news is that the VO_2 max test is an unpleasant affair that entails riding an exercise bike or running on a treadmill at ever greater intensity, while wearing a mask designed to measure oxygen consumption and CO_2 production. The peak amount of oxygen you consume, typically close to the point at which you "fail," meaning the point where you just can't keep going, yields your VO_2 max. We have all our patients do the test at least annually, and they almost all hate it. We then compare their results, normalized by weight, to the population of their age and sex.

Why is this important? Because our VO_2 max is a pretty good proxy measure of our physical capability. It tells us what we can do—and what we cannot do. Take a look at figure 11, which charts low, average, and high VO_2 max levels by age. Two things stand out. First, there's a huge gap in fitness between the top and bottom 5 percent of each age group (the upper vs. lower lines). Second, it's striking how steeply VO_2 max declines with age, and how this decline corresponds to diminished functional capacity. The lower it goes, the less you can do.

For example, a thirty-five-year-old man with average fitness for his age—a VO_2 max in the mid-30s—should be able to run at a ten-minute mile pace (6 mph). But by age seventy, only the very fittest 5 percent of people will still be able to manage this. Similarly, an average forty-five- to fifty-year-old will be able to climb stairs briskly (VO_2 max = 32), but at seventy-five, such a feat

Figure 11. How VO_2 Max Declines with Age

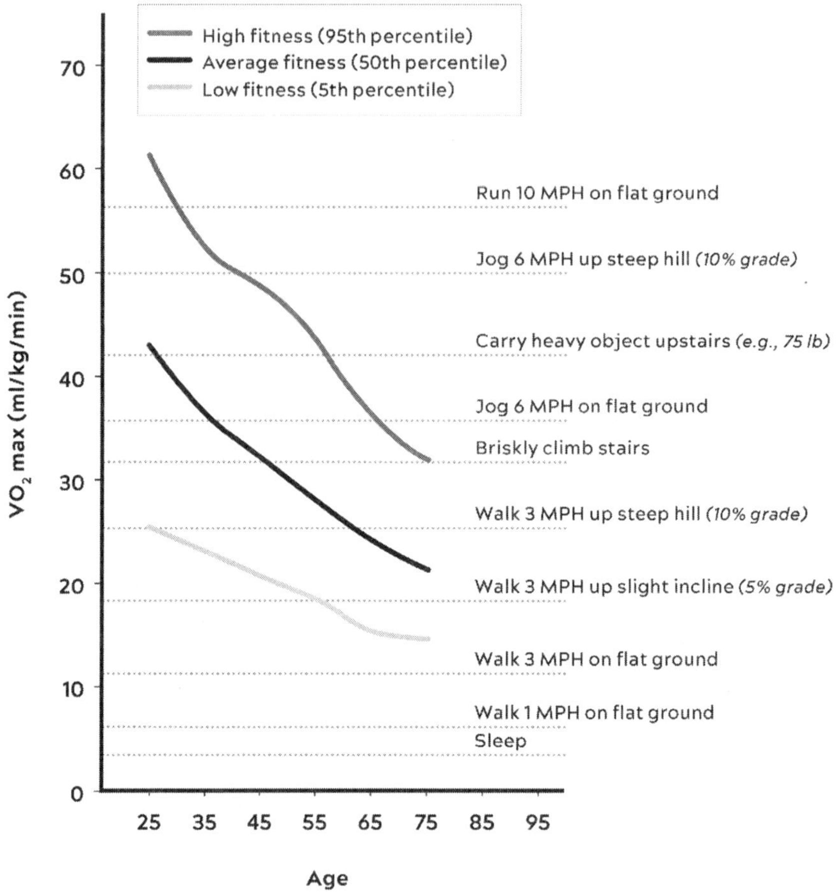

Source: Graph by Jayson Gifford, Brigham Young University, based on data from Ligouri (2020).

demands that a person be in the top tier of their age group. Activities that are easy when we are young or middle-aged become difficult if not impossible as we get older. This explains why so many people are miserable in their Marginal Decade. They simply can't *do* much of anything.

I push my patients to train for as high a VO_2 max as possible, so that they can maintain a high level of physical function as they age. Ideally, I want them to target the "elite" range for their age and sex (roughly the top 2 percent). If they achieve that level, I say good job—now let's reach for the elite level for your sex, but *two decades younger*. This may seem like an extreme goal, but I like to aim high, in case you haven't noticed.

There is a logic to it. Let's say you are a fifty-year-old woman, and you enjoy hiking in the mountains; that's how you want to spend your retirement. That kind of activity would require a VO_2 max of about 30, give or take. Let's also assume, for the sake of argument, that you're in the 50th percentile for your age; that puts you at about 32 ml/kg/min. You can do that hike now!

That seems like good news—but it's really the bad news. Studies suggest that your VO_2 max will decline by roughly 10 percent per decade—and up to 15 percent per decade after the age of fifty. So simply having average or even above-average VO_2 max now just won't cut it. We are planning for you to live for another thirty years, or forty. If you are only starting at 32 ml/kg/min now, at fifty, you can expect to be closer to 21 ml/kg/min at age eighty. These are not abstract numbers; they represent a profound decline in function. It's the difference between walking easily up a flight of stairs versus struggling to even walk on an inclined surface. It's a far cry from hiking in the Dolomites. To arrive in her ninth decade with a sufficient level of fitness to achieve her goal, our fifty-year-old would need to have a VO_2 max of about 45 to 49 right now. This is the top tier for her sex, but two decades younger.

It is important that your goals reflect your own priorities—the activities that you enjoy, and what you want to be able to accomplish in your later decades. The more active you want or plan to be as you age, the more you need to train for it now.

Keep in mind, increasing your VO_2 max *by any amount* is going to improve your life, not only in terms of how long you live but also how well you live, today and in the future. Improving your VO_2 max from the very bottom quartile to the quartile above (i.e., below average) is associated with almost a 50 percent reduction in all-cause mortality, as we saw earlier. I believe that almost anyone is capable of achieving this—and they should, because the alternative is so unacceptable. Once maximal oxygen consumption or VO_2 max drops below a certain level (typically about 18 ml/kg/min in men, and

15 in women), it begins to threaten your ability to live on your own. Your engine is beginning to fail.

This is why it's so essential to train VO_2 max in addition to zone 2. It's a key to maintaining a fulfilling, independent life as you age. But it takes hard work over a long period of time to build it up and keep it up.

How trainable is VO_2 max? The conventional wisdom, reflecting the bulk of the literature, suggests that it's possible to improve elderly subjects' aerobic capacity by about 13 percent over eight to ten weeks of training, and by 17 percent after twenty-four to fifty-two weeks, according to one review. That's a good start, but I think it represents only the beginning of what is possible; as usual with Medicine 2.0, these studies are almost always too short. We are talking about a lifelong training program, not one that lasts only eight weeks. Everyone is different, in terms of their fitness potential and their response to training, but Mike Joyner believes that longer and more focused training can yield much larger gains over extended periods of time—periods measured in years, not weeks. I tell my patients that this is not a two-month project; it's a two-year project.

It is not clear how much upside it is possible to achieve, but the literature suggests that sustained, diligent training can pay off. A small study of nine well-trained octogenarian endurance athletes (cross-country skiers) found that their average VO_2 max was 38, versus 21 for a control group of untrained octogenarian men, a difference of more than 80 percent. That's huge. The athletes had the aerobic capacity of people decades younger than them,* while the men in the control group had declined so far that they were on the verge of losing their ability to live independently. True, the study subjects were lifelong athletes—but that's also part of the point here. Our goal is to become elite athletes of aging.

The payoff is that increasing your VO_2 max makes you functionally younger. One study found that boosting elderly subjects' VO_2 max by

---

* Two of the aging athletes posted VO_2 max scores greater than 40, and the oldest subject, a ninety-one-year-old former Olympian, came very close to that, at 36, which would put them in the top quarter of men in their sixties.

6 ml/kg/min, or about 25 percent, was equivalent to subtracting twelve years from their age. If you are a man in your sixties and you are starting with a VO_2 max of 30, you are more or less average for your age group (see figure 12). (Women typically have a somewhat lower average VO_2 max by age, because of various factors, so an "average" woman in her sixties would be at about 25 ml/kg/min.) If you can boost that up to 35 via training, you will be squarely in the top 25 percent of your age group. Nice work. Now, here's another way to look at it: In your sixties, you will have achieved the aerobic fitness of an average man in his fifties, a decade younger than you. If you can get it still higher, to 38 or 39, you will be the aerobic equivalent of an average thirty-something. This means you will have bought yourself a phase shift, like we talked about with the centenarians: you now have the fitness of someone decades younger than you. So give yourself a pat on the back; you've earned it.

The beauty of this is that VO_2 max can *always* be improved by training, no matter how old you are. Don't believe me? Then let me introduce you to an amazing Frenchman named Robert Marchand, who set an age-group world record in 2012 by cycling 24.25 kilometers in an hour, at the age of 101. Apparently, he wasn't satisfied with that performance, so he decided he needed to train harder. Following a strict program designed by top coaches and physiologists, he managed to boost his VO_2 max from an already-impressive 31 ml/kg/min up to 35 ml/kg/min, which would put him in the elite 2.5 percent of men in their eighties. Two years later, now 103, he came back and broke his own record, riding almost twenty-seven kilometers in an hour. That's impressive, and it shows that it's never too late to improve your VO_2 max.

Even if we are not out to set world records, the way we train VO_2 max is pretty similar to the way elite athletes do it: by supplementing our zone 2 work with one or two VO_2 max workouts per week.

Where HIIT intervals are very short, typically measured in seconds, VO_2 max intervals are a bit longer, ranging from three to eight minutes—and a notch less intense. I do these workouts on my road bike, mounted to a stationary trainer, or on a rowing machine, but running on a treadmill (or a track) could also work. The tried-and-true formula for these intervals is to go

Figure 12. VO_2 Max by Age, Sex, Fitness

	Performance Group by VO_2 max (ml/kg/min)				
Age	**Low**	**Below Average**	**Above Average**	**High**	**Elite**
Women					
18–19	< 35	35–39	40–45	40–52	≥ 53
20–29	< 28	28–35	36–40	41–50	≥ 51
30–39	< 27	27–33	34–38	39–48	≥ 49
40–49	< 26	26–31	32–36	37–46	≥ 47
50–59	< 25	25–28	29–35	36–45	≥ 46
60–69	< 21	21–24	25–29	30–38	≥ 40
70–79	< 18	18–21	22–24	25–35	≥ 36
≥ 80	< 15	15–19	20–22	23–29	≥ 30
Men					
18–19	< 38	38–45	46–49	50–57	≥ 58
20–29	< 36	36–42	43–48	49–55	≥ 56
30–39	< 35	35–39	40–45	46–52	≥ 53
40–49	< 34	34–38	39–43	44–51	≥ 52
50–59	< 29	29–35	36–40	41–49	≥ 50
60–69	< 25	25–29	30–35	36–45	≥ 46
70–79	< 21	21–24	25–29	30–40	≥ 41
≥ 80	< 18	18–22	23–25	26–35	≥ 36

Group comparisons for VO_2 max are Low (bottom 25%), Below Average (26th to 50th percentile), Above Average (51st to 75th percentile), High (75th to 97.6th percentile), and Elite (top 2.3%).

Source: Mandsager et al. (2018).

four minutes at the maximum pace you can sustain for this amount of time—not an all-out sprint, but still a very hard effort. Then ride or jog four minutes easy, which should be enough time for your heart rate to come back down to below about one hundred beats per minute. Repeat this four to six times and cool down.*

* In practice, I've found that my ideal VO_2 max pace works out to about 33 percent more power than my zone 2 pace, if I'm doing four-on/four-off intervals. So if your zone 2 pace represents an output of 150 watts, your VO_2 max training pace should be about 200 watts for four minutes, followed by four minutes of rest. Better yet, if you know your functional

You want to make sure that you get as close to fully recovered as possible before beginning the next set. If you fail to recover sufficiently between sets, you will not be able to reach your peak effort in the working sets and you'll consequently miss the desired adaptation. Also, be sure to give yourself enough time to warm up and then cool down from this intense effort.

The good news, I suppose, is that you don't need to spend very much time in the pain cave. Unless you are training to be competitive in elite endurance sports like cycling, swimming, running, triathlon, or cross-country skiing, a single workout per week in this zone will generally suffice. You'll pretty quickly find that it boosts your performance across the rest of your exercise program as well—and, more importantly, in the rest of your life.

I learned this lesson very vividly not long ago, when my wife and I had a very tight connection at Heathrow Airport in London. Anyone who has connected there knows that getting from Terminal 5 to Terminal 3 is basically a trip within a trip. The only way we were going to catch our connecting flight was to run the equivalent of a mile in less than eight minutes, while each carrying a twenty-pound suitcase. This was not going to be a zone 2 effort; we were going to have to go much harder than that, for eight straight minutes. We needed to be able to produce a burst of power that was much closer to our VO_2 max than to zone 2.

In that moment, we were in a situation not all that different from what our hunter-gatherer ancestors frequently faced (apart from the setting, obviously). Besides being much more fun than traveling through airports, hunting requires 95 percent slow and steady effort, and 5 percent all-out intensity. If you were going to have a chance to kill the antelope or mammoth or whatever else you were tracking, you really needed that extra power to close the deal.

My point is that if you really stop to consider the kind of aerobic fitness that most people actually *need* in the course of their lives, it basically boils down to being really good at going slow for a long time, but also able to go

---

threshold power (FTP), which is the highest power you can sustain for sixty minutes, you should target 120 percent of this for three-minute intervals and 106 percent of this for eight-minute intervals and adjust for everything in between.

hard and fast when needed. Training and maintaining a high level of aerobic fitness, and doing it now, is essential to preserving this range of function in your later years.

In a way, maximum aerobic output is like guitarist Nigel Tufnel's special amplifier, in the classic film *This Is Spinal Tap:* Where most amps only let you turn the volume up to 10, his went up to 11. As he memorably explained, "It's one higher."

Every once in a while, it's nice to have that range. We made our flight with a few seconds to spare.

Strength

Weight training has been a touchstone for me since I was fourteen years old and my best friend John and I, both wannabe prizefighters, first wandered into the gym at the Scarborough campus of the University of Toronto. It was a smelly dungeon two floors underground, inhabited by very sweaty dudes who absolutely lived to lift heavy metal weights. It had no heat, no windows, and no AC, so winters were freezing and summers were so hot it was not uncommon for someone to pass out after a max-effort set. We loved it. It was as mythical to us as Gold's Gym at Venice Beach.

Back then, I went to the gym to pursue my boxing ambitions. I literally had no thought of what my life would look like after about age twenty-three. Now that I'm a middle-aged guy myself, I finally understand the seriousness with which those older guys approached their training. I am chasing a different dream—the Centenarian Decathlon, in case you forgot—but I suspect that I'm on the same page as them now.

The sad fact is that our muscle mass begins to decline as early as our thirties. An eighty-year-old man will have about 40 percent less muscle tissue (as measured by cross section of the *vastus lateralis,* aka the "quad" muscle of the thigh) than he did at twenty-five. But muscle mass may be the least important metric here. According to Andy Galpin, a professor of kinesiology at California State University, Fullerton, and one of the foremost authorities on strength and performance, we lose muscle strength about two to three times more quickly than we lose muscle mass. And we lose power (strength x speed) two to three times faster than we lose strength. This is because the biggest single change in the aging muscle is the atrophy of our fast twitch or type 2 muscle fibers. Ergo, our training must be geared towards improving these with heavy resistance training. Daily life and zone 2 endurance work may be enough to prevent atrophy of type 1 fibers—but unless you are working against significant resistance, your type 2 muscle fibers will wither away.

It takes much less time to lose muscle mass and strength than to gain it, particularly if we are sedentary. Even if someone has been training diligently, a short period of inactivity can erase many of those gains. If that inactivity stems from a fall or a broken bone, and lasts longer than a few days, it can often kick off a steep decline from which we may never fully recover, which is pretty much what happened with Sophie. A study of twelve healthy volunteers with an average age of sixty-seven found that after just ten days of bed rest, which is about what a person would experience from a major illness or orthopedic injury, study participants lost an average of 3.3 pounds of lean mass (muscle). That's substantial, and it shows just how dangerous inactivity can be. If someone is sedentary and consuming excess calories, muscle loss accelerates, because one of the primary destinations of fat spillover is into muscle.

In its most extreme form, this muscle loss is called sarcopenia, as noted in chapter 11. Someone with sarcopenia will have low energy, feelings of weakness, and problems with balance. Sarcopenia is a prime marker for a broader clinical condition called frailty, where a person meets three of these five criteria: unintended weight loss; exhaustion or low energy; low physical activity; slowness in walking; and weak grip strength (about which more soon). It can become difficult to stand or walk, and they are at huge risk of falling and breaking bones.

Regaining that muscle, once we've gotten to this state, is no easy task. One study looked at sixty-two frail seniors (average age seventy-eight) who engaged in a program of strength training and found that even after six months of pure strength training, half of the subjects did not gain *any* muscle mass.

They also didn't lose any muscle mass, likely thanks to the weight training, but the upshot is, it is very difficult to put on muscle mass later in life.

Another metric that we track closely in our patients is their bone density (technically, bone mineral density or BMD). We measure BMD in every patient, every year, looking at both of their hips and their lumbar spine using DEXA. This also measures body fat and lean mass, so it's a useful tool across all of the body-composition domains that we care about.

These three bone regions are typically used to make a diagnosis of osteopenia or osteoporosis. The standard guidelines only recommend screening in women at age sixty-five or men at age seventy—which is classic Medicine 2.0, waiting until someone may be staring danger in the eye before doing anything. We think it's important to get a handle on this much earlier, before any problems arise.

The fact is that bone density diminishes on a parallel trajectory to muscle mass, peaking as early as our late twenties before beginning a slow, steady decline. For women, this decline happens much more quickly once they hit menopause, if they are not on HRT (yet another reason we heavily favor HRT), because estrogen is essential for bone strength—in both men and women. Other risk factors for low bone density include genetics (family history), a history of smoking, long use of corticosteroids (e.g., for asthma or autoimmune conditions), drugs that block estrogen (e.g., women taking such drugs for breast cancer), low muscle mass (again), and being undernourished.

Why do we care so much? Just as with muscle, it comes down to protection. We want to slow this decline, armoring ourselves against injury and physical frailty. The mortality from a hip or femur fracture is staggering once you hit about the age of sixty-five. It varies by study, but ranges from 15 to 36 percent in one year—meaning that up to one-third of people over sixty-five who fracture their hip are dead within a year. Even if a person does not die from the injury, the setback can be the functional equivalent of death in terms of how much muscle mass and, hence, physical capacity is lost during the period of bed rest (recall how quickly people over sixty-five lose muscle mass when bedridden).

Our goal is to try to spot this issue, if it arises, decades before a potential fracture might occur. When we detect low or rapidly declining BMD in a middle-aged person, we use the following four strategies:

1. Optimize nutrition, focusing on protein and total energy needs (see nutrition chapters).
2. Heavy loading-bearing activity. Strength training, especially with heavy weights, stimulates the growth of bone—more than impact sports such as running (though running is better than swimming/cycling). Bones respond to mechanical tension and estrogen is the key hormone in mediating the mechanical signal (weight bearing) to a chemical one telling the body to lay down more bone.
3. HRT, if indicated.
4. Drugs to increase BMD, if indicated.

Ideally, we can solve the problem with the first two, but are not afraid to use the second two methods where appropriate. The takeaway for readers here is that your BMD is important, demanding at least as much attention as muscle mass, so you should at least check your BMD every few years. (Particularly if your primary sports are nonweight-bearing, like cycling or swimming.)

I think of strength training as a form of retirement saving. Just as we want to retire with enough money saved up to sustain us for the rest of our lives, we want to reach older age with enough of a "reserve" of muscle (and bone density) to protect us from injury and allow us to continue to pursue the activities that we enjoy. It is much better to save and invest and plan ahead, letting your wealth build gradually over decades, than to scramble to try to scrape together an individual retirement account in your late fifties and hope and pray that the stock market gods help you out. Like investing, strength training is also cumulative, its benefits compounding. The more of a reserve you build up, early on, the better off you will be over the long term.

Yet unlike some guys in the gym, I'm less concerned with how big my biceps are or how much I can bench press. Those might matter if you're a bodybuilder or a powerlifter, but I'd argue they matter less in the Centenarian Decathlon (or in real life). A far more important measure of strength, I've concluded, is how much heavy stuff you can carry. I say this on the basis of my intuition but also research into hunter-gatherers and human evolution. Carrying is our superpower as a species. It's one reason why we have thumbs, as well as long legs (and arms). No other animal is capable of carrying large objects from one place to another with any efficiency. (And the ones that can, like horses and other livestock, do so only because we bred and trained and harnessed them.) This frames how I view strength training in general. It's largely about improving your ability to carry things.

I've always been a fan of carrying heavy objects with my hands. As a teenager working on a construction site over the summers, I always volunteered to haul tools and materials across the site, and today I still incorporate some kind of carrying, typically with dumbbells, kettlebells, or sandbags, into most of my workouts. I've also become semiobsessed with an activity called *rucking,* which basically means hiking or walking at a fast pace with a loaded pack on your back. Three or four days a week, I'll spend an hour rucking around my neighborhood, up and down hills, typically climbing and descending several hundred feet over the course of three or four miles. The fifty- to sixty-pound pack on my back makes it quite challenging, so I'm strengthening my legs and my trunk while also getting in a solid cardiovascular session. The best part is that I never take my phone on these outings; it's just me, in nature, or maybe with a friend or a family member or a houseguest (for whom rucking is mandatory; I keep two extra rucksacks in the garage).

I was introduced to this pastime by Michael Easter in his eye-opening book *The Comfort Crisis.* His intriguing thesis is that because we have removed all discomfort of any kind from modern life, we have lost touch with the fundamental skills (not to mention the frequent suffering) that once defined what it meant to be human. Carrying stuff over long distances is one of these skills; our ancestors likely had to range far and wide to hunt food for their families and then carry their kills back to camp to feed everyone. But it's so effective that the military has incorporated it into their training.

"Carrying shaped our species," he says. "Our ancestors carried often. It gave them robust functional strength and endurance that was likely very protective. But we've engineered carrying out of our lives, just as we have many other forms of discomfort. Rucking is a practical way to add carrying back into our lives."

The main difference is that instead of carrying sixty pounds of antelope meat in my pack, I'm typically hauling heavy metal weights, which are admittedly less appetizing. One thing I specifically focus on when rucking is the hills. Going uphill gives me a chance to push my VO_2 max energy system; first-time ruckers are amazed at how taxing it is to walk up a 15 percent grade with even twenty pounds on your back—and then walk back down. (A good goal is to be able to carry one-quarter to one-third of your body weight once you develop enough strength and stamina. My daughter and wife routinely carry this much when they join me.)

As great as rucking is, it's not the only thing I rely on to build my strength. Fundamentally I structure my training around exercises that improve the following:

1. *Grip strength,* how hard you can grip with your hands, which involves everything from your hands to your lats (the large muscles on your back). Almost all actions begin with the grip.

2. Attention to both *concentric* and *eccentric loading* for all movements, meaning when our muscles are shortening (concentric) *and* when they are lengthening (eccentric). In other words, we need to be able to lift the weight up and put it back down, slowly and with control. Rucking down hills is a great way to work on eccentric strength, because it forces you to put on the "brakes."

3. *Pulling motions,* at all angles from overhead to in front of you, which also requires grip strength (e.g., pull-ups and rows).

4. *Hip-hinging movements,* such as the deadlift and squat, but also step-ups, hip-thrusters, and countless single-leg variants of exercises that strengthen the legs, glutes, and lower back.

I focus on these four foundational elements of strength because they are the most relevant to our Centenarian Decathlon—and also to living a fulfilling and active life in our later decades. If you can grip strongly, you can open a jar with ease. If you can pull, you can carry groceries and lift heavy objects. If you can do a hip-hinge correctly, you can get up out of a chair with no problem. You're setting yourself up to age well. It's not about how much weight you can deadlift now, but how well you will function in twenty or thirty or forty years.

I put grip strength first because it's something that most people don't really think about. Even I was surprised to discover that there is an enormous body of literature linking better grip strength in midlife and beyond to decreased risk of overall mortality.* The data are as robust as for VO_2 max and muscle mass, in fact. Many studies suggest that grip strength—literally, how hard you can squeeze something with one hand—predicts how long you are likely to live, while low grip strength in the elderly is considered to be a symptom of sarcopenia, the age-related muscle atrophy we just discussed. In these studies, grip strength is likely acting as a proxy for overall muscle strength, but it is also a broader indicator of general robustness *and* the ability to protect yourself if you slip or lose balance. If you have the strength to grab a railing, or a branch, and hold on, you might avoid a fall.

Surprisingly, given the extent to which fitness and gym-going have become so commonplace in our culture in the last few decades, American adults actually seem to have far weaker grip strength—and thus less muscle mass—

* The consensus definition of sarcopenia requires the presence of low skeletal muscle mass and either low muscle strength (e.g., handgrip strength) or low physical performance (e.g., walking speed).

than they did even a generation ago. In 1985, men ages twenty to twenty-four had an average right-handed grip strength of 121 pounds, while in 2015, men of the same age averaged just 101 pounds. This suggests that people now in their thirties are entering midlife with much less strength than their parents, which could lead to problems as they get older.

Grip strength is important at all ages. Every interaction that we have begins with our hands (or feet, as we'll discuss later). Our grip is our primary point of contact in almost any physical task, from swinging a golf club to chopping wood; it is our interface with the world. If our grip is weak, then everything else is compromised.

Training grip strength is not overly complicated. One of my favorite ways to do it is the classic farmer's carry, where you walk for a minute or so with a loaded hex bar or a dumbbell or kettlebell in each hand. (Bonus points: Hold the kettlebell up vertically, keeping your wrist perfectly straight and elbow cocked at ninety degrees, as though you were carrying it through a crowded room.) One of the standards we ask of our male patients is that they can carry half their body weight in each hand (so full body weight in total) for at least one minute, and for our female patients we push for 75 percent of that weight. This is, obviously, a lofty goal—please don't try to do it on your next visit to the gym. Some of our patients need as much as a year of training before they can even attempt this test.

In general, we urge our new patients to begin with far *less* weight than they have lifted in the past, sometimes even dropping down to body weight exercises at first. As we will see in the next chapter, on stability, it is far more important to learn and practice ideal movement patterns than to be pounding heavy weights all the time. That said, a farmer's carry is pretty straightforward (weight in each hand, arms at sides, walk). The most important tip is to keep your shoulder blades down and back, not pulled up or hunched forward. If you are new to strength training, start with light weights, even as low as ten to fifteen pounds, and work up from there.

Another way to test your grip is by dead-hanging from a pull-up bar for as long as you can. (This is not an everyday exercise; rather, it's a once-in-a-while test set.) You grab the bar and just hang there, supporting your body weight. This is a simple but sneakily difficult exercise that also helps strengthen the critically important scapular (shoulder) stabilizer muscles, which we will talk about in the next chapter. Here we like to see men hang for at least two minutes and women for at least ninety seconds at the age of forty. (We reduce the goal slightly for each decade past forty.)

No discussion of strength is complete without mentioning *concentric* and especially *eccentric* loading. Again, eccentric loading means loading the muscle as it is lengthening, such as when you lower a bicep curl. It's more intuitive when lifting something to focus on the concentric phase, such as curling the dumbbell with your biceps. This is the strength of a muscle getting shorter. One of the tests we have our patients perform is stepping onto and off an eighteen-inch block and taking three full seconds to reach the ground (a forward step down, like descending a very tall step). The stepping up part is comparatively easy, but most people initially struggle with a controlled three-second descent. That requires eccentric strength and control. (I'll talk about step-ups and step-downs in detail at the end of chapter 13.)

In life, especially as we age, eccentric strength is where many people falter. Eccentric strength in the quads is what gives us the control we need when we are moving down an incline or walking down a set of stairs. It's really important to keep us safe from falls and from orthopedic injuries. When we can eccentrically load our muscles, it also prevents our joints from taking excess stress, especially our knees. Think about creeping slowly down a very steep hill versus running down in an uncontrolled manner. The difference in force transmission to your knees is dramatic (as is the difference in likely outcomes—safe descent versus faceplant and likely knee injury).

Training eccentric strength is relatively simple. Big picture, it means focusing on the "down" phase of lifts ranging from pull-ups or pull-downs to deadlifts to rows; rucking downhill, carrying a weighted pack, is a great way to build both eccentric strength as well as spatial awareness and control, which are important parts of stability training (next chapter). It also helps protect against knee pain. You don't need to do this for every rep of every set. Sometimes you just want to focus on moving the weight quickly or moving a heavier load, but make sure at some point in each workout that you are taking the time to cue the eccentric phase of your lifts.

Next is pulling, which is closely related to grip strength. Pulling motions are how we exert our will on the world, whether we are hoisting a bag of groceries out of the car trunk or climbing El Capitan. It is an anchor movement. In the gym, it typically takes the form of rows, where you're pulling the weight toward your body, or pull-ups. A rowing machine, something I love to use for VO_2 max training, is another simple and effective way to work on pulling strength.

The final foundational element of strength is hip-hinging, which is what it sounds like: You bend at the hips—*not* the spine—to harness your body's largest muscles, the gluteus maximus and the hamstrings. (I repeat: Do not bend your spine.) It is a very powerful move that is essential to life. Whether you are launching off an Olympic ski jump, picking up a lucky penny off the sidewalk, or simply getting up out of a chair, you are hip-hinging.

Hip-hinging under high axial load, as with a heavy deadlift or squat, should be approached with care because of the risk of injury to the spine. This is why we have our patients work up to weighted hip-hinging very slowly, typically beginning with single-leg step-ups (see description below) and split-stance Romanian deadlift, either without weights or with only very light weights held in the hands.

Normally, this would be the section where I would write a lengthy treatise on the finer points of how to do pull-ups and hip-hinges. I've come to the conclusion that they can't actually be done right without dozens of pictures and thousands of words, in a book that is already too long. There are two reasons why I have decided not to give the details. One, I believe that this type of content is best learned in person, from someone who knows how to cue the movements. For example, the "hard" part about teaching a proper hip-hinge is not illustrating the correct position of the spine relative to the femur and lower leg, or the angle of the hips, in a diagram. The hard part is knowing how to eccentrically load the glutes and hamstrings *before* you hinge, and how to feel your feet pushing into the floor evenly across the full surface of your foot.

If this is hard to follow, you see exactly why I've come to the conclusion that the best way to communicate this information to you, in a way that is actionable, is to show you, as opposed to telling you. And the best way I can show you, shy of working out with you, is having you watch me do these exercises over video, with my colleague Beth Lewis cuing me. (I've placed a link to these videos, and a brief description, at the end of the next chapter.)

The second reason I've opted not to describe all these exercises in detail is that when new patients come to us, we typically have them *stop* strength training, at least with heavy weights. Our first step is to put them through a series of strength and movement tests designed to assess not only their physical condition but also their degree of stability. So before you do anything in the gym, I would urge you to read the next chapter, to begin to understand the crucial and complex concept of stability.

CHAPTER 13

The Gospel of Stability

Relearning How to Move to Prevent Injury

The loftier the building, the deeper the foundation
must be laid.

—THOMAS À KEMPIS

By now it should be clear that it is important to stay in good physical condition as we age. But now consider another, related question: Why don't more people actually pull this off?

A typical seventy-year-old will do less than half as much "moderate to vigorous" physical activity as she did at age forty—and after age seventy the decline accelerates. The fit people in their seventies and eighties are the exception, not the rule.

It is tempting to attribute this to aging itself, the aches and pains that accumulate in middle age and beyond, not to mention the steady loss of aerobic capacity and strength. Other factors such as weight gain and poor sleep can also leave one feeling wiped out. But I think the missing X factor that explains *why* so many people just stop moving is something else: injury. That is, older people tend to exercise less, or not at all, because they simply can't. They have hurt themselves in some way, at some point in their lives, and they just never got back on the horse. So they continued to decline.

This was certainly true of Sophie, my friend Becky's mother, but I too could have easily gone down that path. In my twenties, when I was in medical school and still training hard, lifting weights almost daily, I experienced a mysterious back injury that required two separate surgeries (one of which was botched), followed by a long and very difficult recovery. For several months I was almost unable to function, surviving on large amounts of painkillers. I couldn't even brush my teeth without excruciating back pain, and I spent most of the day just lying on the floor. It got so bad that my mom had to fly out to Palo Alto and take care of me. The thing is, people think it's terrible when someone in their twenties has to go through this (and it is), yet they almost expect it for someone Sophie's age.

Sophie and I were not unique: this type of injury and chronic pain is shockingly widespread. According to the CDC, more than 27 percent of Americans over the age of forty-five report suffering from chronic pain, and about 10 to 12 percent say that pain has limited their activities on "most days or every day" during the previous six months. Most days or every day! Back pain, in particular, is a huge driver of opioid prescriptions and surgical procedures that are often of dubious value. It is a leading cause of disability around the world, and in the United States alone it drains off an estimated $635 billion-with-a-B per year in medical costs and lost productivity.

As I learned, all the aerobic fitness or strength in the world won't help you if you get hurt and have to stop exercising for several months—or forever. Studies of college-age athletes who experience injury in their careers find that they report consistently lower quality of life at middle and older ages. Their injuries continue to affect them not only physically but psychologically as well, for decades into their lives. During my long ordeal, I came to appreciate how important our ability to function physically is to our overall well-being.

All of the above, the research and my own experience, support my first commandment of fitness: *First, do thyself no harm.*

How do we do this? I think stability is the key ingredient. But it also requires a change in our mindset. We have to break out of the mentality that we must crush all our workouts every single time we go to the gym—doing the most reps, with the heaviest weights, day after day. As I learned, pushing oneself so hard all the time, without adequate stability, almost inevitably leads to injury. If you are struggling to get through your workout, then you are likely resorting to your body's own "cheats," your ingrained but potentially dangerous movement patterns.

Instead, we need to change our approach so that we are focused on doing things *right,* cultivating safe, ideal movement patterns that allow our bodies to work as designed and reduce our risk of injury. Better to work smart than to work too hard. But as I would see for myself, relearning these movement patterns is no simple task.

Stability is often conflated with "core," but there is much more to it than having strong abdominal muscles (which isn't what "core" means anyway). In my view, stability is essential to any kind of movement, particularly if our goal is to be able to keep doing that movement for years or decades. It is the foundation on which our twin pillars of cardiovascular fitness and strength must rest. Without it, as we used to say in Canada, you are hosed. Maybe not immediately, but sooner or later you will likely experience an injury that limits your movement, kiboshes your daily activities as you age, and possibly knocks you out of the Centenarian Decathlon for good.

One thing that stability training has taught me is that most "acute" injuries, such as a torn ACL or a hamstring tear, are rarely sudden. While their onset may be rapid—instantaneous back or neck or knee pain—there was likely a chronic weakness or lack of stability at the foundation of the joint that was the true culprit. This is the real iceberg in the water. The "acute" injury is just the part you see, the manifestation of the underlying weakness. So if we are to complete the goals we have set in our own Centenarian Decathlon, we need to be able to anticipate and avoid any potential injuries that lie in our path, like icebergs at sea. This means understanding stability and incorporating it into our routine.

Stability is tricky to define precisely, but we intuitively know what it is. A technical definition might be: stability is the subconscious ability to harness, decelerate, or stop force. A stable person can react to internal or external stimuli to adjust position and muscular tension appropriately without a tremendous amount of conscious thought.

I like to explain stability using an analogy from my favorite sport, auto racing. A few years ago I drove to a racetrack in Southern California to spend a couple of days training with my coach. To warm up, I took a few "sedan laps" in my street car at the time, a modified BMW M3 coupe with a powerful 460+ HP engine. After months of creeping along on clogged Southern California freeways, it was hugely fun to dive into the corners and fly down the straightaways.

Then I switched to the track car we had rented, basically a stripped-down, race-worthy version of the popular BMW 325i. Although this vehicle's engine produced only about one-third as much power (165 HP) as my street car, my lap times in it were several seconds faster, which is an eternity in auto racing. What made the difference? The track car's 20 percent lighter weight played a part, but far more important were its tighter chassis and its stickier, race-grade tires. Together, these transmitted more of the engine's force to the road, allowing this car to go much faster through the corners. Though my street car was quicker in the long straights, it was much slower overall because it could not corner as efficiently. The track car was faster because it had better stability.

Without stability, my street car's more powerful engine was not much use. If I attempted to drive it through the curves as fast as I drove the track car, I'd end up spinning into the dirt. In the context of the gym, my street car is the guy with huge muscles who loads the bar with plates but who always seems to be getting injured (and can't do much else *besides* lift weights in the gym). The track car is the unassuming-looking dude who can deadlift twice his body weight, hit a fast serve in tennis, and then go run up a mountain the next day. He doesn't necessarily look strong. But because he has trained for stability as well as strength, his muscles can transmit much more force across his entire body, from his shoulders to his feet, while protecting his vulnerable back and knee joints. He is like a track-ready race car: strong, fast, stable—and healthy, because his superior stability allows him to do all these things while rarely, if ever, getting injured.

Obviously, my street car would be much more comfortable for a long road trip; no analogy is perfect. But this street car/race car comparison works because it forces us to consider stability in the dynamic setting. Unfortunately, the words *stable* and *stability* are too often lumped in with static terms like *strong* and *in balance*. A tree is more stable than a sapling. A Jenga tower cannot stand without stability. But in the exercise context, we're not as interested in how rigid something is. Instead, we want to think about how efficiently and safely force can be transmitted *through* something.

The key word is *safely*. When stability is lacking, all that extra force has to go somewhere. If my street car's powerful engine is transmitting only part of its power to the road through the tires, the remainder of that energy is leaking out, lost to friction and nonproductive motion, primarily. Parts of the car that should not be moving relative to each other are doing just that. As fun as it might be to drift a car around a corner, that lost energy is ravaging the tires and taking a toll on the suspension. Neither will last long. When this happens in our bodies, this force dissipation (as it's called) leaks out via the path of least resistance—typically via joints like knees, elbows, and shoulders, and/or the spine, any or all of which will give out at some point. Joint injuries are almost always the result of this kind of energy leak.

In sum, stability lets us create the most force in the safest manner possible, connecting our body's different muscle groups with much less risk of injury to our joints, our soft tissue, and especially our vulnerable spine. The goal is to be strong, fluid, flexible, and agile as you move through your world.

In action, stability can be magnificent to behold. Stability lets a skinny pitcher throw a blazing fastball. Stability allows Kai Lenny to surf towering waves at Jaws. But stability is also what enables a seventy-five-year-old woman to continue playing tennis injury-free. Stability is what keeps an eighty-year-old grandmother from falling when she steps off a curb that is unexpectedly high. Stability gives a ninety-five-year-old man the confidence to go walk his beloved dog in the park. It lets us keep doing what we love to do. And when you don't have stability, bad things will inevitably happen—as they did to me, and to Sophie, and to millions of other formerly fit people.

My painful lower-back episode was only the beginning of my injury history. I completed one of my Catalina swims with a torn labrum that was almost certainly exacerbated by spending four hours a day training in the pool and ocean, and continuing to do so even after I started feeling pain.* I still needed surgery to fix the problem more than fifteen years later. That was the price I paid for overdoing it in one specific sport. But it took me another couple of decades to really begin to understand *why* I had injured my back.

This knowledge came courtesy of Beth Lewis, a former professional dancer and powerlifter turned trainer and all-around movement genius who was then based in New York. (I've since talked her into moving to Austin.) We had barely even said hello before she ordered me to take off my shirt and squat. I obeyed, and she was not impressed. I was crestfallen. I had always thought of myself as someone who knew what he was doing in the gym. Now I was being told that I couldn't even do a simple squat correctly.

Her iPhone video told a sorry tale, as you can see from the "before" photo on the left (see figure 13): As I loaded my hips and sank down, I automatically shifted my entire body to the right. I look like I'm about to topple over. My problem, as these photos make painfully clear, was that I lacked stability. It even hurts to look at it now, because it reminds me of the thousands of atrocious, strain-inducing squats I'd committed in this awkward position.

---

* A torn labrum is a pretty common injury, but many people never require surgery to fix it. Though endless swimming is what made it worse, the injury was caused by the frequent subluxations or mild dislocations that I had experienced growing up. Each time the shoulder joint is subluxed it gnaws away at the labrum and increases the odds for further shoulder instability and pain.

Figure 13.

Before (1/29/19) **After** (10/24/19)

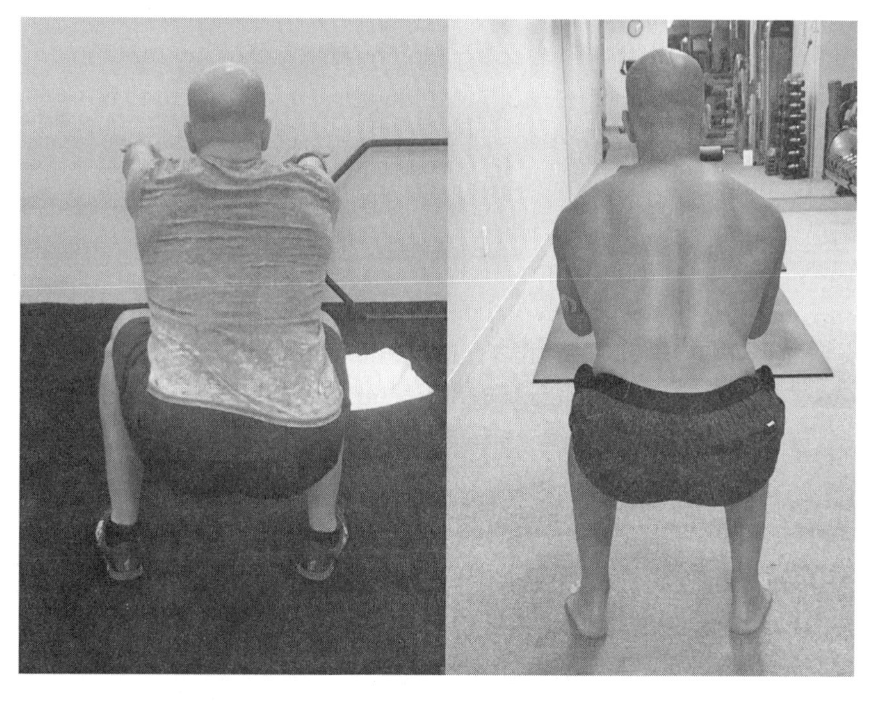

I was not even aware that I was doing this, but I was likely compensating for various injuries and weaknesses that I had accumulated over the years. This is how it works, as I would learn: We try to cheat or work around our existing injuries and limitations and end up creating new problems. This rightward tilt may even explain my back injury when I was only in my twenties; even at that point, I had already been lifting heavy weights for years. Fixing the situation turned out to be a nine-month process, but it ultimately straightened me out, as you can see in the "after" photo on the right. It required retraining not only my body but my brain.

Both Beth and Michael Stromsness, a trainer with whom I'd worked in California and who had introduced me to Beth, were familiar with something I had never heard of called DNS. Short for *dynamic neuromuscular stabilization,* DNS sounds complicated, but it is based on the simplest, most natural movements we make: the way we moved when we were babies.

The theory behind DNS is that the sequence of movements that young children undergo on their way to learning how to walk is not random or accidental but part of a program of neuromuscular development that is essential to our ability to move correctly. As we go through this sequence of motions, our brain learns how to control our body and develop ideal patterns of movement.

DNS originated with a group of Czech neurologists who were working with young children with cerebral palsy in a hospital in Prague in the 1960s. They noticed that because of their illness, these kids did *not* go through the normal infant stages of rolling, crawling, and so forth. Thus they had movement problems throughout their lives. But when the children with cerebral palsy were put through a "training" program consisting of a certain sequence of movements, replicating the usual stages of learning to crawl, sit up, and eventually stand, their symptoms improved and they were better able to control their motions as they matured. The researchers realized that as we grow up, most healthy people actually go through an opposite process—we lose these natural, healthy, almost ingrained movement patterns.

Thus my youngest son, Ayrton, can execute a perfect ass-to-grass squat, dropping his little butt down practically to the ground, bending sharply at the knees yet remaining totally balanced and powerful. It's just a perfect hip-hinge, and it blows my mind every time. He is an absolute master. Yet when I attempted the same movement, I ended up tilted over in the ridiculous half-canted position in the "before" photo, one hip pointed down at the ground, my shoulders askew, my feet rolled outward. My toddler can squat, but apparently I couldn't.

And neither could my fourteen-year-old daughter, Olivia (before Beth got to work on her, too). Flexible as Gumby, skinny but whip-strong, she should be able to squat just as well as, if not better than, her youngest brother. But she couldn't, because even at her young age she had already spent two-thirds of her life in school, mostly sitting in chairs. The ideal movement patterns that she learned as an infant and toddler were erased before she was able to develop the hip stability needed to squat properly. If she spends the next thirty, forty, or fifty years primarily sitting in chairs, as is likely, then she'll be in the same boat as many of my patients, and myself as well: we have essentially forgotten how to move our bodies.

Most adults can't squat correctly, even without any added weight. The only way many of us can come close to matching a toddler's form is to lie on our backs, as Michael Stromsness demonstrated with me in one of our early sessions. Then it becomes much easier to raise our knees into a perfect squat position, with the correct degree of curvature throughout the spine from the base of the skull to the tailbone. This tells us that range of motion per se is not what's stopping most adults from squatting well; it's that when the average adult is under a load, even as little as their own bodyweight, the job of stabilizing his or her own torso becomes too much.

The point of DNS is to retrain our bodies—and our brains—in those patterns of perfect movement that we learned as little kids. As Michael Rintala, a leading American practitioner of DNS, puts it, "DNS beautifully integrates with all the good work you are already doing—it's like a software upgrade for anything you are doing."

My own software was in serious need of an upgrade.

The details of my own journey are too involved to lay out at length here, but in the rest of this chapter I will try to explain at least some of the basic principles that underlie stability training. These may seem a bit strange at first, and if you came to this chapter expecting a high-powered workout program, you may be disappointed. That is part of the point: in my practice, we don't like to push much strength training, including many of the assessments I've discussed, such as dead hangs and weighted step-ups, until we have established some modicum of stability. We don't think it's worth the risk. Just as in engineering, it's worth the extra time to build a solid foundation, even if it delays the project a few months.

A quick caveat: while strength training and aerobic conditioning are relatively straightforward, everyone has very different issues with regard to stability. Thus, it's impossible to give a one-size-fits-all prescription for everyone. My goal in the rest of this chapter is to give you some basic concepts to think about and try out, to help you learn and understand how your own body interacts with the world—which, in the end, is what stability is really about. If you'd like to know more after you've read this chapter, I suggest visiting the websites for DNS (www.rehabps.com) and the Postural Restoration Institute (PRI) (www.posturalrestoration.com), the two leading exponents of what I'm talking about here. Stability is an integral part of my training program. Twice a week, I spend an hour doing dedicated stability training, based on the principles of DNS, PRI, and other practices, with ten to fifteen minutes per day on the other days.

Stability training begins at the most basic level, with the breath.

Breathing is about much more than simple gas exchange or even cardiorespiratory fitness. We exhale and inhale more than twenty thousand times per day, and the way in which we do so has tremendous influence on how we move our body, and even our mental state. How we breathe, as Beth puts it, is who we are.

The link between the body, the mind, and the breath is not new to anyone who has done more than a few Pilates or yoga classes or practiced meditation. In these practices, the breath is our anchor, our touchstone, our timekeeper. It both reflects our mental state and affects it. If our breathing is off, it can disrupt our mental equilibrium, creating anxiety and apprehension; but anxiety can also worsen any breathing issues we might have. This is because deep, steady breathing activates the calming parasympathetic nervous system, while rapid or ragged breathing triggers its opposite, the sympathetic nervous system, part of the fight-or-flight response.

Yet breathing is also important to stability and movement, and even to strength. Poor or disordered breathing can affect our motor control and make us susceptible to injury, studies have found. In one experiment, researchers found that combining a breathing challenge (reducing the amount of oxygen available to study subjects) with a weight challenge reduced the subjects' ability to stabilize their spine. In real-world terms, this means that someone who is breathing hard (and poorly) while shoveling snow is putting themselves at increased risk of a back injury.

It's extremely subtle, but the way in which someone breathes gives tremendous insight to how they move their body and, more importantly, how they stabilize their movements. We run our patients through a series of respiration and movement tests to get the full picture of their respiration strategy and how it relates to their strength and stability issues.

One simple test that we ask of everyone, early on, looks like this: lie on your back, with one hand on your belly and the other on your chest, and just breathe normally, without putting any effort or thought into it. Notice which hand is rising and falling—is it the one on your chest, or your belly, or both (or neither)? Some people tend to flare their ribs and expand the chest on the inhale, while the belly is flat or even goes down. This creates tightness in the upper body and midline, and if the ribs stay flared, it's difficult to achieve a full exhalation. Others breathe primarily "into" the belly, which tilts the pelvis forward. Still others are compressed, meaning they have difficulty moving air in and out altogether, because they cannot expand the rib cage with each inhalation.

Beth identifies three types of breathing styles and associated phenotypes, which she jokingly calls "Mr. Stay Puft," the "Sad Guy," and the "Yogini"—each corresponding to a different set of stability strategies:

Mr. Stay Puft

HYPERINFLATED. This person is an upper-chest breather who tends to pull up into spinal extension for both respiration and stability. Their lumbar spine is in hyperextension, while their pelvis lives in anterior (forward) tilt, meaning their butt sticks out. They are always pulling up into themselves, trying to look like they are in charge. They have a limited sense of grounding in the feet, and limited ability to pronate to absorb shock (the feet turn outward, or supinate). All of the above makes them quite susceptible to lower back pain, as well as tightness in their calves and hips.

Sad Guy

COMPRESSED. Everything about them is sort of scrunched down and tight. Their head juts forward, and so do their shoulders, which kind of roll to the front because they are always pulling forward to try and take in more air. Their midback rolls in an overly flexed or hyperkyphotic posture, and they have limited neck and upper limb motion. Sometimes their lower legs externally rotate, and the feet overpronate. Gravity is weighing them down.

Yogini

UNCONTROLLED. These folks have extreme passive range of motion (i.e., flexibility)—and extremely limited ability to control it. They can often do a toe touch and put their palms flat on the floor, but because of their lack of control, these people are quite prone to joint injuries. They are always trying to find themselves in space, fidgeting and twitching; they compensate for their excessive flexibility by trying to stabilize primarily with their neck and jaw. It is very hard for them to put on lean mass (muscle). Sometimes they have very high anxiety, and possibly also a breathing pattern disorder.

Not everyone fits exactly into one of these three types, but many of us will recognize at least some of these traits in ourselves. There is some overlap as well; it's possible to be a Sad Guy or Mr. Stay Puft and a Yogini at the same time, for example, because the Yogini type is really more about a lack of muscular control.

I was a hyperinflated Mr. Stay Puft, according to Beth: When I inhaled, my ribs would flare out and up, like a rooster thrusting out his chest. This got air into my lungs, but it also pulled my center of mass forward. To balance, my spine would curve into kyphosis, and my butt would stick out (Beth called it "duck butt"). This hyperextended my hamstrings, effectively disconnecting them from the rest of my body, so I was unable to access these muscles. For all those years, before I realized this, I was deadlifting using *only* my back and glutes, with virtually no help from my powerful hamstrings. In terms of breath training, I needed to think about getting air *out*, the exhale—while someone who tends more toward the Sad Guy type should work on getting air *in*, inhaling via the nose rather than the mouth.

The idea behind breath training is that proper breathing affects so many other physical parameters: rib position, neck extension, the shape of the spine, even the position of our feet on the ground. The way in which we breathe reflects how we interact with the world. "Making sure that your breath can be wide and three-dimensional and easy is vital for creating good, efficient, coordinated movement," Beth says.

Beth likes to start with an exercise that builds awareness of the breath and strengthens the diaphragm, which not only is important to breathing but is an important stabilizer in the body. She has the patient lie on their back with legs up on a bench or chair, and asks them to inhale as quietly as possible, with the least amount of movement possible. An ideal inhalation expands the entire rib cage—front, sides, and back—while the belly expands at the same time, allowing the respiratory and pelvic diaphragm to descend. The telltale is that it is quiet. A noisy inhale looks and feels more dramatic, as the neck, chest, or belly will move first, and the diaphragm cannot descend freely, making it more difficult to get air in.

Now, exhale fully through pursed lips for maximum compression and air resistance, to strengthen the diaphragm. Blow all that air out, fully emptying yourself before your shoulders round or your face or jaw gets tense. Very soon, you will see how a full exhale prepares you for a good inhale, and vice versa. Repeat the process for five breaths and do two to three sets. Be sure to pause after each exhale for at least two counts to hold the isometric contraction—this is key, in DNS.

In DNS, you learn to think of the abdomen as a cylinder, surrounded by a wall of muscle, with the diaphragm on top and the pelvic floor below. When the cylinder is inflated, what you're feeling is called *intra-abdominal pressure*, or IAP. It's critical to true core activation and foundational to DNS training. Learning to fully pressurize the cylinder, by creating IAP, is important to safe movement because the cylinder effectively stabilizes the spine.

Here's another quick exercise to help you understand how to create IAP: breathe all the way in, so you feel as if you are inflating the cylinder on all sides and pulling air all the way down into your pelvic floor, the bottom of the cylinder. You're not actually "breathing" there, in the sense that air is actually entering your pelvis; you're seeking maximal lung expansion, which in turn sort of pushes your diaphragm down. With every inhale, focus on expanding the cylinder around its whole diameter and not merely raising the belly. If you do this correctly, you will feel the entire circumference of your shorts expand evenly around your waist, even in the back, not just in the front. When you exhale, the diaphragm comes back up, and the ribs should rotate inward again as your waistband contracts.

This inhale develops tension, and as you exhale, pushing out air, you *keep* that muscular tension all around your cylinder wall. This intra-abdominal pressure is the basic foundation for everything that we do in stability training—a deadlift, squats, anything. It's as if you have a plastic bottle: with the cap off, you can crush the bottle in one hand; with the cap on, there is too much pressure (i.e., stability) and the bottle can't be crushed. I practice this 360-degree abdominal breathing every day, not only in the gym but also while I am at my desk.*

Your "type" also indicates how you should work out, to some extent. The Stay Puft people tend to need more grounding through the feet and more work with weight in front of them so as to pull their shoulders and hips into a more neutral position. Beth typically has someone like me hold a weight in front of my body, a few inches in front of the sternum. This forces my center of mass back, more over my hips. Try it with a light dumbbell or even a milk carton, and you'll see what I mean. It's a subtle but noticeable change of position.

---

* Back when I used to fly every week, I tried a clever trick that Michael Rintala showed me: put two tennis balls in an athletic sock about four to six inches apart, and position them just about at the level of my kidneys, or where my thoracic spine meets my lumbar spine. Then, with every breath I try to make sure I expand fully enough to feel the tennis balls on both sides. The idea is that it cues your breathing. When I did this, I could get off a five-hour flight and feel as if I had not been sitting for longer than about five minutes. (It also kept my seatmates from talking to me when I was trying to work.) It's worth trying on a long flight or drive.

With the Sad Guys and Gals, Beth tends to work more on cross-body rotation, having them swing the arms across the body to open up the chest and shoulders. She is cautious about loading the back and shoulders, preferring to begin with body weight exercises and split-leg work, such as a walking lunge with a reach, either across the body or to the ceiling, on each step.

For the Yoginis, Beth recommends doing "closed-chain" exercises such as push-ups, using the floor or wall for support, as well as using exercise machines with a well-defined and limited range of motion, given their lack of joint control. Machines are important for these folks, and also for people who have not lifted much or at all, because machines keep their movements within safe boundaries. For the Yoginis, as well as for newbies in general, it's important to become more aware of where they are in space, and where they are relative to their range of motion.

The larger point is that someone's breathing style gives us insight into their broader stability strategy, the set of patterns that they have evolved over the years to help them get by in the physical world. All of us have these strategies, and 95 percent of the time, in the course of daily life, they work fine. But once you add different stressors, such as speed, weight, and novelty or unfamiliarity (e.g., stepping off a stair in the dark), then those strategies, those instinctive physical reactions, can create problems. And if our respiration is also taxed, those other problems will be magnified.

If the road to stability begins with the breath, it travels through the feet—the most fundamental point of contact between our bodies and the world. Our feet are literally the foundation for any movement we might make. Whether we're lifting something heavy, walking or running (or rucking), climbing stairs, or standing waiting for a bus, we're always channeling force through our feet. Unfortunately, too many of us have lost basic strength and awareness of our feet, thanks to too much time spent in shoes, especially big shoes with thick soles.

Going back to my race car analogy, our feet are like the tires, the only point of contact between the car and the road. The force of the engine, the stability and stiffness of the chassis, the skill of the driver—all of it is useless if the tires are not firmly gripping the track surface. I would argue that our feet are even more important to us than tires are to a car, as they also play a crucial role in dampening force before it reaches the knees, the hips, and the back (at least a car has suspension rods for that). Failing to pay attention to your feet, as most of us do, is like buying a McLaren Senna (my dream car) and then going to Walmart and getting the cheapest tires you can find. That's what spending years in mushy shoes does to us.

Take another look at my "before" squat. Yes, my hips are obviously askew, but look more closely at my feet. Are they flat on the floor? No, they are not. As you can clearly see, they are rolled out on their outside edges—"supinated," in physiologist-speak. They should be flat, grounded, stable, and strong, to support my weight. But instead they are rolled over and wobbly. No wonder my squat looks so bad.

To help reacquaint us with our feet, Beth Lewis likes to put me, and our patients, through a routine she calls "toe yoga." Toe yoga (which I hate, by the way) is a series of exercises intended to improve the dexterity and intrinsic strength of our toes, as well as our ability to control them with our mind. Toe strength may not be something you think about when you go to the gym, but it should be: Our toes are crucial to walking, running, lifting, and, most importantly, decelerating or lowering. The big toe especially is necessary for the push-off in every stride. Lack of big-toe extension can cause gait dysfunction and can even be a limiting factor in getting up off the floor unassisted as we age. If toe strength is compromised, everything up the chain is more vulnerable—ankle, knee, hip, spine.

Toe yoga is a lot harder than it sounds, which is why I've posted a video demonstration of this and other exercises at www.peterattiamd.com/outlive/videos. First, Beth tells her students to think of their feet as having four corners, each of which needs to be rooted firmly on the ground at all times, like the legs of a chair. As you stand there, try to feel each "corner" of each foot pressing into the ground: the base of your big toe, the base of your pinky toe, the inside and outside of your heel. This is easy, and revelatory; when was the last time you felt that grounded?

Try to lift all ten toes off the ground and spread them as wide as you can. Now try to put just your big toe back on the floor, while keeping your other toes lifted. Trickier than you'd think, right? Now do the opposite: keep four toes on the floor and lift only your big toe. Then lift all five toes, and try to drop them one by one, starting with your big toe. (You get the idea.)*

If you can do this at all, it likely takes a concerted mental effort, your brain *telling* that big toe to drop or rise—which is exactly the point. One of the goals of stability training is to regain mental control, conscious or not, over key muscles and body parts. Because our feet spend so much time crammed into shoes that may or may not fit properly, and likely have a lot of padding in their soles, many of us have lost touch with our feet, or have worked them into unhelpful contortions over time.

In my "before" squatting photo, as noted above, both of my feet are rolled out to the outside, or supinated, a common phenotype. Another common foot strategy is to "pronate" or fold the feet inward—a term you're probably familiar with if you've ever bought running shoes. Beth compares pronation to driving a car with too little air in the tires, meaning you kind of slosh through your movements, unable to transfer force efficiently to the ground. Supination, on the other hand, is like having overinflated tires, so you skid and bounce around. Your feet are unable to absorb shock, and all that bouncing and jarring gets transferred straight to the ankles, hips, knees, and lower back. Both syndromes, pronation and supination, also expose us to risk of plantar fasciitis and knee injury, among other issues. We must be able to move in and out of both supination and pronation to locomote efficiently. Now when I squat, or do any standing lift, my first step is to ground my feet, to be aware of all four "corners," and distribute weight equally. (Also important: I prefer to lift barefoot or in minimal shoes, with little to no cushioning in the soles because it allows me feel the full surface of my feet at all times.)

Feet are also crucial to balance, another important element of stability.

---

* If you really want to go all in on toe yoga, get a set of "toe spacers," which help restore the toes to a more natural, spread position, particularly in people with bunions or other shoe-related issues. I wear these things around the house a lot. I'm typing right now while wearing them. My kids mock me relentlessly.

One key test in our movement assessment is to have our patients stand with one foot in front of the other and try to balance. Now close your eyes and see how long you can hold the position. Ten seconds is a respectable time; in fact, the ability to balance on one leg at ages fifty and older has been correlated with future longevity, just like grip strength. (Pro tip: balancing becomes a lot easier if you first focus on grounding your feet, as described above.)

The structure we most want to protect—and a major focus of stability training in general—is the spine. We spend so much of our time in car seats, in desk chairs, at computers, and peering at our various devices that modern life sometimes seems like an all-out assault on the integrity of our spine.

The spine has three parts: lumbar (lower back), thoracic (midback), and cervical (neck) spine. Radiologists see so much degeneration in the cervical spine, brought on by years of hunching forward to look at phones, that they have a name for it: "tech neck."

This is why it's important to (a) put down the phone, and (b) try to develop some proprioceptive awareness around your spine, so that you really understand what extension (bending back) and flexion (bending forward) feel like, at the level of each single vertebra. The easiest way to start this process is to get on your hands and knees and go through an extremely slowed-down, controlled Cat/Cow sequence, similar to the basic yoga poses of the same names.*

The difference is that you have to *really, really slow down,* moving so slowly and deliberately from one end of your spine to the other that you can feel each individual vertebra changing position, all the way from your tailbone up to your neck, until your spine is bent like a sway-backed cow. Then reverse the movement, tilting your pelvis forward and bending your spine

* Some of these basic DNS stability moves that I am describing have analogues in classic yoga poses, and a top-notch yoga instructor can help you develop the neuromuscular control and awareness that are essential to proper stability, but most yoga classes are too vague and loose for my taste.

one vertebra at a time until your back is arched again, like a really scared cat. (Note: Inhale on Cow, exhale on Cat.)

The point of this exercise is not how much extension or flexion you can reach in extreme Cat or Cow but rather how much segmental control you can achieve, going from one extreme to the other. You should learn to feel the position of each vertebra, which in turn helps you better distribute load and force throughout the spine. Now when I deadlift, this segmental control allows me to maintain a more neutral arc from my thoracic to lumbar spine, spreading the load evenly; before, my spine would have a sharp lordotic bend, meaning I was taking too much force on its hinge points. That's what stability is about: safe and powerful transmission of force through muscles and bones, and not joints or spinal hinge points.

Next we come to the shoulders, which are both complex and evolutionarily interesting. The scapulae (shoulder blades) sit on top of the ribs and have a great ability to move around. The shoulder joint is controlled by a complex set of muscles that attach in various positions to the scapula and the upper portion of the humerus, the long bone in the upper arm (which is why we medical types call it the *glenohumeral* joint). If you compare this ball-and-socket joint to the far more stable and solid one in your hip, it becomes clear that evolution made a huge trade-off when our ancestors began to stand up: we gave up a lot of stability in that shoulder joint in exchange for a much greater range of motion and, in practical terms, the all-important ability to throw a spear. But because there are so many different muscular attachments in the shoulder (no fewer than seventeen), it is much more vulnerable than the hip as I learned in my boxing and swimming careers.

Beth taught me a simple exercise to help understand the importance of scapular positioning and control, a movement known as Scapular CARs, for *controlled articular rotations:* Stand with your feet shoulder-width apart and place a medium to light resistance band under your feet, one handle in each hand (a very light dumbbell also works). Keeping your arms at your sides, raise your shoulder blades, and then squeeze them back and together; this is *retraction,* which is where we want them to be when under load. Then drop them down your back. Finally, bring them forward to the starting point. We start out moving in squares like this, but the goal is to learn enough control that we can move our scapulae in smooth circles. A large part of what we're working on in stability training is this kind of neuromuscular control, reestablishing the connection between our brain and key muscle groups and joints.

Almost everything we do in fitness, and in our daily lives, goes through our hands. If our feet are our contact with the ground, absorbing force, our hands are how we transmit force. They are our interface with the rest of our world. Grip strength—how hard you can squeeze—is only part of the equation. Our hands are quite amazing, actually, in that they are powerful enough to crush the juice out of a lemon yet dexterous enough to play a Beethoven sonata on the piano. Our grip can be firm yet feathery, transmitting force with finesse.

It's all about how you distribute force. If you can transmit and modulate force through your hands, then you can push and pull efficiently. This force originates in the powerful muscles of the trunk and is transmitted down the chain, from rotator cuff to elbow to forearm to wrist. There is a strong correlation between having a weak rotator cuff (shoulder) and weak grip strength.

But it starts with finger strength—which, unfortunately, is another thing that we have sacrificed to comfort and convenience. Back when we carried things, we had to have strong hands to survive. No longer. Many of us don't really even use our hands for much besides typing and swiping. This weakness means pushing and pulling movements bring a higher risk for elbow and shoulder injury.

Because we are not "training" grip in our daily lives, we must be deliberate in our workouts, focusing on initiating movement with the hands and utilizing all the fingers with our upper body movements. Adding carries to your training is a great way to train grip, but it is important always to be mindful of what your fingers are doing and how force is being transmitted through them.

One way that Beth likes to illustrate the importance of this is via a basic bicep curl with a (light) dumbbell. First, try the curl with your wrist bent slightly backward, just a bit out of line with your forearm. Now try the same bicep curl with your wrist straight. Which one felt stronger and more powerful? Which one felt like the fingers were more involved? It's about building awareness of the importance of your fingers, as the last link in the chain.

One last way in which grip is important is in situations requiring reactivity—being able to grab (or let go of) a dog's leash when needed, or gripping a railing to prevent a fall. Our grip and our feet are what connect us to the world, so that our muscles can do what they need to do. Even in a deadlift: one of the key things Beth taught me is that a deadlift is as much about feet and hands as hamstrings and glutes. We're pushing the floor away as we lift with our fingers.

These moves and drills that I've described thus far represent only the very basic elements of stability work. They may seem simple, but they require a great deal of focus; in my practice, we don't even allow our patients to work out with heavy loads until they work on these basic principles for at least six months.

One more note: Trainers can be useful for some purposes, such as basic instruction, accountability, and motivation, but we discourage patients from becoming overly reliant on trainers to tell them exactly what to do every single time they work out. I liken this to learning to swim in a wetsuit. Initially, a wetsuit can help give someone confidence because of the additional flotation it provides. But over the longer term, a wetsuit robs you of the need to figure out your balance in the water. Balance is the real challenge with swimming, because our center of mass is way off from our center of volume, causing our hips to sink. Good swimmers learn to overcome this imbalance with training. But if you never take off the wetsuit, you will never learn how to fix this problem.

Similarly, trainers can be helpful in teaching you the basics of different exercises, and to motivate you to get in the habit of working out. But if you never learn to do the exercises on your own, or never try different ways of doing them, you will never develop the proprioception needed to master your ideal movement patterns. You will rob yourself of the learning progression that is such an important part of stability training—the process of narrowing the gap between what you think you are doing and what you are actually doing.

Everything that we've covered in this last section serves two purposes: as a drill, and as an assessment. I would urge you to film yourself working out from time to time, to compare what you think you are doing to what you are actually doing with your body. I do this daily—my phone on the tripod is one of my most valuable pieces of equipment in the gym. I film my ten most important sets each day and watch the video between sets, to compare what I see to what I think I was doing. Over time, that gap has been narrowing.

It was really difficult, at first, to accept that I wasn't going to be lifting heavy weights anymore, but Beth and Michael Stromsness were persuasive. I couldn't even squat properly or perform a simple pull-up correctly, so doing anything more than that would put me at risk of (further) injury.

I fumed over this for a while. How could I live without weight training? It took several months of work, but eventually I had learned enough that I could deadlift again. Where in the past I'd done four hundred pounds or more, now Beth had me begin at just ninety-five pounds, which seemed like hardly any weight at all.

It helped to recall something that my driving coach, Thomas Merrill, often tells me. He is an incredible driver who in 2022 placed second in the one of the most prestigious motor races in the world, the 24 Hours of Le Mans; he knows what he's talking about. One of his mantras is that in order to go faster, you need to go slower.

Here's what he means: when you "overdrive" a car, as when you're trying too hard to drive as fast as possible, you make mistakes. In driving, mistakes compound. When you spin in turn 5, it's because you probably missed the apex in turn 2 and didn't correct in turn 3. You need to slow down and get the car in the right spot, and it'll take care of the rest.

Slow down, go fast. It's the same, I think, with learning stability.

Hip-Hinging 101: How to Do a Step-Up

Rather than try to describe multiple exercises, I think it's more instructive to provide a deeper explanation of one exercise. I've chosen a step-up, simply stepping up onto a box or a chair, for three reasons. First, it's a hip-hinging movement, one of our core elements of strength training. Second, it's a single-leg exercise that does not require much axial (spine) loading, even with weights in your hands, which means it's very safe, even for beginners (you'll start with just your body weight). Third, it's one of the best exercises to target the eccentric phase of the movement as well as the concentric phase. I also like it because it demonstrates some of the key stability concepts we have been learning in this chapter.

First, find a box or a sturdy chair such that when your foot is on the step your thigh will be parallel to the floor. For most people this is about sixteen to twenty inches, but if that is too difficult start with twelve inches. Place one foot on the box, making sure that the big toe and pinky toe mounds and the entire heel are connected firmly to its surface (I like to do these barefoot). The back foot remains on the floor, roughly twelve inches behind the box, with roughly 40 percent of your weight on the back leg and 60 percent on the front leg. Keep your front hip flexed, spine tall, chest heavy (ribs down), arms relaxed by your sides, and eyes forward.

Now, slightly shift your head, ribs, and pelvis forward at the same time as you quietly but fully inhale through your nose, allowing the diaphragm to descend and creating intra-abdominal pressure. You should feel pressure in the center of the front foot, toward the heel, but keep your toes connected to the box. Glide your front femur back slightly, so that you feel a stretch in both the hamstring and the glute max; they should be very slightly loaded. This sensation is the essence of the hip-hinge. You want to lead with your glutes and hamstrings, not pelvis or ribs. All of your power will come from these muscles working together, and not your back. Keep your knee behind your toes, and your pelvis and ribs in alignment, and load your front foot evenly, not favoring either the toes/forefoot or heel.

With your front foot, push down on the box with intent and with minimal push-off assistance from the back foot. Lift yourself off the floor, exhaling as you initiate the movement, extend the hip, and stand up straight on top of the box. Your head and ribs should finish directly over the pelvis. Bring your rear leg through to finish beside and a little in front of the working leg. Everything should arrive at the same time, as you complete the exhale (feeling the compression in the ribs). Hold this position for a second or two.

On the way down, step the nonworking (now front) foot off the back of the box as your head, ribs, and shoulders shift slightly forward and the hip flexes to (once again) prepare the hamstring and glute to lower your weight. Load the front of the stationary foot, the toes actively flexed into the box. As you lower your body down and back through space, feel the weight shifting from the forefoot into the midfoot, and finally to the heel, in a smooth, coordinated fashion that is controlled by the hamstring (think: slowly rocking backward).

Keep the tempo as slow and even as possible; aim for three seconds from step-off to landing (difficult; two seconds is good). As the back foot lowers, your weight continues to shift back until you "land." Avoid shifting more than 40 percent of your weight to the back foot, to reduce the temptation to use forward momentum to start the next rep. Repeat.

Do five to six reps on each side. Start with body weight only, but once you have the movement and sensation down, you can add weights, ideally a dumbbell or kettlebell in each hand.

(Bonus points: Now you are training grip strength as well as hip-hinging.)

The loaded exercise is essentially the same in terms of sequence and position, with a few caveats:

1. Load is now a function of two things: weight and box height. Box height can be an issue if mobility (flexibility and loading tolerance) is a factor.
2. The weights must hang straight down from the shoulders. The brain will find any way to conserve energy and "cheat," so avoid the subconscious urge to swing the weights forward or lift the shoulders to initiate the step-up (highly likely if the load is too heavy). The glute and hamstring should be doing all the work.
3. If the eccentric phase (step-down) cannot be controlled, the weight is too heavy. You never want to feel as if you are falling back. Try using less weight, or a shorter (two-second) step-down at first.
4. It is crucial to keep the ribs and head above or slightly ahead of the pelvis as you initiate the step-up. If you lead with the pelvis, you will be bending your back and also putting too much pressure on the knee.

You will find more video demonstrations on my website, at www.peterattiamd.com/outlive/videos.

The Power of Exercise: Barry

As a former athlete and lifelong exerciser, I already had a substantial fitness base built up, even if I wasn't necessarily moving or lifting correctly. Many of my problems stemmed from lifting *too much,* cycling *too much,* or swimming *too much.* The vast majority of people have the opposite problem: They're not doing enough. Or they haven't done enough. Or they can't do very much at all. For most people, this is the real challenge. They need a jump start. The good news is that these are the very people who can benefit the most. They have the most to gain.

This is also where we see the true power of exercise—its ability to transform people, to make them functionally younger. It's quite incredible. I mentioned earlier how taking up weight training in her sixties changed my mom's life. But there's no better exemplar, I think, than the amazing, inspiring Barry.

Barry was another client of Beth's (but not a patient of mine), an entrepreneur and executive who had spent his career building a successful business, putting in long hours at work and spending virtually no time on anything else, including his fitness. He took cycling trips occasionally, but that was about it.

I see that a lot among my own patients: they trade health for wealth. Then they reach a certain age and realize they are on a bad path. This was Barry: After spending basically fifty years sitting in a chair, he retired and it dawned on him that he was in terrible shape. Not only was his physical capacity very limited, but he was in almost constant pain. He was then closing in on eighty years old and looking at some painful years ahead—a bad Marginal Decade.

He began to wonder: Why had he worked so hard? In the state he was in, retirement no longer seemed very appealing.

At some point, he had a revelation: instead of retiring, he would give himself a new job. This "job," as he saw it, was to rebuild his neglected body so he could get more enjoyment out of life. He began working with Beth and kept on going even as the pandemic made it impossible to train in person for a while. He was highly motivated. Beth has to remind many of her clients to stick with their workout schedule, but with Barry she had the opposite problem: he wanted to spend *too* much time in the gym. She had to make him take breaks and rest.

Barry's goals are different from mine, obviously, but they went well beyond vaguely wanting to "get healthier." He wanted to be able to do a pull-up—that was his stated fitness goal. What he really wanted was to feel strong, and to be able to move in the world with confidence again, without fear of falling, just as he had done as a younger man. But he was nowhere near that; if Beth had put him on a pull-up bar, he likely would have hurt himself. He could barely walk without pain. So he had to begin at a much more basic level, learning how to do simple movement patterns safely.

Beth started him off with some of the same introductory exercises I'd done: abdominal breathing, progressing into the slowed-down, segmental Cat/Cow. To lessen his risk of falling she had him focus on balance-related movements, beginning with his feet—learning to move and feel his toes again, after decades of having been shoved into shoes. He then progressed into one-leg walking and standing drills. Beth even had him dance, to help him relearn how to move his feet and how to react to visual cues to keep his balance.

They then progressed into building basic strength, beginning with walking lunges to fortify his lower body. His abdominals were still weak from surgery twenty years earlier—it's not uncommon, I've observed, for these things to affect people decades after the fact. So they worked on his abdominal strength, beginning (as I did) with building intra-abdominal pressure. And gradually, they worked toward building his upper and midbody strength—and the scapular stability—he would need. Before long he could do better push-ups than most twenty-something gym bros.

Beth put him through drills designed to improve his ability to react and stay balanced. She had him use an agility ladder, similar to what NFL players and other field-sport athletes use to develop balance, quickness, and footwork. If you're training to be an athlete of life, then you're training to be an athlete, period.

Last, she had Barry work on jumping drills, which is definitely out of the comfort zone of most octogenarians. He was nervous, but eventually he got to the point where he could hop off a pair of yoga blocks and land in a squat—and stick it. The idea was to prepare him for the unexpected, so that if he did find himself stepping off an unexpected stair or curb, he could catch himself and not fall. Most people instinctively brace themselves, out of fear; they don't trust their "brakes," their eccentric strength, and that almost always makes their landing less safe. With stability, you have to be fluid and prepared to react, almost like a dancer.

Another important move that they worked on was simply to get Barry to be able to get up off the ground, using only one arm (or ideally, no arms). This is one of those things that we who are younger take for granted. Of course, we can get up off the ground—until, suddenly, we can't. Children learn to do it without a second thought. But somewhere along the way, adults lose the ability to execute this basic move. Even if we have the requisite physical strength, we might lack neuromuscular control; the message from our brain just doesn't reach our muscles. For someone who is eighty-one, like Barry (at this writing), this is a big deal; it could make the difference between continuing to live independently and having to think about going into a nursing home. So Beth taught him a choreographed sequence of movements that would allow him to stand up from a seated position, and he worked on them until he had mastered it.

The "Barry Get-Up" has become a key part of the fitness assessment that we do with all our patients, as well as one of the key events in the Centenarian Decathlon (it should be in yours, too). It's an important move, whether you're picking yourself up off the ground after a stumble or playing with grandchildren on the floor. (For a video demonstration of the Barry Get-Up, please visit www.peterattiamd.com/outlive/videos.) Everyone should be able to do it.

But I think it's also a metaphor for what's possible with exercise training (and, of course, stability). People like Barry help us to rewrite that narrative of decline that trapped my friend's mom, Sophie, and so many other people. Exercise has the power to change us profoundly, even if we're starting from zero, as Barry was. It gives us the ability to pick ourselves up off the ground—literally and figuratively—and become stronger and more capable. It's not about slowing the decline, it's about getting better, and better, and better.

As Barry puts it, "If you're not pushing ahead, you're going backwards."

CHAPTER 14

Nutrition 3.0

You Say Potato,
I Say "Nutritional Biochemistry"

Religion is a culture of faith;
science is a culture of doubt.

—Richard Feynman

I dread going to parties, because when people find out what I really do for a living (not buying my usual lies about being a shepherd or a race car driver), they always want to talk about the topics I dread most: "diet" and "nutrition."

I will do whatever it takes to get out of that conversation—go and get a drink, even if I'm already holding one, or pretend to answer my phone, or, if all else fails, feign a grand mal seizure. Like politics or religion, it's just not a fit topic of conversation, in my view. (And if I seemed like kind of a jerk to you at a party once, my apologies.)

Diet and nutrition are so poorly understood by science, so emotionally loaded, and so muddled by lousy information and lazy thinking that it is impossible to speak about them in nuanced terms at a party or, say, on social media. Yet most people these days are conditioned to want bullet-point "listicles," bumper-sticker slogans, and other forms of superficial analysis. It reminds me of a story about the great physicist (and one of my heroes) Richard Feynman being asked at a party to explain, briefly and simply, why he was awarded his Nobel Prize. He responded that if he could explain his work briefly and simply, it probably would not have merited a Nobel Prize.

Feynman's rule also applies to nutrition, with one caveat: we actually know far *less* about this subject than we do about subatomic particles. On the one hand, we have made-for-clickbait epidemiological "studies" that make absurd claims, such as that eating an ounce of tree nuts each day will lower your cancer risk by exactly 18 percent (not making this up). On the other, we have clinical trials that tend almost without exception to be flawed. Thanks to the poor quality of the science, we actually don't know that much about how what we eat affects our health. That creates a tremendous opportunity for a multitude of would-be nutrition gurus and self-proclaimed experts to insist, loudly, that only *they* know the true and righteous diet. There are forty thousand diet books on Amazon; they can't all be right.

Which brings us to my final quibble about the world of nutrition and diets, which is the extreme tribalism that seems to prevail there. Low-fat, vegan, carnivore, Paleo, low-carb, or Atkins—every diet has its zealous warriors who will proclaim the superiority of their way of eating over all others until their dying breath, despite a total lack of conclusive evidence.

Once upon a time, I too was one of those passionate advocates. I spent three years on a ketogenic diet and have written and blogged and spoken extensively about that journey. For better and for worse, I'm indelibly associated with low-carb and ketogenic diets. Giving up added sugar—literally, putting down the Coke that I held in my hand, on September 8, 2009, moments after my lovely wife suggested I "work on being a little less not thin"—was the first step on a long, life-changing, but also frustrating journey through the world of diet and nutrition science. The good news is that it reversed my incipient metabolic syndrome and may have saved my life. It also led to me writing this book. The bad news is that it exhausted my patience for the "diet debate."

Consider this chapter my penance.

Overall, I think most people spend either too little or too much time thinking about this topic. Probably more on the "too little" side, as evidenced by the epidemic of obesity and metabolic syndrome. But those on the "too much" side are loud and insistent (check out nutrition Twitter). I was completely guilty of this myself, in the past. Looking back, I now realize that I was too far on the left on the Dunning-Kruger curve, caricatured below in figure 14—my maximal confidence and relatively minimal knowledge having propelled me quite close to the summit of "Mount Stupid."

Figure 14. **Dunning-Kruger Effect**

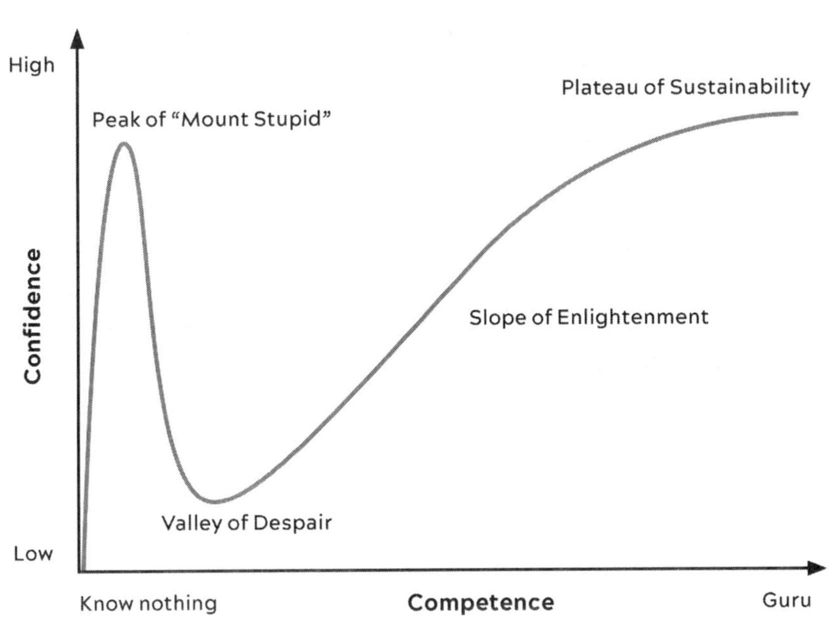

Source: Wikimedia Commons, (2020).

Now I might be halfway up the Slope of Enlightenment on a good day, but one key change I have made is that I am no longer a dogmatic advocate of any particular way of eating, such as a ketogenic diet or any form of fasting. It took me a long time to figure this out, but the fundamental assumption underlying the diet wars, and most nutrition research—that there is one perfect diet that works best for every single person—is absolutely incorrect. More than anything I owe this lesson to my patients, whose struggles have taught me a humility about nutrition that I never could have learned from reading scientific papers alone.

I encourage my patients to avoid using the term *diet* at all, and if I were a dictator, I might ban it entirely. When you eat a slice of prosciutto or a Rice Krispies square, you are ingesting a multitude of different chemical compounds. Just as their chemical makeup differentiates them in terms of taste, the molecules in the foods that we consume affect multiple enzymes and pathways and mechanisms in our bodies, many of which we have discussed in previous chapters. These food molecules—which are basically nothing more than different arrangements of carbon, nitrogen, oxygen, phosphorus, and hydrogen atoms—also interact with our genes, our metabolism, our microbiome, and our physiologic state. Moreover, each of us will react to these food molecules in different ways.

Instead of diet, we should be talking about *nutritional biochemistry.* That takes it out of the realm of ideology and religion—and above all, emotion—and places it firmly back into the realm of science. We can think of this new approach as Nutrition 3.0: scientifically rigorous, highly personalized, and (as we'll see) driven by feedback and data rather than ideology and labels. It's not about telling you what to eat; it's about figuring out what works for your body and your goals—and, just as important, what you can stick to.

What problem are we trying to solve here? What is our goal with Nutrition 3.0?

I think it boils down to the simple questions that we posited in chapter 10:

1. Are you *under*nourished, or *over*nourished?
2. Are you *under*muscled, or *adequately* muscled?
3. Are you metabolically healthy or not?

The correlation between poor metabolic health and being overnourished and undermuscled is very high. Hence, for a majority of patients the goal is to reduce energy intake while adding lean mass. This means we need to find ways to get them to consume fewer calories while also increasing their protein intake, and to pair this with proper exercise. This is the most common problem we are trying to solve around nutrition.

When my patients are undernourished, it's typically because they are not taking in enough protein to sustain muscle mass, which as we saw in the previous chapters is a crucial determinant of both lifespan and healthspan. So any dietary intervention that compromises muscle, or lean body mass, is a nonstarter—for both the under- and overnourished groups.

I used to think that diet and nutrition were the one path to perfect health. Years of experience, with myself and my patients, have led me to temper my expectations a bit. Nutritional interventions can be powerful tools with which to restore someone's metabolic equilibrium and reduce risk of chronic disease. But can they extend and improve lifespan and healthspan, almost magically, the way exercise does? I'm no longer convinced that they can.

I still believe that most people need to address their eating pattern in order to get control of their metabolic health, or at least not make things worse. But I also believe that we need to differentiate between behavior that *maintains* good health versus tactics that *correct* poor health and disease. Wearing a cast on a broken bone will allow it to heal. Wearing a cast on a perfectly normal arm will cause it to atrophy. While this example is obvious, it's amazing how many people fail to translate it to nutrition. It seems quite clear that a nutritional intervention aimed at correcting a serious problem (e.g., highly restricted diets, even fasting, to treat obesity, NAFLD, and type 2 diabetes) might be different from a nutritional plan calibrated to maintain good health (e.g., balanced diets in metabolically healthy people).

Nutrition is relatively simple, actually. It boils down to a few basic rules: don't eat too many calories, or too few; consume sufficient protein and essential fats; obtain the vitamins and minerals you need; and avoid pathogens like *E. coli* and toxins like mercury or lead. Beyond that, we know relatively little with complete certainty. Read that sentence again, please.

Directionally, a lot of the old cliché expressions are probably right: If your great-grandmother would not recognize it, you're probably better off not eating it. If you bought it on the perimeter of the grocery store, it's probably better than if you bought it in the middle of the store. Plants are very good to eat. Animal protein is "safe" to eat. We evolved as omnivores; ergo, most of us can probably find excellent health as omnivores.

Don't get me wrong, I still have a lot to say—that's why these chapters on nutrition are not short. There is so much ideological bickering and utter bullshit out there that I hope to inject at least a little bit of clarity into the discussion. But most of this chapter and the next will be aimed at changing the way you *think* about diet and nutrition, rather than telling you to *eat this, not that.* My goal here is to give you the tools to help you find the right eating pattern for yourself, one that will make your life better by protecting and preserving your health.

What We Sort of Know About Nutritional Biochemistry (and How We Sort of Know It)

One of my biggest frustrations in the area of nutrition—sorry, *nutritional biochemistry*—has to do with how little we actually know about it for certain. The problem is rooted in the poor quality of much nutrition research, which leads to bad reporting in the media, lots of arguing on social media, and rampant confusion among the public. What are we supposed to eat (and not eat)? What is the right diet for *you*?

If all we have to go on is media reporting about the latest big study from Harvard, or the wisdom of some self-appointed diet guru, then we will never escape this state of hopeless confusion. So before we delve into the specifics, it's worth taking a step back to try to understand what we do and don't know about nutrition—what kinds of studies might be worth heeding and which ones we can safely ignore. Understanding how to discern signal from noise is an important first step in coming up with our own plan.

Our knowledge of nutrition comes primarily from two types of studies: epidemiology and clinical trials. In epidemiology, researchers gather data on the habits of large groups of people, looking for meaningful associations or correlations with outcomes such as a cancer diagnosis, cardiovascular disease, or mortality. These epidemiological studies generate much of the diet "news" that pops up in our daily internet feed, about whether coffee is good for you and bacon is bad, or vice versa.

Epidemiology has been a useful tool for sleuthing out the causes of epidemics, including (famously) stopping a cholera outbreak in nineteenth-century London, and (less famously) saving boy chimney sweeps from an epidemic of scrotal cancer that turned out to be linked to their employment.* It has propelled some real public health triumphs, such as the advent of smoking bans and widespread treatment of drinking water. But in nutrition it has proved less insightful. Even at face value, the "associations" that nutritional epidemiologists come up with are often absurd: Will eating twelve hazelnuts every day *really* add two years to my lifespan, as one study suggested?† I wish.

The problem is that epidemiology is incapable of distinguishing between correlation and causation. This, aided and abetted by bad journalism, creates confusion. For example, multiple studies have found a strong association between drinking diet sodas and abdominal fat, hyperinsulinemia, and cardiovascular risk. Sounds like diet soda is bad stuff that causes obesity, right? But that is not what those studies actually demonstrate, because they fail to ask an important question: Who drinks diet soda?

People who are concerned about their weight or their diabetes risk, that's

---

* Back in 1775, Percival Pott, an English surgeon, became the first person on record to demonstrate that cancer may be caused by an environmental factor (now known as a carcinogen). Pott noticed an increase in the number of cases of scrotal warts in young chimney boys, who were given the task of climbing inside chimneys to remove the ash and soot. Pott's investigations led him to the conclusion that the cause of this cancer—a squamous cell carcinoma of the skin—was particles of soot becoming lodged in the ridges of the scrotum.

† According to a 2013 study by Bao et al., people who ate a dozen hazelnuts per day reduced their chances of dying in the next thirty years by 20 percent. (No word on the exact mechanism behind this miraculous outcome.)

who. They may drink diet soda *because* they are heavy, or worried about becoming heavy. The problem is that epidemiology is not equipped to determine the direction of causality between a given behavior (e.g., drinking diet soda) and a particular outcome (e.g., obesity) any more than one of my chickens is able to scramble the egg she has just laid for me.

To understand why, we must consult (again) Sir Austin Bradford Hill, a British scientist whom we met in chapter 11. Hill had helped sleuth out the link between smoking and lung cancer in the early 1950s, and he came up with nine criteria for evaluating the strength of epidemiological findings and determining the likely direction of causality, which we also referenced in regard to exercise.* The most important of these, and the one that can best separate correlation from causation, is the trickiest to implement in nutrition: experiment. Try proposing a study where you would test the effects of a lifetime of eating fast food by randomizing young boys and girls either to Big Macs or a non–fast-food diet. Even if you did somehow receive institutional review board approval for this terrible idea, there are a bunch of different ways in which even a simple experiment can go wrong. Some of the Big Mac kids might secretly go vegetarian, while the controls might decide to frequent the Golden Arches. The point is that humans are terrible study subjects for nutrition (or just about anything else) because we are unruly, disobedient, messy, forgetful, confounding, hungry, and complicated creatures.

This is why we rely on epidemiology, which derives data from observation and often from the subjects themselves. As we saw earlier, the epidemiology around exercise passes the Bradford Hill criteria with flying colors—but using epidemiology to study nutrition often flunks those tests miserably, be-

---

* The Bradford Hill criteria are (1) strength of the association (i.e., effect size), (2) consistency (i.e., reproducibility), (3) specificity (i.e., is it an observation of disease in a very specific population at a specific site, with no other likely explanation?), (4) temporality (i.e., does the cause precede the effect?), (5) dose response (i.e., does the effect get stronger with a higher dose?), (6) plausibility (i.e., does it make sense?), (7) coherence (i.e., does it agree with data from controlled experiments in animals?), (8) experiment (i.e. is there experimental evidence to back up the findings?), and (9) analogy (i.e., the effect of similar factors may be considered).

ginning with the effect size, the power of the association, often expressed as a percentage. While the epidemiology of smoking (like exercise) easily passes the Bradford Hill tests because the effect size is so overwhelming, in nutrition the effect sizes are typically so small that they could easily be the product of other, confounding factors.

Case in point: The claim that eating red meats and processed meats "causes" colorectal cancer. According to a very well-publicized 2017 study from the Harvard School of Public Health and the World Health Organization, eating those kinds of meats raises one's risk of colon cancer by 17 percent (HR = 1.17). That does sound scary—but does it pass the Bradford Hill tests? I don't think so, because the association is so weak. For comparison's sake, someone who smokes cigarettes is at more like 1,000 to 2,500 percent (ten to twenty-five times) increased risk of lung cancer, depending on the population being studied. This suggests that there might actually be some sort of causation at work. Yet very few published epidemiological studies show a risk increase of even 50 percent (HR = 1.50) for any given type of food.

Second, and far more damning, is that the raw data on which these conclusions are typically based are shaky at best. Many nutritional epidemiological studies collect information on subjects via something called a "food frequency questionnaire," a lengthy checklist that asks users to recall everything they ate over the last month, or even the last year, in minute detail. I've tried filling these out and it's almost impossible to recall exactly what I ate two days ago, let alone three weeks.* So how reliable can studies based on such data possibly be? How much confidence do we have in, say, the red meat study?

So do red and processed meats actually cause cancer or not? We don't know, and we will probably never get a more definitive answer, because a clinical trial testing this proposition is unlikely ever to be done. Confusion reigns. Nevertheless, I'm going to stick my neck out and assert that a risk ratio of 1.17 is so minimal that it might not matter that much whether you eat red/

---

* If you'd like to try, Google "Food Frequency Questionnaire," and good luck to you.

processed meats versus some other protein source, like chicken. Clearly, this particular study is very far from providing a definitive answer to the question of whether red meat is "safe" to eat. Yet people have been fighting about it for years.

This is another problem in the world of nutrition: too many people are majoring in the minor and minoring in the major, focusing too much attention on small questions while all but ignoring the bigger issues. Small variations in what we eat probably matter a lot less than most people assume. But bad epidemiology, aided and abetted by bad journalism, is happy to blow these things way out of proportion.

Bad epidemiology so dominates our public discussion of nutrition that it has inspired a backlash by skeptics such as John Ioannidis of the Stanford Prevention Research Center, a crusader against bad science in all its forms. His basic argument is that food is so complex, made up of thousands of chemical compounds in millions of possible combinations that interact with human physiology in so many ways—in other words, nutritional biochemistry—that epidemiology is simply not up to the task of disentangling the effect of any individual nutrient or food. In an interview with the CBC, the normally soft-spoken Ioannidis was brutally direct: "Nutritional epidemiology is a scandal," he said. "It should just go into the waste bin."

The true weakness of epidemiology, at least as a tool to extract reliable, causal information about human nutrition, is that such studies are almost always hopelessly confounded. The factors that determine our food choices and eating habits are unfathomably complex. They include genetics, social influences, economic factors, education, metabolic health, marketing, religion, and everything in between—and they are almost impossible to disentangle from the biochemical effects of the foods themselves.

A few years ago, a scientist and statistician named David Allison ran an elegant experiment that illustrates how epidemiological methods can lead us astray, even in the most tightly controlled research model possible: laboratory mice, which are genetically identical and housed in identical conditions.

Allison created a randomized experiment using these mice, similar to the caloric restriction experiments we discussed in chapter 5. He split them into three groups, differing only in the quantity of food they were given: a low-calorie group, a medium-calorie group, and a high-calorie, *ad libitum* group of animals who were allowed to eat as much as they wanted. The low-calorie mice were found to live the longest, followed by medium-calorie mice, and the high-calorie mice lived the shortest, on average. This was the expected result that had been well established in many previous studies.

But then Allison did something very clever. He looked more closely at the high-calorie group, the mice with no maximum limit on food intake, and analyzed this group separately, as its own nonrandomized epidemiological cohort. Within this group, Allison found that some mice chose to eat more than others—and that these hungrier mice actually lived *longer* than the high-calorie mice who chose to eat less. This was exactly the opposite of the result found in the larger, more reliable, and more widely repeated randomized trial.

There was a simple explanation for this: the mice that were strongest and healthiest had the largest appetites, and thus they ate more. Because they were healthiest to begin with, they also lived the longest. But if all we had to go on was Allison's epidemiological analysis of this particular subgroup, and not the larger and better-designed clinical trial, we might conclude that eating more calories causes *all* mice to live longer, which we are pretty certain is not the case.

This experiment demonstrates how easy it is to be misled by epidemiology. One reason is because general health is a massive confounder in these kinds of studies. This is also known as healthy user bias, meaning that study results sometimes reflect the baseline health of the subjects more than the influence of whatever input is being studied—as was the case with the "hungry" mice in this study.*

One classic example of this, I believe, lies in the vast, well-publicized lit-

---

* I think healthy user bias is also the single biggest confounder in the exercise epidemiology literature. Healthy people tend to do more exercise in part because they are healthy.

erature correlating "moderate" drinking with improved health outcomes. This notion has become almost an article of faith in the popular media, but these studies are also almost universally tainted by healthy user bias—that is, the people who are still drinking in older age tend to do so *because* they are healthy, and not the other way around. Similarly, people who drink zero alcohol often have some health-related reason, or addiction-related reason, for avoiding it. And such studies also obviously exclude those who have already died of the consequences of alcoholism.

Epidemiology sees only a bunch of seemingly healthy older people who all drink alcohol and concludes that alcohol is the cause of their good health. But a recent study in *JAMA*, using the tool of Mendelian randomization we discussed back in chapter 3, suggests that this might not be true. This study found that once you remove the effects of other factors that may accompany moderate drinking—such as lower BMI, affluence, and not smoking—any observed benefit of alcohol consumption completely disappears. The authors concluded that there is *no* dose of alcohol that is "healthy."

Clinical trials would seem like a much better way to evaluate one diet against another: One group of subjects eats diet X, the other group is on diet Y, and you compare the results. (Or, to continue the alcohol example, one group drinks moderately, one group drinks heavily, and the control group abstains altogether.)

These are more rigorous than epidemiology, and they offer some ability to infer causality thanks to the process of randomization, but they too are often flawed. There's a trade-off between sample size, study duration, and control. To do a long study in a large group of subjects, you essentially have to trust that they are following the prescribed diet, whether the Big Mac diet in our hypothetical example from above, or a simple low-fat diet. If you want to ensure that your subjects are actually eating the diet, you need to feed each subject, observe them eating, and keep them locked in the metabolic ward of a hospital (to be sure they are not eating anything else). All of this is doable, but only for a handful of subjects for a few weeks at a time, which is not nearly a large enough sample or long enough duration to infer anything beyond mechanistic insights about nutrients and health.

These studies make pharmaceutical studies seem simple. Determining whether pill X lowers blood pressure enough to prevent heart attacks requires only that study subjects remember to take their pill every day for however many months or years, and even that simple compliance poses a challenge. Now imagine trying to ensure that study subjects lower their dietary fat content to no more than 20 percent of total calories and consume at least five servings of fruits and vegetables daily for a year. In fact, I'm convinced that compliance is the *key* issue in nutrition research, and with diets in general: Can you stick to it? The answer is different for almost everyone. This is why it's so difficult for experiments to answer the central questions about the relationship between diet and disease, no matter how big and ambitious they are.

One classic example of a well-intended nutrition study that created more confusion than clarity is the Women's Health Initiative (WHI), an enormous randomized controlled trial that was meant to test a low-fat, high-fiber diet in nearly fifty thousand women. Begun in 1993, it lasted eight years and cost nearly $750 million (and if it sounds familiar, that is because of the study's highly publicized other arm, discussed earlier, which looked at the effects of hormone replacement therapy on older women). In the end, despite all this effort, the WHI found no statistically significant difference between the low-fat and control diet groups in terms of incidence of breast cancer, colorectal cancer, cardiovascular disease, or overall mortality.*

Many people, myself included, argued that the results of this study demonstrated the lack of efficacy of low-fat diets. But in reality, it probably told us nothing about a low-fat diet because the "low-fat" intervention group con-

---

* While this study did not find a statistically significant difference in death from breast cancer at either the 8.5 or 16.1 year follow-up, it did find a statistically significant reduction in deaths from any cause in those women diagnosed with breast cancer, but the difference in absolute risk was insignificant. At 8.5 years the reduction in deaths was 0.013% and at 16.5 years it was a mere 0.025%.

sumed around 28 percent of their calories from fat, while the control group got about 37 percent of their calories from fat. (And that's even assuming the investigators were able to be remotely accurate in their assessment of what the subjects actually ate over the years, a big assumption.) So this study compared two diets that were pretty similar, and found that they had pretty similar outcomes. Big surprise. Nevertheless, flawed as it was, the WHI study has been fought over for years by partisans of different ways of eating.

Just as an aside, the WHI study does provide a clear example of why it is so important to evaluate any intervention, nutritional or otherwise, through the lens of *efficacy* versus *effectiveness*. Efficacy tests how well the intervention works under perfect conditions and adherence (i.e., if one does everything exactly as prescribed). Effectiveness tests how well the intervention works under real-world conditions, in real people. Most people confuse these and therefore fail to appreciate this nuance of clinical trials. The WHI was not a test of the efficacy of a low-fat diet for the simple reasons that (a) it failed to test an actual low-fat diet, and (b) study subjects did not adhere to the diet perfectly. So it can't be argued from the WHI that low-fat diets do not improve health, only that the *prescription* of a low-fat diet, in this population of patients, did not improve health. See the difference?

That said, some clinical trials have provided some useful bits of knowledge. One of the best, or least bad, clinical trials ever executed seemed to show a clear advantage for the Mediterranean diet—or at least, for nuts and olive oil. This study also focused on the role of dietary fats.

The large Spanish study known as PREDIMED (PREvención con DIeta MEDiterránea) was elegant in its design: rather than telling the nearly 7,500 subjects exactly what they were supposed to eat, the researchers simply gave one group a weekly "gift" of a liter of olive oil, which was meant to nudge them toward other desired dietary changes (i.e., to eat the sorts of things that one typically prepares with olive oil). A second group was given a quantity of nuts each week and told to eat an ounce per day, while the control group was simply instructed to eat a lower-fat diet, with no nuts, no excess fat on the meat they did eat, no *sofrito* (a garlicky Spanish tomato sauce with onions and peppers that sounds delicious), and weirdly, no fish.

The study was meant to last six years, but in 2013 the investigators announced that they had halted it prematurely, after just four and a half years, because the results were so dramatic. The group receiving the olive oil had about a one-third lower incidence (31 percent) of stroke, heart attack, and death than the low-fat group, and the mixed-nuts group showed a similar reduced risk (28 percent). It was therefore deemed unethical to continue the low-fat arm of the trial. By the numbers, the nuts-or-olive-oil "Mediterranean" diet appeared to be as powerful as statins, in terms of number needed to treat (NNT), for primary prevention of heart disease—meaning in a population that had not yet experienced an "event" or a clinical diagnosis.*

It looked like a slam dunk; it's rare when investigators can report hard outcomes like death or heart attack, as opposed to simple weight loss, in a mere dietary study. It did help that the subjects already had at least three serious risk factors, such as type 2 diabetes, smoking, hypertension, elevated LDL-C, low HDL-C, overweight or obesity, or a family history of premature coronary heart disease. Yet despite their elevated risk, the olive oil (or nuts) diet had clearly helped them delay disease and death. A post hoc analysis of PREDIMED data also found cognitive improvement in those allocated the Mediterranean-style diet(s), versus cognitive decline in those allocated the low-fat diet.

But does that mean a Mediterranean diet is right for everyone, or that extra-virgin olive oil is the healthiest type of fat? Possibly—but not necessarily.

To me, perhaps the most vexing issue with diet and nutrition studies is the degree of variation between *individuals* that is found but often obscured. This

---

* In secondary prevention, statins tend to show a somewhat lower NNT. The PREDIMED study was later retracted and reanalyzed to correct for errors in randomization (that is, the subjects were not assigned a particular intervention on a truly random basis); the new analysis did not materially change the study's conclusions, however. In my opinion, the biggest problem with PREDIMED is something called *performance bias*, meaning subjects in the two treatment arms may have changed their behavior due to having more interaction with the investigators than the controls.

is especially true in studies looking mostly or entirely at weight loss as an end point. The published studies report average results that are almost always underwhelming, subjects losing a few pounds on average. In reality, some individuals may have lost quite a bit of weight on the diet, while others lost none or even gained weight.

There are two issues at play here. The first is compliance: how well can you stick to the diet? That differs for everyone; we all have different behaviors and thought patterns around food. The second issue is how a given diet affects you, with your individual metabolism and other risk factors. Yet these are too often ignored, and we end up with generalizations about how diets "don't work." What that really means is that diet X or diet Y doesn't work for *everyone.*

Our goal in the next chapter is to help you figure out the best eating plan for *you,* as an individual. To do that, we must move beyond labels and dive into nutritional biochemistry.

CHAPTER 15

Putting Nutritional Biochemistry into Practice

How to Find the Right Eating Pattern for You

My doctor told me to stop having intimate dinners for four. Unless there are three other people.

—Orson Welles

Most of my patients are already on some sort of "diet" when they come to me. One thing that almost all of them have in common is that they are dissatisfied with the results.

I can empathize. During residency, when I was actually even fatter than Not-Thin Peter, I tried a vegan diet for a while. Theoretically, going vegan should have made it easy to lose weight, simply because you have to chew your way through an awful lot of salad to match the caloric content of a ribeye. But in reality, I was taking most of my meals in the hospital, so that meant a lot of chips and other snacks, and a veggie sub every day for lunch. I didn't lose a single freaking pound in six months. Looking back, the problem should have been obvious. While I was technically following a virtuous "vegan" diet, I was basically eating a bunch of junk food that just didn't happen to contain animal products. In other words, I was on a vegan version of the SAD, the Standard American Diet.

Even going vegan is not enough to free you from the clutches of the SAD. It is our default food environment, occupying the middle of the grocery store: the boxed and frozen and bagged bounty of an agricultural system that produces subsidized corn, flour, sugar, and soybeans by the megaton. On one level, it's brilliant, a solution to four problems that have plagued humanity since the beginning: (1) how to produce *enough* food to feed almost everyone; (2) how to do so *inexpensively;* (3) how to *preserve* that food so it can be stored and transported safely; and (4) how to make it highly *palatable.* If you optimize for all four of these characteristics, you're pretty much guaranteed to end up with the SAD, which is not so much a diet as a business model for how to feed the world efficiently. Two cheers for modern industrial food systems.

But notice that a fifth criterion is missing: how to *make it harmless.* The SAD was not specifically intended to do harm, of course. The fact that it does harm most of us, if consumed in excess, is a consequence of the four points above colliding with millions of years of evolution that have optimized us to be highly efficient fat-storage vehicles. It is an unfortunate externality of its business model, sort of like with cigarettes. Tobacco manufacturers set out to make a lot of money from a plentiful agricultural commodity, but the solution they devised, the cigarette, had an unfortunate side effect: it slowly killed the customer.

The elements that constitute the SAD are almost as devastating to most people as tobacco when consumed in large quantities, as intended: added sugar, highly refined carbohydrates with low fiber content, processed oils, and other very densely caloric foods. I should note that this does not mean *all* "processed" foods are bad. Almost everything we eat, aside from fresh vegetables, is processed to some degree. For example, cheese is a processed food, invented as a way to preserve milk, which would otherwise spoil quickly without refrigeration. What we're really talking about, when we talk about the SAD, is *junk* food.

The basic problem we face is that, for perhaps the first time in human history, ample calories are available to many if not most people on the planet. But evolution has not prepared us for this situation. Nature is quite happy for us to be fat and frankly doesn't care if we get diabetes. Thus, the SAD foils our key objectives with regard to nutrition: it induces us to eat more than we need to, becoming overnourished, while its preponderance of low-quality, ultra-processed ingredients tends to displace other nutrients that we need, such as protein, to maintain optimal health.

The SAD disrupts the body's metabolic equilibrium. It places enormous strain on our ability to control our blood glucose levels, and causes us to store fat when we should be utilizing it. The leading source of calories that Americans consume is a category called "grain-based desserts," like pies, cakes, and cookies, according to the US Department of Agriculture. That is our *number one* "food group." If we consume a bunch of grain-based desserts in a Cheesecake Factory binge, our blood glucose levels will surge. And if we do it over and over and over again, as we saw in previous chapters, we will eventually overwhelm our ability to handle all those calories in a safe way. The SAD essentially wages war on our metabolic health, and, given enough time, most of us will lose the war.

The farther away we get from the SAD, the better off we will be. This is the common goal of most "diets"—to help us break free of the powerful gravitational pull of the SAD so that we can eat less, and hopefully eat better. But eating less is the primary aim. Once you strip away the labels and the ideology, almost all diets rely on at least one of the following three strategies to accomplish this:

1. CALORIC RESTRICTION, or CR: eating less in total, but without attention to what is being eaten or when it's being eaten
2. DIETARY RESTRICTION, or DR: eating less of some particular element(s) within the diet (e.g., meat, sugar, fats)
3. TIME RESTRICTION, or TR: restricting eating to certain times, up to and including multiday fasting

In other words, if you are overnourished, and statistically speaking about two-thirds of us are, you will need to apply at least one of these methods of caloric reduction: deliberately tracking (and reducing) what you eat; cutting out certain foods; and/or giving yourself less time in which to eat. That's it. Breaking down our approach to nutrition to these three strategies allows us to speak about dietary interventions more objectively, instead of relying on labels such as "low-fat" or "Mediterranean" that don't tell us very much. If we modify none of these variables—eating whatever we want, whenever we want, in as great a quantity as we want—we end up right back at the SAD.

Each of these approaches has its pros and cons, as I've observed over a decade of working on nutrition issues with countless patients. These will be covered in more detail below, but here is the tl;dr:

1. From the standpoint of pure efficacy, CR or caloric restriction is the winner, hands down. This is how bodybuilders shed weight while holding on to muscle mass, and it also allows the most flexibility with food choices. The catch is that you have to do it perfectly—tracking every single thing you eat, and not succumbing to the urge to cheat or snack—or it doesn't work. Many people have a hard time sticking with it.
2. DR or dietary restriction is probably the most common strategy employed for reducing energy intake. It is conceptually simple: pick a type of food, and then don't eat that food. It only works, obviously, if that food is both plentiful and significant enough that eliminating it will create a caloric deficit. Saying you're going on the "no lettuce" diet is pretty much doomed to fail. And you can still overeat while adhering perfectly to a particular DR, as I found out when I attempted to go vegan.
3. TR or time restriction—also known as intermittent fasting—is the latest trend in ways to cut calories. In some ways I think it's the easiest. When I was a cyclist, and I was trying to drop that six final pounds from my already very light (for me) frame, this became my

jam. I would allow myself only one meal per day, despite doing about three hours per day of training. But this can still backfire if you overeat. I have, much to my amusement, watched patients gain weight on a one-meal-a-day approach by turning their meal into a contest to see who could eat the most pizza and ice cream. But the more significant downside of this approach is that most people who try it end up very protein deficient (we'll cover protein needs later in this chapter). One not uncommon scenario that we see with TR is that a person loses weight on the scale, but their body composition alters for the worse: they lose lean mass (muscle) while their body fat stays the same or even increases.

These three approaches are what we'll spend the rest of the chapter exploring, beginning with the most important: how much we eat.

CR: Calories Matter

I may be starting to sound like a broken record, but it should be obvious by now that many of the problems we want to address or avoid stem from consuming calories in excess of what we can use or safely store. If we take in more energy than we require, the surplus ends up in our adipose tissue, one way or another. If this imbalance continues, we exceed the capacity of our "safe" subcutaneous fat tissue, and excess fat spills over into our liver, our viscera, and our muscles, as we discussed in chapter 6.

How many calories you consume has a huge impact on everything else we're talking about in this book. If you're ingesting one thousand extra calories a day, of anything, you're going to have problems sooner or later. In prior chapters, we've seen how excess calories contribute to many chronic diseases, not only metabolic disorders but also heart disease, cancer, and Alzheimer's disease. We also know from decades of experimental data (chapter 5) that eating *fewer* calories tends to lengthen lifespan, at least in lab animals such as rats and mice—although there is debate over whether this represents true lifespan extension or an elimination of the known hazards of overfeeding, the default state of the control animals in most of these experiments. (And also, of many modern humans.)

In human beings, as opposed to laboratory animals, caloric restriction typically goes by a different name: calorie *counting*. There is plenty of research showing that people who count their calories and limit them can and do lose weight, the primary end point of such studies. This is how Weight Watchers works. The biggest obstacles to doing this successfully are first of all hunger and second the requirement that you track what you eat in meticulous detail. The apps that help you do this today are better than ten years ago, but it's still not easy. For the right person, this approach works incredibly well—it's a favorite of bodybuilders and athletes—but for many, the requirement of constant tracking makes it unfeasible.

One slight advantage is that calorie counting is agnostic to food choices; you can eat whatever you want so long as you stay within your daily allowance. But if you make too many poor decisions, you will be very hungry, so buyer beware. You can lose weight on a restricted-calorie diet consisting only of Snickers bars, but you will feel much better if you opt for steamed broccoli and chicken breasts instead.

There has been a long-running controversy over whether caloric restriction could or should be applied to humans as a tool to enhance longevity. It did seem to work for Luigi Cornaro, the dieting Italian gentleman from the sixteenth century—he claimed to have lived to one hundred, although he was probably actually in his eighties when he died. This supposed longevity benefit is, obviously, a difficult proposition to study in human beings over the long term, for some of the reasons I've just outlined. So the hypothesis was tested in monkeys, in two long-running primate studies. The results were so surprising that they are still being debated.

In July of 2009, a study published in *Science* found that rhesus monkeys that had been fed a reduced-calorie diet for more than two decades had lived markedly longer than those who were allowed to eat freely. "Dieting Monkeys Offer Hope for Living Longer," declared the headline on the front page of *The New York Times*. The accompanying photos told the story: on the left was a monkey named Canto, looking trim and spry at the relatively advanced age of twenty-seven, while on the right sat Owen, just two years older but looking like Canto's flabby, dissipated uncle. Canto had been on a calorically restricted diet for most of his life, while Owen had eaten just about as much food as he wanted.

Owen and Canto were two of seventy-six monkeys in this study, begun two decades earlier at the University of Wisconsin–Madison. Half of the monkeys (the control group) were fed *ad libitum*, meaning they could eat as much food as they wanted, while the other half were placed on a "diet" allowing them about 25 percent fewer calories than the controls. They then lived out their lives as the researchers observed them growing older.

Aging studies tend to be about as exciting as watching paint dry, but the end results were pretty dramatic. In the end, the calorically restricted monkeys lived significantly longer and proved far less likely to die of age-related diseases than the *ad libitum*–fed control monkeys. They were healthier by many other measures, such as insulin sensitivity. Even their brains were in better shape than those of the controls, retaining more gray matter as they aged. "These data demonstrate that caloric restriction slows aging in a primate species," the study authors concluded.

Case closed, or so it seemed.

Three years later, in August 2012, another monkey study made the front page of the *Times*, but with a markedly different headline: "Severe Diet Doesn't Prolong Life," the paper declared grimly, adding, "At Least in Monkeys." This study, also begun in the 1980s, was conducted under the auspices of the National Institute on Aging, one of the National Institutes of Health, and the study design was nearly identical to the Wisconsin study, with one group of monkeys being fed about 25 to 30 percent less than the other. Yet the NIH researchers found that their calorically restricted monkeys had *not* lived longer than the controls. There was no statistically significant difference in the lifespans of the two groups. From a headline writer's point of view, caloric restriction had not "worked."

Journalists love it when a study contradicts whatever the last well-publicized study said. In the small world of people who study aging, the NIH results caused consternation. Everyone had expected that the NIH monkey study would confirm the results seen in Wisconsin. Now it appeared as if the two research teams had spent tens of millions of dollars in federal grant money to demonstrate that caloric restriction lengthens monkey lifespans in Wisconsin but not in Maryland, where the NIH monkeys were kept.

But sometimes science tells us more when an experiment "fails" than when it yields the expected results, and so it was with the monkeys. When examined side by side, the two monkey studies had some seemingly minor differences that turned out to be hugely significant—and very pertinent to our strategy as well. Together, the dueling monkey studies constitute one of the most rigorous experiments ever done about the complex relationship between nutrition and long-term health. And like many of the best scientific experiments, this one happened at least partly by accident.

The most profound difference between the two studies was also the most fundamental, for a diet study: the food that the monkeys ate. The Wisconsin animals ate an off-the-shelf commercial monkey chow that was "semipurified," meaning its ingredients were highly processed and rigorously titrated. The NIH monkeys were fed a diet that was similar in its basic macronutrient profile, but their chow was "natural" and less refined, custom formulated from whole ingredients by an in-house primate nutritionist at NIH. The most glaring contrast: while the NIH monkey chow contained about 4 percent sugar, the Wisconsin diet comprised an astonishing 28.5 percent sucrose, by weight. That is a greater proportion of sugar than you'll find in vanilla Häagen-Dazs ice cream.

Could that alone have explained the difference in survival outcomes? Possibly: more than 40 percent of the Wisconsin control monkeys, the ones not subject to calorie limitations, developed insulin resistance and prediabetes, while just one in seven of the NIH controls became diabetic.* And the Wisconsin control monkeys proved far more likely to die from cardiovascular

---

* The Wisconsin researchers recorded markers of diabetes such as insulin resistance, while the NIH researchers only noted the diagnosis of type 2 diabetes.

causes and cancer than monkeys from any other group. This could suggest that caloric restriction was eliminating early deaths because of the bad Wisconsin diet more than it was actually slowing aging—which is still useful information, as avoiding diabetes and related metabolic disorders is important to our strategy.

The Wisconsin researchers defended their diet as more similar to what Americans actually eat, which is a fair point. The comparison is not exact by any means, but in human terms the Wisconsin monkeys were more or less living on fast food, while the NIH monkeys were eating at the salad bar. The Wisconsin control monkeys ate the most calories, of the worst food, and their health suffered. Makes sense; if your diet consists mostly of cheeseburgers and milkshakes, then eating fewer cheeseburgers and milkshakes is going to help you.

The NIH diet was much higher in quality. Instead of ultraprocessed ingredients like corn oil and cornstarch (another 30 percent of the Wisconsin diet), the NIH monkey chow contained ground whole wheat and corn, and thus more phytochemicals and other possibly beneficial micronutrients like those typically found in fresh food. While not exactly natural, it was at least closer to what rhesus monkeys would actually eat in the wild. So giving the NIH monkeys more or less of that feed may have had less of an impact because the diet was not as harmful to begin with. Upshot: the *quality* of your diet may matter as much as the *quantity*.

Taken together, then, what do these two monkey studies have to tell us about nutritional biochemistry?

1. Avoiding diabetes and related metabolic dysfunction—especially by eliminating or reducing junk food—is very important to longevity.

2. There appears to be a strong link between calories and cancer, the leading cause of death in the control monkeys in both studies. The CR monkeys had a 50 percent lower incidence of cancer.

3. The *quality* of the food you eat could be as important as the *quantity*. If you're eating the SAD, then you should eat much less of it.

4. Conversely, if your diet is high quality to begin with, and you are metabolically healthy, then only a slight degree of caloric restriction—or simply not eating to excess—can still be beneficial.

I think this last point is key. These two studies suggest that if you are eating a higher-quality diet—and are metabolically healthy to begin with—then severe caloric restriction may not even be necessary. The NIH control monkeys ate as much as they wanted of their better diet and *still* lived nearly as long as the CR monkeys in both studies. Interestingly, the post facto analyses also revealed that the NIH control monkeys naturally consumed about 10 percent fewer calories per day than the Wisconsin controls, likely because their higher-quality diet left them feeling less hungry. The researchers speculated that even this very slight degree of caloric reduction may have been significant—certainly, it supports our thesis that it is better to avoid being overnourished.

Note that these study results do *not* suggest that everyone needs to undertake a drastic, severe reduction in caloric intake. Limiting calories can be helpful for people who are metabolically unhealthy and/or overnourished. But I'm not convinced that whatever longevity boost long-term, deep caloric restriction may confer is worth some of the trade-offs—including potentially weakened immunity and greater susceptibility to cachexia and sarcopenia (muscle loss), not to mention constant hunger. These unwanted side effects would accelerate some of the negative processes that already go along with aging, suggesting that in older people especially, caloric restriction might do more harm than good.

The monkeys teach us that if you are metabolically healthy and not overnourished, like the NIH animals, then avoiding a crap diet may be all you need. Some of the NIH CR monkeys ended up with some of the longest lifespans ever recorded in rhesus monkeys. It seems quite clear, then, that even for monkeys, limiting caloric intake *and* improving diet quality "works"—it's how you pull it off that is tricky. As we'll see in the next section, there are many other strategies we can adopt to limit the calories we consume and to tailor our food consumption to suit our metabolism and way of life.

DR: The Nutritional Biochemistry "Diet"

Dietary restriction (DR) represents the land of conventional "diets," where 90 percent of the attention—and research funding, and energy, and anger, and, of course, arguing—over nutritional biochemistry is focused. But it is pretty simple, when you get down to it: identify one or more bogeymen in your nutrition world, such as wheat gluten (for example), and exclude it. The more ubiquitous the bogeyman, the more restrictive the "diet," and the more likely you are to reduce your overall caloric intake. Even if you decided to eat nothing but potatoes, you would still lose weight, because a human being can only choke down so many potatoes in a day. I've seen people do this, and it works. The hard part is figuring out *what* foods to eliminate or restrict.

This wasn't an issue for our ancestors. There is ample evidence to suggest that they were opportunistic omnivores, out of necessity. They ate anything and everything they could get their hands on: lots of plants, lots of starch, animal protein whenever they could, honey and berries whenever possible. They also seemed to be, at least on the basis of the study of the few remaining hunter-gatherer societies, very metabolically healthy.

Should we do the same? Should we be opportunistic omnivores eating anything and everything we can get our hands on? That's how evolution has formed us, but our modern food environment makes it a little too easy to find food. Thus, being overnourished and metabolically unhealthy is now commonplace. We have too many choices and too many delicious ways to take calories into our body. Hence the need for dietary restriction. We need to erect walls around what we can and cannot (or should not) eat.

The advantage of DR is that it is highly individualized; you can impose varying degrees of restriction, depending on your needs. For example, you could decide to eliminate all sugar-sweetened beverages, and that would be a great first step (and a relatively easy one). You could go a step further and quit drinking sweet fruit juices as well. You could quit eating other foods with added sugar. Or you could go as far as reducing or eliminating carbohydrates in general.

One reason carbohydrate restriction is so effective for many people is that it tends to reduce appetite as well as food choices. But some people have a harder time maintaining it than others. (Even I am pretty sure I could never go back to a ketogenic diet for more than a few days.) While fat restriction also limits food choices, it can be less effective at reducing appetite if you pick the wrong low-fat foods to eat (e.g., high-carb junk food). If you consume most of your carbohydrates in the form of Fruit Loops, for example, you will still be very hungry all the time.

A major risk with DR is that you can still easily end up overnourished if you are not deliberate about it. People tend to (erroneously) assume you can't eat too much if you're just restricting *fill-in-the-blank* (e.g., carbohydrates). This is incorrect. Even if done correctly and strictly, DR can still result in overnutrition. If you cut out carbohydrates altogether but overdo it on the Wagyu steaks and bacon, you will fairly easily find yourself in a state of caloric excess. The key is to pick a strategy to which you can adhere but that also helps achieve your goals. This takes patience, some willpower, and a willingness to experiment.

We also want to be sure we're not compromising our other goals along the way. Any form of DR that restricts protein, for example, is probably a bad idea for most people, because it likely also impairs the maintenance or growth of muscle. Similarly, replacing carbohydrates with lots of saturated fats can backfire if it sends your apoB concentration (and thus your cardiovascular disease risk) sky-high.

A more significant issue with DR is that everyone's metabolism is different. Some people will lose tremendous amounts of weight and improve their metabolic markers on a low-carbohydrate or ketogenic diet, while others will actually gain weight and see their lipid markers go haywire—on the exact same diet. Conversely, some people might lose weight on a low-fat diet, while others will gain weight. I have seen this happen time and again in my own practice, where similar diets yield very different outcomes, depending on the individual.

For example, when my patient Eduardo came to see me a few years ago with what turned out to be a case of full-blown type 2 diabetes, cutting his carbohydrate intake was clearly the way to go. Type 2 diabetes is a condition of impaired carbohydrate metabolism, after all. From the outside, Eduardo seemed like a pretty healthy guy, with a soccer player's build and a physical job in construction. He certainly did not fit the (bogus) stereotype of the lazy, gluttonous diabetic. But tests showed that he had almost no ability to store excess sugar that he consumed. His hemoglobin A1c was 9.7 percent, well into the diabetic red zone. Being Latino meant that Eduardo was at a higher risk of NAFLD and diabetes to begin with, thanks to his genes. He wasn't even forty, but unless we did something drastic, he more than likely was going to die a painful and early death.

The obvious first step was to wean Eduardo off carbohydrates almost entirely. No more tortillas, rice, or starchy beans—and no Gatorade, either. Because he worked outside in the heat, he was pounding about three or four liters of "sports drinks" each day. I never once described this diet as "ketogenic," and Eduardo certainly wasn't telling the guys on the job site about his new trendy keto diet. He just wasn't drinking Gatorade anymore. (I also put him on the diabetes drug metformin, which is cheap as well as effective.) Within about five months, Eduardo's markers had all normalized, and his diabetes seemed to have been reversed; his hemoglobin A1c was now a completely normal 5.3 percent, just thanks to dietary changes and metformin. And along the way he lost about twenty-five pounds. I'm not saying this diet was the only possible path to this result, but this relatively simple and achievable form of DR created enough of an energy imbalance that he lost weight, and everything else improved in lockstep.

In the past, I was a huge proponent of ketogenic diets, finding them particularly useful to manage or prevent diabetes in patients like Eduardo. I also like that they have a strict definition, unlike "low carb" or "low fat." A ketogenic diet means restricting carbohydrates to such an extent that the body begins metabolizing fat into "ketone bodies" that the muscles and brain can utilize as fuel. A ketogenic diet helped fix Not-Thin Peter, and it had likely saved Eduardo's life. I thought it was the medicine that every metabolically unhealthy person needed.

But my patients brought me back to earth, as they so often do. As a physician, one often receives feedback in a very direct, personal way. If I give someone a medication or a recommendation, I will find out pretty quickly whether it is working. It's not "data" in the strict sense, but it can be equally powerful. I've had more than one patient for whom a ketogenic diet has completely failed. They didn't lose weight, and their liver enzymes and other biomarkers failed to improve. Or they found it impossible to sustain. I've had other patients who were able to stick to the diet, but then their lipid numbers (especially their apoB) went through the roof, probably because of all the saturated fats they were eating.

At the time, this confused me. What was wrong with them? Why couldn't they just follow the diet correctly? I had to remind myself of what Steve Rosenberg used to say when a patient's cancer progressed despite treatment: *The patient has not failed the treatment; the treatment has failed the patient.*

These patients needed a different treatment.

The real art to dietary restriction, Nutrition 3.0–style, is not picking which evil foods we're eliminating. Rather, it's finding the best mix of macronutrients for our patient—coming up with an eating pattern that helps them achieve their goals, in a way that they can sustain. This is a tricky balancing act, and it requires us (once again) to forget about labels and viewpoints and drill down into nutritional biochemistry. The way we do this is by manipulating our four macronutrients: alcohol, carbohydrates, protein, and fat. How well do you tolerate carbohydrates? How much protein do you require? What sorts of fats suit you best? How many calories do you require each day? What is the optimal combination *for you*?

Let's now look at each of the four macronutrients in more detail.

Alcohol

It's easy to overlook, but alcohol should be considered as its own category of macronutrient because it is so widely consumed, it has such potent effects on our metabolism, and it is so calorically dense at 7 kcal/g (closer to the 9 kcal/g of fat than the 4 kcal/g of both protein and carbohydrate).

Alcohol serves no nutritional or health purpose but is a purely hedonic pleasure that needs to be managed. It's especially disruptive for people who are overnourished, for three reasons: it's an "empty" calorie source that offers zero nutrition value; the oxidation of ethanol delays fat oxidation, which is the exact opposite of what we want if we're trying to lose fat mass; and drinking alcohol very often leads to mindless eating.

While I certainly enjoy an occasional glass of my favorite Belgian beer, Spanish red wine, or Mexican tequila (never in the same sitting, obviously), I also believe that drinking alcohol is a net negative for longevity. Ethanol is a potent carcinogen, and chronic drinking has strong associations with Alzheimer's disease, mainly via its negative effect on sleep, but possibly via additional mechanisms. Like fructose, alcohol is preferentially metabolized in the liver, with well-known long-term consequences in those who drink to excess. Last, it loosens inhibitions around other kinds of food consumption; give me a few drinks, and the next thing you know I'm elbow-deep in the Pringles can as I pace around the pantry looking for my next snack.

There have been numerous well-publicized studies suggesting that moderate levels of alcohol consumption can be beneficial, for example by improving endothelial function and reducing clotting factors, both of which would reduce cardiovascular disease risk. But heavier drinking tends to reverse those effects. And as shown by the Mendelian randomization study in *JAMA* that we talked about in the previous chapter, "moderate drinking" is so confounded by healthy user bias that it is impossible to put much faith in these studies purporting to show a health benefit for drinking.

Nevertheless, for many of my patients, the lifestyle around moderate drinking (e.g., a nice glass of wine with a non-SAD dinner) helps them dissipate stress. My personal bottom line: if you drink, try to be mindful about it. You'll enjoy it more and suffer fewer consequences. Don't just keep drinking because they're serving it on the plane. I strongly urge my patients to limit alcohol to fewer than seven servings per week, and ideally no more than two on any given day, and I manage to do a pretty good job adhering to this rule myself.

Carbohydrates

The balance of our nonalcohol diet consists of carbohydrates, protein, and fat, and it's largely a job of finding the right mix for you as an individual. In the days of labeled diets, we would assemble our macronutrients and sort out the different types of foods, using rules and arbitrary boundaries—you can eat this, but not that; these, but not those. We would basically be guessing at the right mix. And then we would wait to see whether it "worked," typically defined in terms of whether the person lost weight over a period of weeks or months. Now we have more sophisticated ways of looking at macronutrients, beginning with the most abundant: carbohydrates.

Carbs probably create more confusion than any other macro. They are neither "good" nor "bad"—although some types are better than others. Overall, it's more a question of matching dose to tolerance and demand, which is much less tricky than it used to be. Thanks to advancements in technology, we no longer need to guess; we now have data.

Carbohydrates are our primary energy source. In digestion, most carbohydrates are broken down to glucose, which is consumed by all cells to create energy in the form of ATP. Excess glucose, beyond what we need immediately, can be stored in the liver or muscles as glycogen for near-term use or socked away in adipose tissue (or other places) as fat. This decision is made with the help of the hormone insulin, which surges in response to the increase in blood glucose.

We already know that it's not good to consume excessive calories. In the form of carbohydrates, those extra calories can cause a multitude of problems, from NAFLD to insulin resistance to type 2 diabetes, as we saw in chapter 6. We know that elevated blood glucose, over a long enough period of time, amplifies the risk of all the Horsemen. But there is also evidence suggesting that repeated blood glucose spikes, and the accompanying rise(s) in insulin, may have negative consequences in and of themselves.

Each person will respond differently to an influx of glucose. Too much glucose (or carbohydrate) for one person might be barely enough for another. An athlete who is training or competing in high-level endurance events might easily take in—and burn up—six hundred or eight hundred grams of carbohydrates per day. If I consumed that much now, day to day, it would probably render me a diabetic within a year. So how much is too much? And what about quality? Obviously that piece of pie is going to affect an endurance athlete differently from a sedentary person—and the pie will also have a different effect than a baked potato or french fries.

Now we have a tool to help us understand our own individual carbohydrate tolerance and how we respond to specific foods. This is called continuous glucose monitoring, or CGM, and it has become a very important part of my armamentarium in recent years.*

The device consists of a microscopic filament sensor that is implanted in the upper arm, attached to a fingertip-sized transmitter that sends data† to the patient's phone in real time. As its name suggests, CGM gives continuous, real-time information on blood glucose levels, which is extraordinary: the patient can see, moment by moment, how their blood sugar levels are responding to whatever they eat, whether a doughnut, a steak, or a handful of Raisinets. More importantly, it also keeps track of glucose levels over time, capturing historical averages and variance, and registering each and every time that blood glucose spikes upward or crashes downward.

CGM represents a huge improvement over the Medicine 2.0 standard of one fasting glucose test per year, which in my opinion tells you almost nothing of value. Think back to my self-driving cars analogy from Part I: fasting blood glucose, annually, does tell us something, but it's not too far from strapping a brick to the gas pedal. With CGM, you start to approximate the sensors currently found on cars with elaborate driver-assistance tools.

The power of CGM is that it enables us to view a person's response to carbohydrate consumption in real time and make changes rapidly to flatten the curve and lower the average. Real-time blood glucose serves as a decent proxy

---

* Disclosure: I have used CGM periodically since 2015, and in 2021 I was a paid adviser to a company (Dexcom) that manufactures and sells CGM devices, though my work with them focused on the measurement of other (nonglucose) analytes.

† The filament does not actually touch the patient's blood but measures glucose levels in the interstitial fluid and extrapolates blood glucose levels from that.

for the insulin response, which we also look to minimize. And, last, I find that it is much more accurate, and more actionable, than HbA1c, the traditional blood test used to estimate average blood glucose over time.

At the moment, CGM is available only by prescription and is most commonly worn by patients diagnosed with type 1 or type 2 diabetes, who need to monitor their glucose levels from moment to moment. For these people, CGM is an essential tool that can protect them from life-threatening swings in blood glucose. But I think nearly every adult could benefit from it, at least for a few weeks, and it will likely be available to consumers without a prescription in the not-too-distant future.* It's currently fairly easy for a nondiabetic to obtain a CGM from one of several online metabolic health start-ups.

Yet some experts and evidence-based medicine types have criticized the growing use of CGM in nondiabetic people. They argue, as these sorts of people always do, that the "cost" is excessive. CGM costs about $120 a month, which is not insignificant—but I would argue that even this is still far cheaper than allowing someone to slide into metabolic dysfunction and eventually type 2 diabetes. Insulin treatment alone can cost hundreds of dollars a month. Also, as CGM becomes more common, and more readily available without a prescription, the cost is sure to come down. Typically, my healthy patients need to use CGM only for a month or two before they begin to understand what foods are spiking their glucose (and insulin) and how to adjust their eating pattern to obtain a more stable glucose curve. Once they have this knowledge, many of them no longer need CGM. It's a worthwhile investment.

The second argument against using CGM in healthy patients is also pretty typical: There are no randomized clinical trials showing a benefit from the technology. This is true, strictly speaking, but it is also a weak argument. For one thing, use of CGM is growing so fast and the technology is advancing so much that by the time you are reading this there may very well be published

---

* In the meantime, you can approximate your own CGM with a simple drugstore glucose monitor, simply by taking a reading every hour on the hour and plotting out the results (noting mealtimes and snacks, as well). It's also enlightening to take glucose measurements before and after a meal, at thirty-minute intervals up to two hours postprandial, and to observe how different foods and combinations of foods affect your glucose "curve."

RCTs (assuming a study can be designed to test the metrics that matter most over a long enough period of time).

I am confident that such studies will show a benefit, if done correctly, because there are already ample data showing how important it is to keep blood glucose low and stable. A 2011 study looking at twenty thousand people, mostly *without* type 2 diabetes, found that their risk of mortality increased monotonically with their average blood glucose levels (measured via HbA1c). The higher their blood glucose, the greater their risk of death—even in the nondiabetic range of blood glucose. Another study in 2019 looked at the degree of variation in subjects' blood glucose levels and found that the people in the highest quartile of glucose variability had a 2.67 times greater risk of mortality than those in the lowest (most stable) quartile. From these studies, it seems quite clear that we want to lower average blood glucose *and* reduce the amount of variability from day to day and hour to hour. CGM is a tool that can help us achieve that. We use it in healthy people in order to help them stay healthy. That shouldn't be controversial.

When I've put my patients on CGM, I've observed that there are two distinct phases to the process. The first is the insight phase, where you learn how different foods, exercise, sleep (especially lack thereof), and stress affect your glucose readings in real time. The benefit of this information can't be overstated. Almost always, patients are stunned to see how some of their favorite foods send their glucose soaring, then crashing back to earth. This leads to the second phase, which is what I call the behavior phase. Here you mostly know how your glucose is going to respond to that bag of potato chips, and that knowledge is what prevents you from mindlessly eating it. I've found that CGM powerfully activates the Hawthorne effect, the long-observed phenomenon whereby people modify their behavior when they are being watched. (The Hawthorne effect is also what makes it difficult to study what people actually eat, for the same reason.)

Typically, the first month or so of using CGM is dominated by insights. Thereafter, it's really dominated by behavior modification. But both are quite powerful, and even after my patients stop using CGM, I find that the Hawthorne effect persists, because they know what that bag of potato chips will do to their glucose levels. (Those who need more "training" to break their snacking habit will typically need to use CGM for longer.) CGM has proved especially useful in patients with *APOE e4,* where we often see big glucose spikes, even in relatively young people. In these patients, the behavior modification that CGM prompts is an important part of their Alzheimer's disease prevention strategy.

The real beauty of CGM is that it allows me to titrate a patient's diet while remaining flexible. No longer do we need to try to hit some arbitrary target for carbohydrate or fat intake and hope for the best. Instead, we can observe in real time how their body handles the food they are eating. Is their average blood glucose a little bit high? Are they "spiking" above 160 mg/dL more often than I would like? Or could they perhaps tolerate a little bit *more* carbohydrate in their diet? Not everyone needs to restrict carbohydrates; some people can handle more than others, and some have a hard time sticking to severe carbohydrate restriction. Overall, I like to keep average glucose at or below 100 mg/dL, with a standard deviation of less than 15 mg/dL.* These are aggressive goals: 100 mg/dL corresponds to an HbA1c of 5.1 percent, which is quite low. But I believe that the reward, in terms of lower risk of mortality and disease, is well worth it given the ample evidence in nondiabetics and diabetics alike.

All of this takes experimentation and iteration; dietary restriction has to be adaptive, changing with the patient's lifestyle, age, exercise habits, and so on. It's always interesting to see which specific foods cause elevated CGM readings in some patients but not in others. The SAD sends most people's CGM readings through the roof, as all the sugar and processed carbohydrates dump into the bloodstream at once, provoking a strong insulin response,

---

* Standard deviation, a statistical calculation that indicates the extent of variation within a group (or within an individual), gives us an idea how much the patient's glucose levels are fluctuating *around* that average and also serves as a poor man's proxy for how much insulin they are likely secreting to accomplish the job of glucose disposal. A higher standard deviation means there are greater fluctuations, and probably much more insulin is required to bring their glucose under control. This, to me, is a key early warning sign of hyperinsulinemia.

which is what we don't want. But seemingly "healthy" meals, for example certain kinds of vegetarian tacos, can also send glucose levels soaring in some people but not others. It also depends on when those carbs are eaten. If you eat 150 grams of carbohydrates as a serving of rice and beans in one sitting, that has a different effect than eating the same amount of rice and beans spread out over the day (and, obviously, much different from ingesting 150 grams of carbs in the form of Frosted MiniWheats). Also, everyone tends to be more insulin sensitive in the morning than in the evening, so it makes sense to front-load our carb consumption earlier in the day.

One thing CGM pretty quickly teaches you is that your carbohydrate tolerance is heavily influenced by other factors, especially your activity level and sleep. An ultraendurance athlete, someone who is training for long rides or swims or runs, can eat many more grams of carbs per day because they are blowing through those carbs every time they train—and they are also vastly increasing their ability to dispose of glucose via the muscles and their more-efficient mitochondria.* Also, sleep disruption or reduction dramatically impairs glucose homeostasis over time. From years of experience with my own CGM and that of my patients, it still amazes me how much even one night of horrible sleep cripples our ability to dispose of glucose the next day.

Another surprising thing I've learned thanks to CGM is about what happens to a patient's glucose levels during the night. If she goes to bed at, say, 80 mg/dL, but then her glucose ramps up to 110 for most of the night, that tells me that she is likely dealing with psychological stress. Stress prompts an elevation in cortisol, which in turn stimulates the liver to drip more glucose into circulation. This tells me that we need to address her stress levels and probably also her sleep quality.

This doesn't need to be an exercise in deprivation: one patient of mine gleefully confessed that his CGM, which he had only reluctantly agreed to wear, had given him a "superpower" to cheat. By eating certain "forbidden" types of carbohydrates only at certain times, either mixed with other foods or

* As we saw in the previous chapter, this glucose disposal takes place both with and without insulin.

after exercising, he had figured out how he could hit his average glucose goals while still enjoying all the foods he loved. He was gaming his CGM, but he had also unwittingly discovered another rule of nutrition, which is that timing is important: If you scarf a large baked potato before working out, it will leave much less of a footprint on your daily glucose profile than if you eat it right before bedtime.

It is important to remember the limitations of CGM—chiefly, that it measures *one* variable. This variable happens to be very important, but it is not the only one. Thus, CGM data alone are not going to help you find the ideal diet. Eating bacon for breakfast, lunch, and dinner might give you a great CGM tracing, even though it's obviously not an optimal diet. Similarly, a bathroom scale will suggest that smoking is good for you because you lost weight. This is why I monitor my patients' other biomarkers closely as well, to ensure that their CGM-driven choices are not increasing their risk of something else, such as cardiovascular disease. We also monitor other variables that are relevant to diet, beginning with weight (obviously) but continuing with body composition, the ratios of lean mass and fat mass, and how they change. We can also look at biomarkers such as lipids, uric acid, insulin, and liver enzymes. All of these taken together start to give us a better way to evaluate our progress than any one in isolation.

Lessons from Continuous Glucose Monitoring

In the years that I have used CGM, I have gleaned the following insights—some of which may seem obvious, but the power of confirmation cannot be ignored:

1. Not all carbs are created equal. The more refined the carb (think dinner roll, potato chips), the faster and higher the glucose spike. Less processed carbohydrates and those with more fiber, on the other hand, blunt the glucose impact. I try to eat more than fifty grams of fiber per day.

2. Rice and oatmeal are surprisingly glycemic (meaning they cause a sharp rise in glucose levels), despite not being particularly refined; more surprising is that brown rice is only slightly less glycemic than long-grain white rice.
3. Fructose does *not* get measured by CGM, but because fructose is almost always consumed in combination with glucose, fructose-heavy foods will still likely cause blood-glucose spikes.
4. Timing, duration, and intensity of exercise matter a lot. In general, aerobic exercise seems most efficacious at removing glucose from circulation, while high-intensity exercise and strength training tend to *increase* glucose transiently, because the liver is sending more glucose into the circulation to fuel the muscles. Don't be alarmed by glucose spikes when you are exercising.
5. A good versus bad night of sleep makes a world of difference in terms of glucose control. All things equal, it appears that sleeping just five to six hours (versus eight hours) accounts for about a 10 to 20 mg/dL (that's a lot!) jump in peak glucose response, and about 5 to 10 mg/dL in overall levels.
6. Stress, presumably, via cortisol and other stress hormones, has a surprising impact on blood glucose, even while one is fasting or restricting carbohydrates. It's difficult to quantify, but the effect is most visible during sleep or periods long after meals.
7. Nonstarchy veggies such as spinach or broccoli have virtually no impact on blood sugar. Have at them.
8. Foods high in protein *and fat* (e.g., eggs, beef short ribs) have virtually no effect on blood sugar (assuming the short ribs are not coated in sweet sauce), but large amounts of lean protein (e.g., chicken breast) will elevate glucose slightly. Protein shakes, especially if low in fat, have a more pronounced effect (particularly if they contain sugar, obviously).

9. Stacking the above insights—in both directions, positive or negative—is very powerful. So if you're stressed out, sleeping poorly, *and* unable to make time to exercise, be as careful as possible with what you eat.
10. Perhaps the most important insight of them all? Simply tracking my glucose has a positive impact on my eating behavior. I've come to appreciate the fact that CGM creates its own Hawthorne effect, a phenomenon where study subjects change their behavior *because* they are being observed. It makes me think twice when I see the bag of chocolate-covered raisins in the pantry, or anything else that might raise my blood glucose levels.

Protein

Why is protein so important? One clue lies in the name, which is derived from the Greek word *proteios,* meaning "primary." Protein and amino acids are the essential building blocks of life. Without them, we simply cannot build or maintain the lean muscle mass that we need. As we saw in chapter 11, this is absolutely critical to our strategy, because the older we get, the more easily we lose muscle, and the more difficult it becomes to rebuild it.

Remember the study we discussed in chapter 11 that looked at the effect of strength training in sixty-two frail seniors? The subjects who did only strength training for six months gained no muscle mass. What I didn't mention there was that another group of subjects was given protein supplementation (via a protein shake); those subjects added an average of about three pounds of lean mass. The extra protein likely made the difference.*

Unlike carbs and fat, protein is not a primary source of energy. We do

* Similar results have been found in multiple other studies, although it remains unclear whether protein supplementation helps to improve muscle *strength* as well as muscle mass.

not rely on it in order to make ATP,* nor do we store it the way we store fat (in fat cells) or glucose (as glycogen). If you consume more protein than you can synthesize into lean mass, you will simply excrete the excess in your urine as urea. Protein is all about structure. The twenty amino acids that make up proteins are the building blocks for our muscles, our enzymes, and many of the most important hormones in our body. They go into everything from growing and maintaining our hair, skin, and nails to helping form the antibodies in our immune system. On top of this, we *must* obtain nine of the twenty amino acids that we require from our diet, because we can't synthesize them.

The first thing you need to know about protein is that the standard recommendations for daily consumption are a joke. Right now the US recommended dietary allowance (RDA) for protein is 0.8 g/kg of body weight. This may reflect how much protein we need to stay alive, but it is a far cry from what we need to thrive. There is ample evidence showing that we require more than this—and that consuming less leads to worse outcomes. More than one study has found that elderly people consuming that RDA of protein (0.8 g/kg/day) end up *losing* muscle mass, even in as short a period as two weeks. It's simply not enough.

On a related note, some of you may have the impression that low-protein diets are helpful for longevity purposes. Certainly, a number of mouse studies have suggested that restricting protein can improve mouse lifespan. I am not convinced that these results are applicable to humans, however. Mice and human beings respond very differently to low protein, and numerous studies suggest that low protein in the elderly leads to low muscle mass, yielding greater mortality and worse quality of life. I am more persuaded by this human data than I am by studies in mice, who are simply not the same as us.

How much protein do we actually need? It varies from person to person.

---

* Although it can. The liver can turn amino acids into glucose via a process known as gluconeogenesis. This is not a primary source of glucose, nor is it a preferred use for protein.

In my patients I typically set 1.6 g/kg/day as the minimum, which is twice the RDA. The ideal amount can vary from person to person, but the data suggest that for active people with normal kidney function, one gram per *pound* of body weight per day (or 2.2 g/kg/day) is a good place to start—nearly triple the minimal recommendation.

So if someone weighs 180 pounds, they need to consume a minimum of 130 grams of protein per day, and ideally closer to 180 grams, especially if they are trying to add muscle mass. This is a lot of protein to eat, and the added challenge is that it should not be taken in one sitting but rather spread out over the day to avoid losing amino acids to oxidation (i.e., using them to produce energy when we want them to be available for muscle protein synthesis). The literature suggests that the ideal way to achieve this is by consuming four servings of protein per day, each at ~0.25 g/lb of body weight. A six-ounce serving of chicken, fish, or meat will provide about 40 to 45 grams (at about 7 grams of actual protein per ounce of meat), so our hypothetical 180-pound person should eat four such servings a day.

Most people don't need to worry about consuming too much protein. It would require an overwhelming effort to eat more than 3.7 g/kg/day (or ~1.7 g/lb of body weight), defined as the safe upper limit of protein consumption (too much stress on the kidneys, for one). For someone my size, that maximum amount would be nearly 300 grams per day, or the equivalent of seven or eight chicken breasts.

How much protein you need depends on your sex, body weight and lean body mass, activity level, and other factors, including age. There is some evidence that older people might require more protein because of the anabolic resistance that develops with age—that is, their greater difficulty in gaining muscle. Unfortunately, there's no CGM for protein, so it becomes a bit of a process of trial and error. I try to consume enough to maintain muscle mass as I train. If I find that I'm losing muscle mass, then I endeavor to eat more. Older people in particular should try to keep track of their lean mass, such as via a body-composition-measuring scale (or better yet, DEXA scan), and adjust their protein intake upwards if lean mass declines. For me and my patients, this works out to four servings, as described, with at least one of them being a whey protein shake. (It's very difficult for me to consume four actual meals. Typically, I will consume a protein shake, a high-protein snack, and two protein meals.)

Now, a word on plant protein. Do you need to eat meat, fish, and dairy to get sufficient protein? No. But if you choose to get all your protein from plants, you need to understand two things. First, the protein found in plants is there for the benefit of the plant, which means it is largely tied up in indigestible fiber, and therefore less bioavailable to the person eating it. Because much of the plant's protein is tied to its roots, leaves, and other structures, only about 60 to 70 percent of what you consume is contributing to your needs, according to Don Layman, professor emeritus of food science and human nutrition at the University of Illinois Urbana-Champaign, and an expert on protein.

Some of this can be overcome by cooking the plants, but that still leaves us with the second issue. The distribution of amino acids is not the same as in animal protein. In particular, plant protein has less of the essential amino acids methionine, lysine, and tryptophan, potentially leading to reduced protein synthesis. Taken together, these two factors tell us that the overall quality of protein derived from plants is significantly lower than that from animal products.

The same is true of protein supplements. Whey protein isolate (from dairy) is richer in available amino acids than soy protein isolate. So if you forgo protein from animal sources, you need to do the math on your protein quality score. In truth, this can get pretty complicated pretty quick, because you get wrapped around the axle of something called the Digestible Indispensable Amino Acid Score (DIAAS) and the Protein Digestibility-Corrected Amino Acid Score (PDCAAS). These are great if you have the time to comb through databases all day, but for those of us with day jobs, Layman suggests focusing on a handful of important amino acids, such as leucine, lycine, and methionine. Focus on the absolute amount of these amino acids found in each meal, and be sure to get about three to four grams per day of leucine and lycine and at least one gram per day of methionine for maintenance of lean mass. If you are trying to increase lean mass, you'll need even more leucine, closer to two to three grams per serving, four times per day.

Multiple studies suggest that the more protein we consume, in general, the better. A large prospective study called the Healthy Aging and Body Composition Study, with more than two thousand elderly subjects, found that those who ate the most protein (about 18 percent of caloric intake) kept more of their lean body mass over three years than those in the lowest quintile of protein consumption (10 percent of calories). The difference was significant: the low-protein group lost 40 percent more muscle than the high-protein group.

You could make the case that protein is a performance-enhancing macronutrient. Other studies have found that boosting protein intake even moderately above the RDA can slow the progressive loss of muscle mass in older people, including patients with heart failure and cachexia (wasting). Adding thirty grams of milk protein to the diet of frail elderly people, in another study, significantly improved their physical performance.

Beyond its role in building muscle, protein may have beneficial effects on our metabolism. One study found that giving elderly people supplements containing essential amino acids (that is, mimicking some effects of increasing dietary protein) lowered their levels of liver fat and circulating triglycerides. Another study in men with type 2 diabetes found that doubling their protein intake from 15 to 30 percent of total calories, while cutting carbohydrates by half, improved their insulin sensitivity and glucose control. Eating protein also helps us feel satiated, inhibiting the release of the hunger-inducing hormone ghrelin, so we eat fewer calories overall.

In case my point here isn't clear enough, let me restate it: don't ignore protein. It's the one macronutrient that is absolutely essential to our goals. There's no minimum requirement for carbohydrates or fats (in practical terms), but if you shortchange protein, you will most certainly pay a price, particularly as you age.

Fat

The balance of our diet is composed of fat—or rather fats, plural. Fat is essential, but too much can be problematic both in terms of total energy intake and also metabolically. It should be relatively straightforward, but dietary fat has a sordid past that also creates a lot of confusion.

Fats have long had a bad rap, on two counts: their high caloric content (9 kcal/g) and their role in raising LDL cholesterol and thus heart disease risk. Like carbohydrates, fats are often labeled "good" or "bad" on the basis of one's tribal or political stripes; in actuality, of course, it's not that black and white. Fats have an important place in any diet, and therefore it's important to understand them.

While carbohydrates are primarily a source of fuel and amino acids are primarily building blocks, fats are both. They are very efficient fuel for oxidation (think: slow-burning logs) and also the building blocks for many of our hormones (in the form of cholesterol) and cell membranes. Eating the right mix of fats can help maintain metabolic balance, but it is also important for the health of our brain, much of which is composed of fatty acids. On a practical level, dietary fat also tends to leave one feeling more satiated than many types of carbohydrates, especially when combined with protein.

There are (broadly) three types of fats: saturated fatty acids (SFA), monounsaturated fatty acids (MUFA), and polyunsaturated fatty acids (PUFA).* The differences between these have to do with differences in their chemical structure; a "saturated" fat simply has more hydrogen atoms attached to its carbon chain.† Within PUFA, we make one more important distinction,

---

* There are also the dreaded trans fats, but they have largely been removed from our diet, so I'll omit them from this discussion.

† The differences between types of fats all come down to organic chemistry. Fatty acids are essentially chains of carbon atoms of various lengths. That's why we refer to some fats as medium-chain fatty acids versus long-chain fatty acids, for example. A saturated fat gets its name from the fact that it is fully "saturated" with hydrogen atoms attached to that carbon chain. A "monounsaturated" fat refers to the fact that the chain is not fully saturated with hydrogens, and in this case, the reason is that there is one (i.e., mono) double bond in the chain of carbons rather than a single bond. With polyunsaturated fats, there is more than

which is to separate the omega-6 from the omega-3 variants (also a chemical distinction having to do with the position of the first double bond). We can further subdivide omega-3 PUFA into marine (EPA, DHA) and nonmarine sources (ALA). Salmon and other oil-rich seafood provide the former, nuts and flaxseed the latter.

The key thing to remember—and somehow this is almost always overlooked—is that virtually no food belongs to just one group of fats. Olive oil and safflower oil might be as close as you can get to a pure monounsaturated fat, while palm and coconut oil might be as close as you can get to a pure saturated fat, but all foods that contain fats typically contain *all three* categories of fat: PUFA, MUFA, and SFA. Even a ribeye steak contains a lot of monounsaturated fats.

So it's not really possible or feasible to try to eliminate certain categories of fatty acids from the diet entirely; instead, we try to tweak the ratios. The default fat state of most of my patients (i.e., their baseline fat consumption when they come to me) works out to about 30–40 percent each of MUFA and SFA, and 20–30 percent PUFA—and within that PUFA group, they are generally consuming about six to ten times more omega-6 than omega-3s and usually scant amounts of EPA and DHA.

From our empirical observations and what I consider the most relevant literature, which is less than perfect, we try to boost MUFA closer to 50–55 percent, while cutting SFA down to 15–20 percent and adjusting total PUFA to fill the gap. We also boost EPA and DHA, those fatty acids that are likely important to brain and cardiovascular health, with marine fat sources and/or supplementation. We titrate the level of EPA and DHA in our patients' diets by measuring the amount of each found in the membranes of their red

one double bond (confused yet?). Double bonds cause bends in the carbon chain and make the fatty acid more prone to oxidation. Saturated fats are more stable and do not easily react with other molecules. Since saturated fats are linear and can be densely packed together, they can be more solid at room temperature. Because unsaturated fats have kinks in their structure, they are more likely to be liquid at room temperature.

blood cells (RBC), using a specialized but readily available blood test.* Our target depends on a person's *APOE* genotype and other risk factors for neurodegenerative and cardiovascular disease, but for most patients the range we look for is between 8 and 12 percent of RBC membrane composed of EPA and DHA.

Putting all these changes into practice typically means eating more olive oil and avocados and nuts, cutting back on (but not necessarily eliminating) things like butter and lard, and reducing the omega-6-rich corn, soybean, and sunflower oils—while also looking for ways to increase high-omega-3 marine PUFAs from sources such as salmon and anchovies.†

But once again, this is where the SAD, our modern food environment, comes in to complicate things. A hundred years ago, our ancestors would have gotten all their fat from animals, in the form of butter, lard, and tallow, and/or fruits, such as olives, coconuts, and avocados. They would have done so mostly by consuming these foods in their relatively natural state, and achieving a reasonable balance of fatty acids would have come fairly easily. Over the course of the twentieth century, advances in food-processing technology enabled us to chemically and mechanically extract oil from vegetables and seeds that otherwise would have been impossible to get. These new technologies suddenly allowed vast quantities of oils high in polyunsaturated fats, such as corn and cottonseed oil (aka linoleic acid, a PUFA), to flood into the food supply. Our per capita consumption of soybean oil, for example, has increased over a thousand-fold since 1909; meanwhile, studies have found that levels of linoleic acid found in human fat tissue have also increased, by 136 percent over the last half century.

This industrial fat revolution also helped create trans fats, listed on ingredient labels as "partially hydrogenated vegetable oils" (think: margarine),

---

* The fancy version of this test can also determine a person's omega-6/omega-3 ratio as well as the levels of all fatty acids in their blood.

† Interestingly, the baseline composition of human fat tissue, made up of roughly 55 percent MUFA, 30 percent SFA, and 15 percent PUFA (Seidelin 1995), falls right in line with the *dietary* fat distribution that works well in most of my patients.

which in turn helped enable the proliferation of the SAD, in part because they allowed foods to remain shelf stable for longer periods. But trans fats also contributed to atherosclerosis (by raising apoB) and have been banned by the FDA.

It is tempting to indict this massive proliferation of soybean and other seed oils as the dietary bad guy responsible for our obesity and metabolic syndrome epidemic. Anything that goes up by a thousand-fold in the same few decades in which our health goes to hell in a handbasket can't be good, right? Even just a few years ago, I used to think this was the case. But the closer and closer I look at the data, the less and less sure I am that we can say much in this regard.

In fact, the most comprehensive review of this topic, *Polyunsaturated Fatty Acids for the Primary and Secondary Prevention of Cardiovascular Disease*, published by the Cochrane Collaboration in 2018—a 422-page summation of all relevant literature from forty-nine studies, randomizing over twenty-four thousand patients—drew the following conclusion: "Increasing PUFA probably makes little or no difference *(neither benefit nor harm)* to our risk of death, and may make little or no difference to our risk of dying from cardiovascular disease. However, increasing PUFA probably slightly reduces our risk of heart disease events and of combined heart and stroke events (moderate-quality evidence)."

Slight advantage to increasing PUFA, noted. A more recent publication by the Cochrane Collaboration, published in 2020 as a 287-page treatise titled *Reduction in Saturated Fat Intake for Cardiovascular Disease*, looked at fifteen RCTs in over fifty-six thousand patients and found, among other things, that "reducing dietary saturated fat reduced the risk of combined cardiovascular events by 17%." Interesting. But the same review also found "little or no effect of reducing saturated fat on all-cause mortality or cardiovascular mortality." Furthermore, "There was little or no effect on cancer mortality, cancer diagnoses, diabetes diagnosis, HDL cholesterol, serum triglycerides or blood pressure, and small reductions in weight, serum total cholesterol, LDL cholesterol and BMI."

Slight disadvantage to saturated fats, but no observed effect on mortality.

Last, yet another recent review, published in late 2020, titled *Total Dietary Fat Intake, Fat Quality, and Health Outcomes: A Scoping Review of Systematic Reviews of Prospective Studies,* examined fifty-nine systematic reviews of RCTs or prospective cohort studies and found "mainly no association of total fat, monounsaturated fatty acid (MUFA), polyunsaturated fatty acid (PUFA), and saturated fatty acid (SFA) with risk of chronic diseases."

I could go on, but I think you get the point. The data are very unclear on this question, at least at the population level. As we discussed in the introduction to Medicine 3.0 and earlier in this chapter, any hope of using broad insights from evidence-based medicine is bound to fail when it comes to nutrition, because such population-level data cannot provide much value at the individual level when the effect sizes are so small, as they clearly are here. All Medicine 2.0 has to offer is broad contours: MUFA seems to be the "best" fat of the bunch (based on PREDIMED and the Lyon Heart study), and after that the meta-analyses suggest PUFA has a slight advantage over SFA. But beyond that, we are on our own.

Medicine 3.0 asks, what is the "best" mix of fats for our patient? I use an expanded lipid panel to keep track of how changes in fatty acid consumption may affect my patients' cholesterol synthesis and reabsorption, and their overall lipid and inflammatory response. Subtle changes in fat intake, particularly of saturated fats, can make a significant difference in lipid levels in some people, as I have learned over and over again—but not in others. Some people (like me)* can consume saturated fats with near impunity, while others can hardly even look at a slice of bacon without their apoB number jumping to the 90th percentile.

Medicine 2.0 says this proves that nobody should eat saturated fats, period. Medicine 3.0 takes these data and says, "While it is obviously not good that our patient's apoB has gone up this much, it now presents us with a choice: Should we consider medication to lower their apoB, or reduce their

---

* In my keto days I was consuming about 250 to 350 grams of fat per day, easily 40 to 50 percent of which was SFA, yet I had perfectly normal lipids and unmeasurable inflammatory markers. I have zero idea why, other than perhaps I was also exercising about three to four hours per day.

intake of saturated fats? Or both?" There is no obvious or uniform answer here, and addressing this not-too-uncommon situation comes down to a judgment call.

In the final analysis, I tell my patients that on the basis of the least bad, least ambiguous data available, MUFAs are probably the fat that should make up most of our dietary fat mix, which means extra virgin olive oil and high-MUFA vegetable oils. After that, it's kind of a toss-up, and the actual ratio of SFA and PUFA probably comes down to individual factors such as lipid response and measured inflammation. Finally, unless they are eating a lot of fatty fish, filling their coffers with marine omega-3 PUFA, they almost always need to take EPA and DHA supplements in capsule or oil form.

TR: The Case for (and Against) Fasting

Fasting, or time-restricted (TR) eating (regulating *when* you eat), presents us with a tactical conundrum. On the one hand, it is a powerful tool for accomplishing some of our goals, large and small. On the other, fasting has some potentially serious downsides that limit its usefulness. While intermittent fasting and eating "windows" have become popular and even trendy in recent years, I've grown skeptical of their effectiveness. And frequent longer-term fasting has enough negatives attached to it that I am reluctant to use it in all but the most metabolically sick patients. The jury is still out on the utility of infrequent (e.g., yearly) prolonged fasts. Overall, I've come to believe that fasting-based interventions must be utilized carefully and with precision.

There is no denying that some good things happen when we are not eating. Insulin drops dramatically because there are no incoming calories to trigger an insulin response. The liver is emptied of fat in fairly short order. Over time, within three days or so, the body enters a state called "starvation ketosis," where fat stores are mobilized to fulfill the need for energy—yet at the same time, as I often noticed when I was undergoing regular lengthy fasts, hunger virtually disappears. This paradoxical phenomenon is likely due to the ultrahigh levels of ketones that this state produces, which tamp down feelings of hunger.

Fasting over long periods also turns down mTOR, the pro-growth and pro-aging pathway we discussed in chapter 5. This would also be desirable, one might think, at least for some tissues. At the same time, lack of nutrients accelerates autophagy, the cellular "recycling" process that helps our cells become more resilient, and it activates *FOXO*, the cellular repair genes that may help centenarians live so long. In short, fasting triggers many of the physiological and cellular mechanisms that we want to see. So why don't I recommend it to all my patients?

It's a tricky question, because the scientific literature on fasting is still relatively weak, notwithstanding the many popular books that have been written about fasting in its various forms. I have recommended (and practiced) different forms of fasting myself, from time-restricted feeding (eating in a defined time period each day) to water-only fasting for up to ten days. Because my thinking about fasting has evolved considerably, I feel that I need to address the topic here. I still think it can be useful sometimes, in some patients—typically the ones with the most severe metabolic dysfunction—but I am less persuaded that it is the panacea that some believe it to be.

There are really three distinct categories of time-restricted feeding, and we'll look at each of them in order. First, we have the short-term eating windows that we've mentioned previously, where someone will limit their consumption of food to a specific time frame, such as six or eight hours out of the day. In practice, that could mean skipping breakfast, eating a first meal at 11 a.m. and finishing dinner by 7 p.m. every evening; or someone could eat breakfast at 8 a.m., another meal at 2 p.m., and nothing thereafter.

There are an almost infinite number of variations on this, but the trick is that it works only provided you make the feeding window small enough. The standard 16/8 (sixteen hours of fasting, eight hours to eat) is barely enough for most people, but it can work. Usually a narrower window, such as 18/6 or 20/4, is needed to eke out enough of a caloric deficit. For a time, I was experimenting with a two-hour eating window, which basically meant that I would eat one huge meal per day. I always enjoyed the look on a waiter's face when I would order multiple entrees.

In my experience, most people find this to be the easiest way to reduce their caloric intake, by focusing on *when* they are eating rather than *how much* and/or *what* they are eating. But I am not convinced that short-term time-restricted feeding has much of a benefit beyond this.

The original 16/8 model came from a study conducted in mice. This study found that mice fed in only eight hours out of the day, and fasted for the other sixteen, were healthier than mice fed continuously. The time-restricted mice gained less weight than the mice that ate whenever they wanted, even though the two groups consumed the same number of calories. This study gave birth to the eight-hour diet fad, but somehow people lost sight of the fact that this is a big extrapolation from research *in mice.* Because a mouse lives for only about two to three years—and will die after just forty-eight hours without food—a sixteen-hour fast for a mouse is akin to a multiday fast for a human. It's just not a valid comparison.

Human trials of this eating pattern have failed to find much of a benefit. A 2020 clinical trial by Ethan Weiss and colleagues found no weight loss or cardiometabolic benefits in a group of 116 volunteers on a 16/8 eating pattern. Two similar studies also found minimal benefit. One other study did find that shifting the eating window to early in the day, from 8 a.m. to 2 p.m., actually did result in lower twenty-four-hour glucose levels, reduced glucose excursions, and lower insulin levels compared to controls. So perhaps an early-day feeding window could be effective, but in my view sixteen hours without food simply isn't long enough to activate autophagy or inhibit chronic mTOR elevation, or engage any of the other longer-term benefits of fasting that we would want to obtain.

Another drawback is that you are virtually guaranteed to miss your protein target with this approach (see "Protein," above). This means that a person who needs to gain lean body mass (i.e., undernourished or undermuscled), should either abandon this approach completely or consume a pure protein source outside their feeding window (which more or less defeats the purpose of time-restricted feeding). Also, it's very easy to fall into the trap of overindulgence during your feeding window and mindlessly consume, say, a half gallon of ice cream in one sitting. Taken together, this combo of too little protein and too many calories can have the exact opposite effect we want:

gaining fat and losing lean body mass. In my clinical experience, this result is quite common.

As I said, I will sometimes put certain patients on a time-restricted eating pattern because I've found it helps them reduce their overall caloric intake with minimal hunger. But it's more of a disciplinary measure than a diet. Setting time limits around food consumption helps foil a key feature of the SAD, which is that it's difficult to *stop* eating it. Time-restricted feeding is a way of putting the brakes on snacking and late-night meals—the type of mindless eating-just-to-eat that the Japanese call *kuchisabishii,* for "lonely mouth." But beyond that, I don't think it's particularly useful.

Next, we have alternate-day fasting (ADF), which has also become popular. This is where you eat normally or even a bit more than normal one day, and then very little (or nothing) the next. There is more extensive research into this eating pattern in humans—and, of course, there have been books written about it too—but the results are not particularly appealing. Some studies have found that subjects can indeed lose weight on alternate-day fasting diets, but more nuanced research suggests there are some significant downsides. One small but revealing study found that subjects on an alternate-day fasting diet did lose weight—but they also lost more lean mass (i.e., muscle) than subjects who simply ate 25 percent fewer calories every day.

This study was limited because of its small size and short duration, but it suggests that fasting might cause some people, especially lean people, to lose too much muscle.* On top of this, the ADF group had much lower activity levels during the study, which suggests that they were not feeling very good on the days they were not eating. With longer-term fasting, these effects only become more pronounced, particularly the loss of muscle mass. Therefore, I

* I experienced something like this in my cycling phase. At my peak I was doing very strict time-restricted feeding to the tune of about 20/4 every day. Lunch was a basically a chicken salad at 2 p.m. and dinner was normal size at 6 p.m., and I was twenty pounds lighter than I am today—mostly because I had less muscle. It was great for cycling, where light weight is an advantage, but bad for upper-body muscle mass.

am inclined to agree with lead investigator James Betts: "If you are following a fasting diet, it is worth thinking about whether prolonged fasting periods [are] actually making it harder to maintain muscle mass and physical activity levels, which are known to be very important factors for long-term health."

As a result of this and other research, I have become convinced that frequent, prolonged fasting may be neither necessary nor wise for most patients. The cost, in terms of lost lean mass (muscle) and reduced activity levels, simply does not justify whatever benefits it may bring. My rule of thumb for any eating pattern, in fact, is that you *must* eat enough to maintain lean mass (muscle) and long-term activity patterns. That is part of what makes any diet sustainable. If we are going to use a powerful tool like fasting, we must do so carefully and deliberately.

But fasting can still prove useful sometimes, in some patients—generally, patients for whom no other dietary intervention has worked. Case in point: my friend Tom Dayspring, the lipidologist whom we met in chapter 7. Tom became a patient of mine a few years ago because I was so concerned about his metabolic health. Then in his midsixties, he was carrying 240 pounds on his five-eight frame, giving him a BMI of 36.5, well into the obese range. Blood tests revealed that he was also working on a serious case of NAFLD, if not outright NASH. Over the years, I had nagged him constantly until he finally agreed to try to do something about it. Given his issues, a ketogenic diet was the obvious place to start. If we limited his carb intake, I figured, he'd lose weight and hopefully his NAFLD would have a chance to dissipate, and his biomarkers and weight would also come under control.

But they didn't. After Tom struggled for six months to stay on the diet, his liver enzymes, and his weight, had failed to budge. A year later, same story. Two years, three years, nothing had changed. In the meantime, his health continued to deteriorate, to the point where he had difficulty walking a single city block. He eventually required both a hip replacement and spinal fusion. The problem was that Tom was simply unable to stay on the strict ketogenic diet for very long. He would be fine for two weeks or so, but then he would break down and eat a sandwich or a plate of pasta. It simply wasn't sustainable for him.

Tom clearly required some sort of stronger medicine, and I concluded that he needed to try fasting. Unfortunately, like many SAD-trained North Americans, Tom hated the very thought of hunger. This was why he had difficulty adhering to the strict ketogenic diet for very long—he felt hungry, and he craved his old familiar carb-heavy foods. Thus, he was never able to switch his metabolism into ketosis and drive down his hunger. Because of his persistently high insulin, his fat cells were refusing to give up the energy they had stored. So he felt hungry all the time, and he could not lose any fat. Clearly, he needed to break out of this vicious cycle.

At first, Tom was horrified by the very notion of fasting. But he is also a scientist, and after delving into some of the research on nutrient deprivation and putting that together with what he already knew about lipids and metabolism and disease risk, he agreed to give it a try. His scientific mind was persuaded, but I think at some point he also realized that he was staring down the barrel of what might be the last five years of his life unless he made some drastic changes. We came up with an aggressive plan, at the limit of what he thought he could tolerate: one week per month, Monday through Friday, Tom would subsist on a drastically reduced diet of about seven hundred calories per day, comprising mostly fat, with a little protein and almost no carbohydrates.

This kind of fasting is called "hypocaloric" because you are not truly fasting in the sense of eating no food at all. You are eating just enough to quell the worst hunger pangs, but not so much that your body thinks you are fully fed. For twenty-five days out of each month, Tom ate a "normal" diet (though in his case, very starch- and sugar-restricted), and only between noon and 8 p.m.; during his fasting week, a typical day's menu might consist of a salad with light dressing, an avocado, and some macadamia nuts or olives. He was surprised at how good he felt. "It wasn't as horrific as I thought it was going to be," he told me later. "After day three, the hunger disappears."

It did not take long for his blood biomarkers to improve dramatically: where his complete blood chemistry report used to be largely yellow and red—meaning, most of his values were borderline to "bad"—it is now almost entirely green. His lipids are under control, and his liver enzymes have plummeted back to safe, normal ranges. After several cycles of this, he was able to do things like climb a flight of steps or walk several city blocks again without feeling out of breath. His blood pressure is lower, and he has been able to stop taking many of the countless medications he was on. Last, he now weighs sixty-seven pounds less than he used to, a sign that his metabolic health really is back on track, and a powerful incentive for him to keep at it. "The weight just poured off," he told me.

Fasting had effectively reset or rebooted his crashed metabolism in a way that no other dietary intervention was able to achieve. Because it has such deleterious effects on muscle mass, I only use it in hard-to-fix patients like Tom. Tom was so overweight to begin with that he could tolerate the loss of muscle because he was losing so much fat at the same time. But most people can't safely lose muscle mass, so fasting is a tool that we can only really use in extremis, when there are no other viable options.

Conclusion

In the last two chapters, we have explored the impact of what we eat—and sometimes what we do not eat—on our health, and the importance of moving our thinking toward a Nutrition 3.0 mindset, based on feedback and data rather than labels and trends and ideology.

I once believed that diet and nutrition could cure almost all ills, but I no longer feel that strongly about it. Nutritional biochemistry is an important component of our tactics, but it is not the only path to longevity, or even the most powerful one. I see it more as a rescue tactic, particularly for patients like Eduardo and Tom, with really severe metabolic problems such as NAFLD and type 2 diabetes. It is also essential for older people who need to build or maintain muscle mass. But its power to leverage increased lifespan and healthspan is more limited. Bad nutrition can hurt us more than good nutrition can help us. If you're already metabolically healthy, nutritional interventions can only do so much.

I know this seems hard to believe, after all we've been conditioned to think and given all the grandstanding that goes into promoting this diet versus that one. But in reality, the first-, second-, and third-order terms in this problem come down to energy balance. CR, DR, and TR are just tools to reduce energy intake, to correct the state of being overnourished and/or metabolically unhealthy.

The bad news is that most Americans are *not* metabolically healthy, so they need to pay attention to nutrition. In most cases, addressing the problem means reducing overall energy intake—cutting calories—but in a way that is sustainable for the individual person. We also have to focus on eliminating those types of foods that raise blood glucose too much, but in a way that also does not compromise protein intake and lean body mass.

This is where it can get tricky. Protein is actually the most important macronutrient, the one macro that should not be compromised. Remember, most people will be overnourished—but also undermuscled. It is counterproductive for them to limit calories at the expense of protein and hence muscle mass.

This is also where other tactics can play a role. As we saw in chapter 12, zone 2 aerobic training can have a huge impact on our ability to dispose of glucose safely, and also on our ability to access energy we have stored as fat. And the more muscle mass we have, the more capacity we have to use and store excess glucose, and utilize stored fat. In the next chapter, we will see how important good sleep can be to maintaining metabolic balance.

If your issues fall more in the domain of lipoproteins and cardiovascular risk, then it makes sense to focus on the fats side of the equation as well, meaning mostly saturated fats, which raise apoB in some people, although this is relatively easy to control pharmacologically. Excessive carbohydrate intake can also have spillover effects on apoB, in the form of elevated triglycerides. (If there is one type of food that I would eliminate from everyone's diet if I could, it would be fructose-sweetened drinks, including both sodas *and* fruit juices, which deliver too much fructose, too quickly, to a gut and liver that much prefer to process fructose slowly. Just eat fruit and let nature provide the right amount of fiber and water.)

In the end, the best nutrition plan is the one that we can sustain. How you manipulate the three levers of diet—calorie restriction, dietary restriction, and time restriction—is up to you. Ideally, your plan improves or maintains all the parameters we care about—not only blood glucose and insulin but also muscle mass and lipid levels, and possibly even weight—while reducing your risk of your most proximate Horseman or Horsemen. Your nutrition goals depend on your individual risk profile: Are you more at risk of metabolic dysfunction, or cardiovascular disease? There is no one right answer for everyone; each patient finds their own balance, their own best approach. Hopefully, in this chapter I've given you some tools with which to come up with a plan that works for you.

And, one last thing. If, after reading this chapter, you're upset because you don't quite agree with some detail I've covered—be it the ratio of MUFA to PUFA to SFA, or the exact bioavailability of soy protein, the role of seed oils and lectins, or the ideal target for average blood glucose levels—or if I have offended your sensibilities because I didn't say *your* diet is the *best* diet, I have one final piece of advice. Stop overthinking nutrition so much. Put the book down. Go outside and exercise.

CHAPTER 16

The Awakening

How to Learn to Love Sleep, the Best Medicine for Your Brain

Each night, when I go to sleep, I die.
And the next morning, when I wake up, I am reborn.

—Mahatma Gandhi

There's a reason why medical residency is called "residency": You're basically living at the hospital, day and night, for the duration. At one point, I was averaging nearly 120 hours per week at work, often for more than thirty hours at a stretch. That left a grand total of about 48 hours a week for eating, sleeping, working out, going on dates (mostly first-and-last), and everything else in life. A friend who had been a year ahead of me in medical school offered what seemed like sage advice: "Even if you spend every single one of those free hours sleeping, you will still be tired—and if you only work and sleep, you will be miserable. So live a little. Sleep can be sacrificed."

One summer evening during my internship, after an unusually long stint at work, I got a taste of what acute sleep deprivation could do. One of my co-residents was sick and I volunteered to take his call shift, which was the night before my own scheduled call. That meant I was at work from 5:30 a.m. Monday until 6 p.m. on Wednesday. When I left the hospital, I got into my car and headed for the freeway to drive home. As I sat at a traffic light, my head suddenly snapped upright. *Holy shit,* I said to myself. *I just fell asleep behind the wheel.* At the next light, it happened again, and this time my left foot slipped off the clutch and the engine stalled.

To this day I thank the stars that, despite having gone more than sixty hours without sleep, I was at least able to muster the vestigial good judgment required to save my own life. I pulled over to the curb along Eastern Avenue and got out of the car for some fresh air. There was a nice warm breeze, and the low-hanging sun felt good on my face. There happened to be a park right there, and I decided to set the alarm on my pager (yes, my pager) for thirty minutes later and lie down on the grass to "rest my eyes."

Six hours later, I woke up in the middle of Baltimore's Patterson Park, then an open-air heroin market and thriving hub for prostitution. Our ER had patched up quite a few of the locals. It was now the middle of the night, and I was sprawled out on my back in my bright green scrubs, with a puddle of drool at my neck. I had mysterious bite marks on my forearms, and there were a few syringes scattered around. Otherwise, I was fine. Apparently, nobody had dared mess with the crazy guy sleeping on the ground in his hospital scrubs.

I would like to be able to say that I learned my lesson from this scary incident and that I immediately recognized the importance of sleep. I didn't. In truth, it took almost another decade for the message of that episode to sink in, in part because such extreme examples of acute sleep debt were easy to dismiss as artifacts of residency. It was just part of the deal. This was not the only time this kind of thing had happened to me: on another occasion, I fell asleep in my car in the parking lot of the gym with the radio on, and Jill had to come give me a jump start at 2 a.m., after only a few months of dating. (I'm a lucky man.)

At the time of my unplanned nap in the park, there was a huge debate over residents' working hours, and I am embarrassed to admit that I was ardently opposed to reducing them. The proposal was to limit the maximum number of hours we could work to just 80, down from more than 110. I thought it would make us all soft, and many of my senior colleagues agreed.

Looking back, it is shocking that such a cavalier disregard for sleep was tolerated, even cultivated, in a medical setting. It's almost as if they had encouraged us to smoke and drink heavily while on the job. This is not an idle analogy: We now know that even one sleepless night can create a state that is the functional equivalent of being legally drunk. Studies have found that sleep-deprived medical personnel, in particular, commit many more errors and cause many more deaths than those who are well rested. Count me among them: one of my worst moments as a sleep-starved resident came during yet another absurdly long shift (more than forty-eight hours), when I face-planted into the drapery of a patient on whom I was about to perform a "lap chole," a laparoscopic gallbladder removal. Luckily nothing bad happened to the patient, but the memory still makes me cringe.

Even then, less than twenty years ago, we knew relatively little about why we sleep, what happens *while* we are asleep, and the importance of sleep to both short-term performance and long-term health. We now know that chronic sleep debt is a far more insidious killer than the acute sleep deprivation that results in falling asleep at stop signs. Many studies have found powerful associations between insufficient sleep (less than seven hours a night, on average) and adverse health outcomes ranging from increased susceptibility to the common cold to dying of a heart attack. Poor sleep dramatically increases one's propensity for metabolic dysfunction, up to and including type 2 diabetes, and it can wreak havoc with the body's hormonal balance. Looking back, I now suspect that at least some of my own health issues, in my thirties, had their roots in my cavalier disregard of sleep.

As important as sleep is for the body, it may even be more so for the brain. Good sleep, in terms of not only quantity but quality, is critical to our cognitive function, our memory, and even our emotional equilibrium. We feel better, in every way, after a night of good sleep. Even while we are unconscious, our brain is still working, processing thoughts and memories and emotions (hence, dreams). It even cleans itself, in a manner similar to a city sweeping the streets. Relatedly, there is a growing body of evidence that sleeping well is essential to preserving our cognition as we age and staving off Alzheimer's disease.

These conclusions are mainly based on observational studies, which I questioned in chapter 14 as they pertain to nutrition, and they share some of the same flaws—notably, that subjects' recollections of how much they sleep might not be terribly accurate. (Do you know exactly how long, and how well, you slept last night? Likely not.) But these studies differ from nutritional epidemiology because there is only one input, sleep; several of their key findings have been confirmed in more rigorous clinical studies; and the data are more uniform, consistently pointing in the same direction.

The long and the short of it is that poor sleep can take a wrecking ball to both your long-term health and your ability to function day-to-day. When you look at the ripple effects of this, across a society that places as little value on sleep as I once did, a devastating picture emerges.

"[The] decimation of sleep throughout industrialized nations is having a catastrophic impact on our health, our life expectancy, our safety, our productivity, and the education of our children," declares Matthew Walker, director of the Center for Human Sleep Science at the University of California at Berkeley, in his book *Why We Sleep.* I've found that my own patients' health problems can often be traced to poor sleep—and that fixing their sleep issues makes our other tactics more effective.

Luckily, it did not take another near disaster to awaken me to the importance of sleep. Rather, it was a pointed question from my friend Kirk Parsley, a former Navy SEAL who later tended SEALs as a naval physician. Over dinner one night in 2012, I had been arguing to Kirk that five to six hours a night was more than enough sleep, and if I didn't feel tired, then I didn't need more sleep. In fact, I went so far as to declare that it was a pity that we needed to waste time in bed at all. Imagine how much more we could accomplish if we just cut out sleep entirely!

There I was again, bravely ascending the flanks of Mount Stupid. But Kirk stopped me short with a simple, Socratic question. If sleep is so unimportant, he asked, then why hasn't evolution gotten rid of it?

His logic was inarguable. When we are asleep, we are accomplishing nothing useful: we are not reproducing, gathering food, or protecting our family. Even worse, in that slumbering state we are extremely vulnerable to predators and enemies, as I had been in Patterson Park. This, he argued, demonstrates precisely why sleep is so important. Why would evolution allow us to spend up to a third of our lives in a state of unconsciousness, where we could easily be killed or eaten? He pressed the issue: Don't you think natural selection would have eliminated the need to sleep hundreds of millions of years ago—unless, somehow, it was absolutely essential?

He was so right that it was as if he had struck a gong inside my brain. Every animal engages in some form of sleep; scientists have found no exceptions, so far. Horses can do it standing up; dolphins sleep one half of their brain at a time; and even great white sharks, who never stop moving, spend time in a sleep-like, restful state. Elephants sleep only four hours per day, while the brown bat snoozes for nineteen hours per twenty-four, which strikes me as perhaps a bit too much, but the point is that every animal that has been carefully studied to date sleeps in some way. Kirk was correct: evolutionarily, sleep is non-negotiable.

I was not alone in ignoring or dismissing the importance of sleep; it had long been shortchanged by science and in Western, industrialized society. Decades ago, sleep was considered merely a blank state, a period of unconsciousness during which nothing of any importance happened. Nowadays, our high-achieving culture still seems to regard sleeping as wasted time, something that only babies, dogs, and lazy people need. But the science of sleep has taken off in the last three decades, and newer findings suggest that this attitude is exactly wrong. We now know that sleep is as fundamental to our health as stability is fundamental to strength.

Now that I've made sleep a priority in my own life, I reap the benefits every day. There's no feeling more powerful than waking up after I've slept really, really well. My brain is brimming with new ideas, I feel like crushing my workout, and I am a genuinely better person to those around me. But when was the last time you woke up feeling that way? This morning? Last week? Last month? You can't remember?

If this is you, then you need to be taking stock of your sleep patterns and sleep quality, and working to fix them—just as much as you should address your lipoproteins, your metabolic health, or your markers of physical fitness. It's that important.

How long do we need to sleep? This question is tricky, because our sleep cycles are powerfully influenced by external cues such as sunlight, noise, and artificial lighting, not to mention our own emotions and stresses. Also, we are quite good at adapting to inadequate sleep, at least for a while. But many, many studies have confirmed what your mother told you: We need to sleep about seven and a half to eight and a half hours a night. There is even some evidence, from studies conducted in dark caves, that our eight-ish-hour sleep cycle may be hard-wired to some extent, suggesting that this requirement is non-negotiable. Getting significantly less sleep than this, or significantly more, will almost inevitably cause problems in the long run.

Even a single night of bad sleep has been found to have deleterious effects on our physical and cognitive performance. Athletes who sleep poorly the night before a race or a match perform markedly worse than when they are well rested. Endurance drops, VO_2 max drops, and one-rep-max strength drops. Even our ability to perspire is impaired. And we are more likely to be injured: A 2014 observational study found that young athletes who slept less than six hours per night were more than two and a half times more likely to experience an injury than their peers who slept eight hours or more.

Good sleep is like a performance-enhancing drug. In one study, Stanford basketball players were encouraged to strive for ten hours of sleep per day, with or without naps, and to abstain from alcohol or caffeine. After five weeks, their shooting accuracy had improved by 9 percent, and their sprint times had also gotten faster.* LeBron James makes sleep a key part of his recovery routine, always trying for nine and sometimes ten hours of sleep per night, plus a daily nap. "When you get in that good sleep, you just wake up, and you feel fresh," he has said. "You don't need an alarm clock. You just feel like, okay, I can tackle this day at the highest level that you can get to."

For those of us who are not professional athletes, sleep is still essential to performance in more mundane—and dangerous—tasks, such as driving. One study found that after a night of sleep deprivation, a group of professional drivers displayed far worse reaction time in situations such as braking to avoid a crash. Unfortunately, there's no breathalyzer for sleep-impaired driving, so it's harder to capture precise statistics. But a survey conducted by AAA found that nearly one in three drivers (32 percent) reported that in the past thirty days they had driven when they were so tired they had a hard time keeping their eyes open.

Yet we are often unaware of the devastating effect that poor sleep is having on our energy levels and our performance. Research has found that people who are sleep deprived almost always underestimate its effects on them, because they adapt to it. As anyone who has had infant children knows, we come to accept the resulting state of mild exhaustion and mental fog as a new normal, a process called "baseline resetting." I know I did. I assumed I was sleeping sufficiently, as a resident and then as a consultant, because I didn't have anything to compare it to. Now that I sleep better, I'm amazed that I survived for as long as I did in that state. It's like, a regular TV looks fine if that's all you've ever seen. But once you see a 4K screen, you realize that your old cathode-ray tube TV was not very clear after all. The difference is that dramatic.

---

* It's not just a matter of getting enough sleep; the timing also matters. Studies have looked at winning percentages of teams in the NBA/NFL/NHL, and there is a clear circadian disadvantage for teams who have to travel westward (Roy and Forest, 2018).

Old-Man Blood

Scary as it can be in some situations, the short-term harm done by a night or three of poor sleep pales in comparison to the damage that we do to ourselves if this situation continues. Kirk Parsley observed this when he was a physician to the SEALs. Outwardly, these men appeared to be prime physical specimens, finely honed by their rigorous training. But when Parsley analyzed their blood tests, he was shocked: many of these young guys had the hormone levels and inflammatory markers of men several decades older than them—"old-man blood," Parsley called it. Because their training exercises and missions often began at odd hours of the night and required them to stay awake for twenty-four hours or more at a stretch, they were chronically sleep deprived, their natural sleep-wake cycles utterly disrupted.

When Kirk told me that story, I experienced a jolt of recognition: I, too, had had "old-man blood," during my Not-Thin Peter phase, with elevated insulin, high triglycerides, and a testosterone level in the bottom 5 percent of men in the United States. I had always attributed my poor health and hormone imbalance at that point to my lousy diet, and diet alone, but I had also spent at least a decade in a state of severe sleep deprivation, in residency and afterward. Belatedly, I realized that not sleeping had actually caught up to me as well. It was probably even evident in my face: studies have found that people who sleep less chronically tend to have older-looking, flabbier skin than people their same age who sleep more.

Now I recognize that sleep, diet, and risk of long-term diseases are all intimately connected to each other. Knowing what I do now, I would bet that a few months of perfect sleep could have fixed 80 percent of my problems back then, even on a crappy diet.

This may come as a surprise to you, as it did to me, but poor sleep wreaks havoc on our metabolism. Even in the short term, sleep deprivation can cause profound insulin resistance. Sleep researcher Eve van Cauter of the University of Chicago subjected healthy young people to severely restricted sleep, just 4.5 hours a night, and found that after four days they had the elevated insulin levels of obese middle-aged diabetics and, worse yet, approximately a

50 percent reduction in their capacity for glucose disposal. This turns out to be one of the most consistent findings in all of sleep research. No fewer than nine different studies have found that sleep deprivation increases insulin resistance by up to a third. Very rarely in medicine do we see such consistent findings, with experimental evidence confirming the epidemiology so powerfully, so it's worth paying attention. It seems quite clear that poor or inadequate sleep can help tilt us into metabolic dysfunction.

Unfortunately, similar longer-term trials haven't been done, but observational studies suggest a clear link between short sleep and long-term metabolic disturbances. Multiple large meta-analyses of sleep studies have revealed a close relationship between sleep duration and risk of type 2 diabetes and the metabolic syndrome. But it cuts both ways: *long* sleep is also a sign of problems. People who sleep eleven hours or more nightly have a nearly 50 percent higher risk of all-cause mortality, likely because long sleep = poor quality sleep, but it may also reflect an underlying illness. Similar risk associations have been found between poor or short sleep and hypertension (17 percent), cardiovascular diseases (16 percent), coronary heart diseases (26 percent), and obesity (38 percent). Taken together, these findings all suggest that the long-term effects of inadequate sleep parallel what we would expect from the short-term studies: increased insulin resistance and more of the diseases that accompany it, from NASH and type 2 diabetes to heart disease. If your sleep is chronically compromised, then your metabolism might be too.

This association between sleep and metabolic health seems puzzling at first, but I think the missing link here is stress. Higher stress levels can make us sleep poorly, as we all know, but poor sleep also makes us more stressed. It's a feedback loop. Both poor sleep and high stress activate the sympathetic nervous system, which—despite its name—is the opposite of calming. It is part of our fight-or-flight response, prompting the release of hormones called glucocorticoids, including the stress hormone cortisol. Cortisol raises blood pressure; it also causes glucose to be released from the liver, while inhibiting the uptake and utilization of glucose in the muscle and fat tissues, perhaps in order to prioritize glucose delivery to the brain. In the body, this manifests as elevated glucose due to stress-induced insulin resistance. I see this often, in myself and some of my patients: high overnight glucose on CGM is almost always a sign of excessive cortisol, sometimes exacerbated by late-night eating and drinking. If it persists, this elevated blood glucose can lead to type 2 diabetes.

Compounding the problem, poor sleep also changes the way we behave around food. Studies by Eve van Cauter's group have found that limiting subjects' sleep to four or five hours a night suppresses their levels of leptin, the hormone that signals to us that we are fed, while increasing levels of ghrelin, the "hunger" hormone. When we sleep poorly, we can be desperately, irrationally hungry the next day, and more likely to reach for high-calorie and sugary foods than their healthy alternatives. Studies show that people who are more sleep deprived tend to have a higher likelihood of indulging in a fourth meal late in the evening. Follow-up studies by van Cauter's group have found that short-sleeping subjects ate about three hundred extra calories' worth of food the following day, compared with when they were well rested. Taken together, this all adds up to a perfect recipe for the beginnings of NAFLD and insulin resistance.

Sleep and Cardiovascular Disease

The sympathetic nervous system may also help explain why poor sleep is so strongly associated with cardiovascular disease and heart attacks. When we perceive a threat, it takes over, mobilizing stress hormones such as cortisol and adrenaline, which raise our heart rate and blood pressure. Unfortunately, poor sleep has much the same effect, putting the sympathetic nervous system on permanent alert; we get stuck in fight-or-flight mode, and our blood pressure and heart rate remain elevated. This, in turn, multiplies the stress placed on our vasculature. I've noticed this myself, via some of the self-tracking devices that I like to play with: During a night of poor sleep, my resting heart rate will be higher (bad), and my heart rate variability will be lower (also bad).

This may explain why inadequate sleep over long periods is associated with an increased risk of cardiac events. This is something that is difficult to study definitively, as in a randomized controlled trial. Two large meta-analyses have found that short sleep (defined as less than six hours a night) is associated with about a 6 to 26 percent increase in cardiovascular disease. This does not tell us about causality. Surely, some of the reasons why people are sleeping poorly may also be contributing to their risk of heart disease: longer work hours, less income, more stress, et cetera. But one particularly interesting study compared observational and Mendelian randomization data in people with previously identified genetic variants that either increase or decrease their lifelong exposure to longer or shorter sleep duration. The MR data confirmed the observational findings, that sleeping less than six hours a night was associated with about a 20 percent higher risk of a heart attack. Even more noteworthy, the researchers found that sleeping six to nine hours a night (i.e., adequately, by the researchers' definition) was associated with a reduction in heart attack risk—even among individuals with a high genetic predisposition for coronary artery disease.

Translation: good sleep may help mitigate some of the genetic risk of heart disease faced by people like me. All of the above has convinced me to make sleep a top priority in my own life, and to pay attention to my patients' sleep habits.

Sleep and the Brain

What is really striking about most of what we've discussed so far in this chapter—the crucial role that sleep plays in metabolic health and cardiovascular health—is how much of this effect is mediated through the brain. Sleep plays a major role in brain health, especially as we get older, not only in terms of daily cognitive function but also in terms of our long-term cognitive health, a crucial pillar of healthspan.

We have all felt groggy and slow after a restless night; our brain simply is not as sharp as it should be. A good night's sleep or even a solid nap usually restores us. But sleep researchers are revealing myriad ways in which good sleep is essential to long-term brain health—and how bad sleep inflicts major damage. Poor sleep was long considered to be one of the first symptoms of incipient Alzheimer's disease. Subsequent research, however, has pointed to chronic bad sleep as a powerful potential *cause* of Alzheimer's disease and dementia. Sleep, it turns out, is as crucial to maintaining brain health as it is to brain function.

When we get into bed and close our eyes, a series of physiological changes begins to take place as we descend into sleep. Our heart rate slows, our core temperature drops, and our breathing becomes regular as we wait for sleep to overtake us. Meanwhile, our brain is embarking on its own journey.

Researchers now know that we sleep in a series of well-defined stages, each of which has a specific function and a specific electrical brain wave "signature," which is how researchers initially identified the different sleep stages. To visualize these stages, imagine that when you go to bed and close your eyes, you are embarking on a deep-ocean dive in a submarine. As your body relaxes, you fall asleep, hopefully within a few minutes, and your metaphorical vessel slips beneath the waves and begins its descent.

Normally, our descent is quite rapid: we plunge into the depths, passing through a period of light sleep before dropping into deep sleep. This sleep stage is called non-REM, or NREM, sleep, and it comes in two strengths, light NREM and deep NREM. The latter is the more important of the two, especially for neurological health. In our submersible analogy, this is when we descend into the lightless depths of the sea, where our brain is immune from external stimuli. But that does not mean that there is nothing going on. As we descend into deep sleep, our brain waves slow until they reach an extremely low frequency, a chanting rhythm of about one to four cycles per second. This deep sleep dominates the first half of the night, although we typically cycle back and forth between deep and lighter NREM.

Later in the night, typically, our "submersible" rises back up toward the surface, into a zone called rapid eye movement (REM) sleep. In this state, our eyeballs really will dart around behind our eyelids. We are "seeing" things, but only in our mind. This is where most of our dreaming occurs, as our mind processes images and events that seem familiar but are also strange or dislocated from their typical context. Interestingly, the electrical signature of REM sleep is very similar to that when we are awake; the main difference is that our body is paralyzed, which is probably not accidental, since it prevents us from acting on our bizarre dream thoughts. It wouldn't be good if we could just get up and run around while we were in REM sleep. (This probably also explains those dreams where we are trying to run away from something and our body just won't seem to cooperate.)

During a typical night, we will cycle between these sleep stages. These sleep cycles last about ninety minutes, and we may even wake up momentarily in between them—which is likely evolution's way of protecting us from getting eaten by a lion or attacked by enemies during the night, notes Dr. Vikas Jain, a Stanford-trained sleep physician who works with me on my patients' sleep issues.

Both REM and deep NREM sleep (which we'll call "deep sleep" for convenience) are crucial to learning and memory, but in different ways. Deep sleep is when the brain clears out its cache of short-term memories in the hippocampus and selects the important ones for long-term storage in the cortex, helping us to store and reinforce our most important memories of the day. Researchers have observed a direct, linear relationship between how much deep sleep we get in a given night and how well we will perform on a memory test the next day.

When we are young, REM sleep is important in helping our brains grow and develop. Even while we are asleep, our brain is forming new connections, expanding our neural network; this is why younger people spend more time in REM. In adulthood, our REM sleep time tends to plateau, but it remains important, especially for creativity and problem solving. By generating seemingly random associations between facts and memories, and by sorting out the promising connections from the meaningless ones, the brain can often come up with solutions to problems that stumped us the previous day. Research has also found that REM sleep is especially helpful with what is called procedural memory, learning new ways of moving the body, for athletes and for musicians.

Another very important function of REM sleep is to help us process our emotional memories, helping separate our emotions from the memory of the negative (or positive) experience that triggered those emotions. This is why, if we go to bed upset about something, it almost always seems better in the morning. We remember the event but (eventually) forget the pain that accompanied it. Without this break for emotional healing, we would live in a state of constant anxiety, every memory triggering a renewed surge of the emotions around that event. If this sounds like PTSD, you are correct: studies of combat veterans found that they are less able to separate memories from emotions, precisely due to their lack of REM sleep. It turned out that the veterans put out high levels of noradrenaline, the fight-or-flight hormone that effectively prevented their brains from relaxing enough to enter REM.*

Perhaps most intriguing, REM sleep helps us maintain our emotional awareness. When we are deprived of REM, studies have found, we have a more difficult time reading others' facial expressions. REM-deprived study subjects interpreted even friendly or neutral expressions as menacing. This is not trivial: our ability to function as social animals† depends on our ability to understand and navigate the feelings of others. In short, REM sleep seems to protect our emotional equilibrium, while helping us process memories and information.

Deep sleep, on the other hand, seems to be essential to the very health of our brain as an organ. A few years ago, researchers in Rochester discovered that while we are in deep sleep, the brain activates a kind of internal waste disposal system that allows cerebrospinal fluid to flood in between the neurons and sweep away intercellular junk; while this happens, the neurons themselves pull back to allow this to happen, the way city residents are sometimes required to move their cars to allow street sweepers to pass through. This cleansing process flushes out detritus, including both amyloid-beta and tau, the two proteins linked to neurodegeneration. But if we do not spend enough time in deep sleep, the system cannot work as effectively, and amyloid and tau build up among the neurons. Broader studies have found that people

---

* Noradrenaline can be lowered by the blood pressure drug prazosin.

† What's interesting is that REM sleep appeared relatively late in the game of evolution; all animals display NREM sleep, but only birds and nonaquatic mammals experience REM, although recent studies suggest that a REM sleep–like state may exist in nonavian reptiles. (Aquatic mammals need to surface periodically to breathe, so they do not enter deep sleep.)

who have generally slept less than seven hours per night, over decades, tend to have much more amyloid-beta and tau built up in their brains than people who sleep for seven hours or more per night. Tau, the protein that collects in "tangles" inside unhealthy neurons, is itself correlated to sleep disturbances in cognitively normal people and in those with MCI, or mild cognitive impairment, an early stage of dementia.

This can become a vicious cycle. If someone has Alzheimer's disease, they are likely to experience sleep disturbances. People with Alzheimer's disease spend progressively less time in deep sleep and REM sleep, and they may also experience dramatic changes in their circadian rhythm (i.e., sleep-wake cycle). Also, up to half of people with Alzheimer's disease develop sleep apnea.

But sleep disturbances, in turn, may help create conditions that allow Alzheimer's to progress. Insomnia affects 30 to 50 percent of older adults, and there is ample research showing that sleep disturbances often precede the diagnosis of dementia by several years; they may even appear before cognitive decline. One study linked poor sleep quality in cognitively normal people with the onset of cognitive impairment—just one year later.

Meanwhile, superior sleep quality in older adults is associated with a lower risk of developing MCI and Alzheimer's disease, and with maintaining a higher level of cognitive function. Successfully treating sleep disturbance may delay the age of onset into MCI—by about eleven years, according to one study—and may improve cognitive function in patients already diagnosed with Alzheimer's disease.

Clearly, sleep and cognitive health are deeply intertwined; this is why one of the pillars of Alzheimer's disease prevention, particularly for our high-risk patients, is improving their sleep. It is not enough merely to spend time in bed; *good-quality* sleep is essential to long-term brain health. This is the crucial distinction. Sleep that is irregular, or fragmented, or not deep enough will not allow the brain to enjoy any of these benefits.

Unfortunately, our ability to obtain deep sleep declines with age, beginning as soon as our late twenties or early thirties, but worsening as we enter middle age. It's not entirely clear how much this decline in sleep quality is due to growing older itself, versus the increased likelihood of health conditions that result in poor sleep as we age. One analysis suggests that the bulk of the changes in adult sleep patterns occur between the ages of nineteen and sixty and only minimally decline after that, if one remains in good health (a big if).

One possible contributor to this age-related reduction in deep sleep is changes in growth hormone secretion. Growth hormone is typically released in a pulse about one hour after we begin to sleep at night, or around the time at which we're likely to enter deep sleep. On the other hand, inhibiting growth hormone reduces deep sleep, so it's not clear which is cause or effect here. Growth hormone reaches its peak during adolescence, rapidly declines between young adulthood to middle age, and then declines more slowly after that. This pattern parallels the changes in the amount of deep sleep we get as we age.

More research points to the forties and sixties as the decades of life when deep sleep is especially important for the prevention of Alzheimer's disease. People who have slept less during those decades seem to be at higher risk of developing dementia later on. Thus, good sleep in middle age appears to be especially important for maintaining cognitive health.

What I realize now is that for all those years when I was sleeping five or six hours a night and thinking I was on top of my game, I was in fact likely performing far below my potential, thanks to my lack of sleep. And at the same time, I was probably putting myself at risk of long-term illness—metabolic, cardiac, and cognitive. I always bragged, "I'll sleep when I'm dead."

Little did I know that my not sleeping was doing much to hasten that day.

Assessing Your Sleep

It would be nice if science could locate some sort of "sleep switch," some brain pathway that could be triggered, or inhibited, to make us fall asleep instantly and cycle smoothly in and out of deep sleep and REM sleep all night, until we wake up feeling refreshed. But that hasn't happened yet.

It's not for lack of effort on the part of big pharma. Sleep is such a problem for so many people that there are about a dozen FDA-approved sleep medications on the market. The first real blockbuster sleeping medication, Ambien (zolpidem), generated $4 billion in revenues within the first two years after it was approved in the 1990s. The demand was huge, but the phenomenon goes back much further than that. The drug morphine, first isolated from the opium poppy in 1806, was named for Morpheus, the god of dreams, because it put people to sleep quite effectively. This was fitting, in that sleeping and dreaming can be a refuge from both physical and emotional pain. But morphine, being addictive, is obviously not ideal as a sleep drug.

Currently, the US sleep medication market is estimated to be worth about $28 billion a year. But the number of prescriptions has actually been declining recently, perhaps because consumers are catching on to the fact that, by and large, these medicines do not actually work very well. They can be good at inducing unconsciousness, but then again, so was Muhammad Ali's right cross. Sleep medications such as Ambien and Lunesta do not promote healthy, long-lasting sleep so much as they tend to promote a sleep-like state of unconsciousness that does not really accomplish much if any of the brain-healing work of either REM or deep sleep. One study found that Ambien actually decreased slow-wave sleep (deep sleep) without increasing REM, meaning people who take it are basically trading high-quality sleep for low-quality sleep. Meanwhile, Ambien has the well-publicized side effect that some users have been known to walk around and do things while "sleeping," leading to all sorts of problems.

The pharmaceutical industry then came up with a new class of sleep drugs that purportedly solved the sleepwalking problem by blocking a wake-promoting brain chemical called orexin. Interestingly, orexin was initially thought to be more relevant to appetite, which it also regulates (by increasing hunger). But so-called orexin antagonist inhibitors such as Dayvigo (lemborexant) and Quviviq (daridorexant) have been approved for treating insomnia, and they appear to be promising—not least because users have better ability to respond to auditory stimuli at night (e.g., a parent who wants to sleep but still be able to respond if a child is crying). They are, however, quite expensive.

Then there are the older benzodiazepine drugs, such as Valium (diazepam) and Xanax (alprazolam), which remain very popular—almost ubiquitous in our society—and are also sometimes used to treat insomnia. These typically induce unconsciousness without improving sleep quality. Somewhat worryingly, their use has also been associated with cognitive decline, and they are generally not recommended for older adults beyond very short-term time horizons (nor is Ambien, by the way).

When new patients come into our practice, it is not uncommon for them to be relying on one of these sleep aids. If they are using Ambien or Xanax once a month, or only with travel, or to help them sleep during a time of emotional distress, it's not alarming. But if they are using such drugs regularly, it becomes our highest priority to get them off these sleep aids and have them begin to learn to sleep correctly without them.

One drug that we do find helpful for assisting with sleep is trazodone, a fairly old anti-depressant (approved in 1981) that never really took off. At the doses used to treat depression, two hundred to three hundred milligrams per day, it had the unwanted side effect of causing users to fall asleep. But one man's trash is another man's treasure. That side effect is what we want in a sleep medication, especially if it also improves sleep architecture, which is exactly what trazodone does—and most other sleep meds do not.* We typically use it at much lower doses, from one hundred milligrams down to fifty milligrams or even less; the optimal dosing depends on the individual, but the goal is to find the amount that improves their sleep quality without next-day grogginess. (We have also had good results with the supplement ashwagandha.)

There is still no pharmacological magic bullet for sleep, but there are some fairly effective things you can do to improve your ability to fall asleep and stay asleep—and, hopefully, sleep well enough to avoid all the scary stuff that we've been talking about in this chapter. Keep in mind, however, that these

* The use of trazodone for sleep is becoming more common but is still considered an "off-label" use by the FDA. It appears especially helpful for enabling patients to stay asleep and not wake up during the night.

tips and strategies are not going to be effective if you have a true sleep disorder, such as insomnia or sleep apnea (see below for questionnaire assessments that you can take to your doctor for discussion).

The first step in this process echoes the first step in a recovery program: we must renounce our "addiction" to chronic sleep deprivation and admit that we need more sleep, in sufficient quality and quantity. We are giving ourselves permission to sleep. This was actually rather difficult for me at first, as I had spent decades practicing just the opposite. I hope by now I've convinced you of the importance of sleep, across multiple dimensions of health.

The next step is to assess your own sleep habits. There are numerous sleep trackers out there that can give you a pretty good idea about how well you are actually sleeping. They work by measuring variables such as heart rate, heart rate variability (HRV), movement, breathing rate, and more. These inputs are used to estimate sleep duration and stage and do so with reasonable (but not perfect) accuracy. While I've found these to be quite helpful in optimizing my own sleep, some people get worked up over poor sleep scores—which can further impair their sleep. In these situations I insist that my patients take a tracker holiday for a few months. It's also worth reiterating that *long* sleep is also often a sign not only of poor sleep quality, but other potential health problems.

In parallel, you should make a longer-term assessment of your sleep quality over the last month. Probably the best-validated sleep questionnaire is the Pittsburgh Sleep Quality Index, a four-page document that asks questions about your sleep patterns over the last month: for example, how often you have had trouble falling asleep within thirty minutes, have woken up during the night, have had difficulty breathing (i.e., snoring), have had trouble staying awake during the day (such as while driving), or have "felt a lack of enthusiasm for getting things done."

It's easy to find the questionnaire and scoring key online,* and I often find that it helps persuade my patients that it's time to take sleep seriously and

---

* The Pittsburgh Sleep Quality Index questionnaire is available at www.sleep.pitt.edu/instruments/#psqi; for a detailed guide to scoring, see Buysse et al. (1989).

make it a priority in their lives. There's another, even simpler quiz called the Epworth Sleepiness Scale, which asks users to rate how likely they are to fall asleep in certain situations, on a scale of 0 (not likely) to 3 (very likely):

- Sitting and reading
- Watching TV
- Sitting in a meeting or other public place
- As a passenger in a car for an hour
- Lying down to rest in the afternoon
- Sitting and talking to someone
- Sitting after lunch (without alcohol)
- In a car, stopped for a few minutes in traffic

A total score of 10 or more indicates excessive sleepiness and likely points to an issue with sleep quality.*

Yet another helpful screening tool is the Insomnia Severity Index, which provides an opportunity to reflect on and report your experience of sleep problems and their impact on your functioning and well-being.†

One important but often-ignored factor in sleep assessment is that different people may have widely differing "chronotypes," which is a fancy way of saying that someone is a "morning person" or "not a morning person." We all have different relationships to the circadian cycle, and much of that relationship is genetic: a morning person and a night owl will have different circadian genes.‡ Studies have found that some individuals are genetically predisposed to leap out of bed first thing in the morning, while others naturally tend to wake up later (and go to sleep later), not really hitting their stride until some-

---

* The Epworth Sleepiness Scale and its scoring can be viewed at www.cdc.gov/niosh/emres/longhourstraining/scale.html.

† The Insomnia Severity Index and information on its scoring and interpretation are available at www.ons.org/sites/default/files/InsomniaSeverityIndex_ISI.pdf.

‡ To figure out your sleep chronotype, take the Morningness/Eveningness Questionnaire (MEQ) at https://reference.medscape.com/calculator/829/morningness-eveningness-questionnaire-meq.

time in the afternoon. The latter are not "lazy," as was long assumed; they may simply have a different chronotype.

Like so much else in biology, this has a possible basis in evolution: if all members of a clan or a tribe adhered to the exact same sleep schedule, the entire group would be vulnerable to predators and enemies for several hours every night. Obviously not ideal. But if their sleep schedules were staggered, with some individuals going to bed early while others were more inclined to stay up late and tend the fire, the group as a whole would be much less vulnerable. This may also explain why teenagers want to go to bed late and then sleep in: Our chronotype appears to undergo a temporary shift in adolescence toward late sleeping and later rising. School start times, unfortunately for both teens and for those of us who are parents, remain stubbornly fixed at very early hours—but there is a growing nationwide movement to push school times later, to better suit adolescent sleep schedules.

Last, it is important to rule out—or rule in—the possibility of obstructive sleep apnea, which is surprisingly prevalent yet underdiagnosed. It is possible to get a formal test for this, in a sleep lab or at home, but there is yet another questionnaire, called STOP-BANG, that correlates pretty strongly with the formal apnea test.* If you snore, have high blood pressure, feel tired most days, or if your partner has observed that you stop breathing occasionally during the night, even for a moment, you are a candidate for further sleep apnea testing by a medical professional. (Other risk factors include having a BMI greater than thirty and being male.) Sleep apnea is a serious medical problem that can have implications for cardiovascular health and dementia risk.

Sleeping Better

Once you have ruled out (or addressed) serious problems like sleep apnea, there are concrete steps that you can take to improve your sleep, or at least improve your chances of getting good sleep.

---

* The STOP-BANG Questionnaire is available at www.stopbang.ca/osa/screening.php.

Most important, you must create an environment for yourself that is conducive to sleeping well. The first requirement for good sleep is darkness. Light is the enemy of sleep, full stop. Thus, you want to make your bedroom itself as dark as possible—installing room-darkening curtains if you live somewhere with a lot of outdoor evening light, and removing all light sources in the bedroom, even down to electronic equipment like TVs and cable boxes and such. Their little pinpoint LEDs are more than bright enough to keep you from sleeping well. Digital clocks are especially deadly, not only because of their bright numerals but also because if you wake up and see that it's 3:31 a.m., you might start worrying about your 7 a.m. flight and never fall back asleep.

This is easier said than done, because it essentially amounts to evicting the twenty-first century from your bedroom. Modern life almost systematically destroys our ability to sleep properly, beginning with the ubiquity of electric light. Not only does non-natural lighting interfere with our natural circadian rhythm, but it also blocks the release of melatonin, the darkness-activated hormone that tells our brain that it is time to fall asleep. It's similar to the way that the SAD interferes with satiety hormones that normally tell us that we are full and we can stop eating.

Even worse is the relatively recent advent of LED household lighting, which is predominantly on the blue end of the spectrum, meaning it more resembles daylight. When our brain detects that blue light, it thinks it is daytime and that we should be awake, so it tries to block us from falling asleep. Therefore, you should also reduce the amount of bright, LED light that you're exposed to in the evening. A couple of hours before you go to bed, begin turning off unnecessary lights in your house, gradually reducing your light exposure from there. Also, try to swap out blue-intensive LED bulbs for those on the warmer end of the spectrum.

The devices we stare at before bed—phones, laptops, video games—are even worse for our sleep. Not only do they bombard us with more blue light, but they also activate our minds in ways that impede our ability to sleep. One large-scale survey found that the more interactive devices subjects used during the hour before bedtime, the more difficulties they had falling asleep and staying asleep—whereas passive devices such as TV, electronic music players, and, best of all, books were less likely to be associated with poor sleep. This may partially explain why watching TV before bed does not seem to affect sleep quite as negatively as playing video games or scrolling social media does, according to research by Michael Gradisar, a sleep researcher and professor of psychology at Flinders University in Australia.

I am increasingly persuaded that our 24-7 addiction to screens and social media is perhaps our most destructive habit, not only to our ability to sleep but to our mental health in general. So I banish those from my evenings (or at least, I try to). Turn off the computer and put away your phone at least an hour before bedtime. Do NOT bring your laptop or phone into bed with you.

Another very important environmental factor is temperature. Many people associate sleep with warmth, but in fact the opposite is true: One of the signal events as we are falling asleep is that our body temperature drops by about one degree Celsius. To help that happen, try to keep your bedroom cool—around sixty-five degrees Fahrenheit seems to be optimal. A warm bath before bed may actually help with this process, not only because the bath itself is relaxing but also because when we get out of the bath and climb into our cool bed, our core temperature drops, which signals to our brain that it is time to fall asleep. (There are also a variety of cooling mattresses and mattress toppers out there that could help people who like to sleep cool.)

Our internal "environment" is just as important to good sleep. The first thing I tell my patients who are having difficulty sleeping is to cut back on alcohol—or better yet, give it up entirely. It's counterintuitive, because alcohol initially acts as a sedative, so it can help us fall asleep more quickly. But as the night wears on, alcohol turns from friend of sleep to foe, as it is metabolized into chemicals that impair our ability to sleep. Depending on how much we've had to drink, during the second half of the night we may have a harder time entering REM sleep and be more prone to waking up and lingering in unproductive light sleep.

The effects of alcohol on memory and cognition are apparent even in moderate drinkers. Studies have found that young people who drink heavily are more likely to forget even basic tasks like locking the door or mailing a letter. Students who averaged nine drinks per week (not much, by collegiate standards) performed worse on a word-based memory test. And, in a finding that will surprise no one, students who drank more slept later and felt sleepier in the daytime, as well as performing worse on tests. More alarming is the finding that students who drank heavily two days *after* a bout of learning or study forgot or failed to retain most of what they had learned.

Note that these are all findings in young people, students who are presumably at their cognitive peak. If you extrapolate to those of us in middle and older age, who may have a lower tolerance for alcohol and a greater propensity to forget things, the implications are worrisome. I find that my own threshold is two drinks in an evening: any more than that, my sleep goes sideways, and my work performance suffers the next day, no matter how much coffee I drink.

Coffee is not a solution to the problem of poor sleep, especially if consumed to excess or (especially) at the wrong time. Most people think of caffeine as a stimulant that somehow gives us energy, but actually it functions more as a sleep blocker. It works by inhibiting the receptor for a chemical called adenosine, which normally helps us go to sleep every night. Over the course of the day, adenosine builds up in our brain, creating what scientists call "sleep pressure," or the drive to sleep. We may be tired and needing sleep, but if we ingest caffeine it effectively takes the phone off the hook, so our brain never gets the message.

This is obviously helpful in the morning, particularly if our "chronotype" is telling us we should still be asleep at 6 a.m. But the half-life of caffeine in the body is up to six hours, so if we drink a cup of coffee at noon, we will still have half a cup's worth of caffeine in our system at 6 p.m. Now multiply this by the number of cups of coffee you drink in a day and work forward from the time of your last cup. If you down one last double espresso at 3 p.m., you will still have a full shot's worth of caffeine in your system at 9. What you won't have, most likely, is much of an urge to fall asleep anytime soon.

Everyone differs in their caffeine tolerance, based on genes and other factors (23andMe tests for one common caffeine-related gene). I'm a very fast metabolizer, so I can handle that afternoon espresso without it affecting my sleep too much; I can even drink coffee after dinner, and it seems to have no impact (unlike alcohol). Someone who metabolizes caffeine slowly should probably stop at one or two cups, before noon.

This concept of *sleep pressure,* our need or desire for sleep, is key to many of our sleep tactics. We want to cultivate sleep pressure, but in the right amounts, at the right times—not too much, not too little, and not too soon. This is why one of the primary techniques that doctors use to treat patients with insomnia is actually sleep restriction, limiting the hours when they are "allowed" to sleep to six, or less. This basically makes them tired enough that they fall asleep more easily at the end of the day, and (hopefully) their normal sleep cycle is restored. Their sleep pressure builds up to the point where it overwhelms whatever is causing their insomnia. But this also helps explain why napping can be counterproductive. Taking a nap during the day, while sometimes tempting, can also relieve too much of that sleep pressure, making it harder to fall back asleep at night.

Another way to help cultivate sleep pressure is via exercise, particularly sustained endurance exercise (e.g., zone 2), ideally not within two or three hours of bedtime. My patients often find that a thirty-minute zone 2 session can do wonders for their ability to fall asleep. Even better is exercise that entails some exposure to sunlight (i.e., outdoors). While blue light late in the evening can interfere with sleep, a half-hour dose of strong daylight, during the day, helps keep our circadian cycle on track, setting us up for a good night of sleep.

It is also important to mentally prepare ourselves for sleeping. For me, this means avoiding anything that might create stress or anxiety, such as reading work emails or especially checking the news. This activates the sympathetic nervous system (the fight-or-flight one) at a time when we want to be destressing and generally winding down. I have to force myself to step away from the computer in the evening; that queue of emails will still be there in the morning. If there's a burning issue that I can't get off my mind, I'll write a few lines about it, creating a plan of action for the next morning. Another way to turn down the sympathetic nervous system and prepare the brain for sleep is through meditation. There are several very good apps that can help with guided meditations, including some that are focused entirely on sleep.

The overarching point here is that a good night of sleep may depend in part on a good day of wakefulness: one that includes exercise, some outdoor time, sensible eating (no late-night snacking), minimal to no alcohol, proper management of stress, and knowing where to set boundaries around work and other life stressors.

How to Improve Your Sleep

The following are some rules or suggestions that I try to follow to help me sleep better. These are not magic bullets but are mostly about creating better conditions for sleeping and letting your brain and body do the rest. The closer you can come to these operating conditions, the better your sleep will be. Of course, I'm not suggesting that it's necessary to do all t hese things—in general, it's best not to obsess over sleep. But the more of these you can check off, the better your odds or a good night of sleep.

1. Don't drink any alcohol, period-and if you absolutely, positively must, limit yourself to one drink before about 6 p.m. Alcohol probably impairs sleep quality more than any other factor we can control. Don't confuse the drowsiness it produces with quality sleep.
2. Don't eat anything less than three hours before bedtime—and ideally longer. It's best to go to bed with just a little bit of hunger (although being ravenous can be distracting.)
3. Abstain from stimulating electronics, beginning two hours before bed. Try to avoid anything involving a screen if you're having trouble falling asleep. If you must, use a setting that reduces the blue light from your screen.
4. For at least one hour before bed, if not more, avoid doing anything that is anxiety-producing or stimulating, such as reading work email or, God help you, checking social media.

These get the ruminative, worry-prone areas of our brain humming, which is not what you want.

5. For folks who have access, spend time in a sauna or hot tub prior to bed. Once you get into the cool bed, your lowering body temperature will signal to your brain that it's time to sleep. (A hot bath or shower works too.)
6. The room should be cool, ideally in the midsixties. The bed should be cool too. Use a "cool" mattress or one of the many bed-cooling devices out there. These are also great tools for couples who prefer different temperatures at night, since both sides of the mattress can be controlled individually
7. Darken the room completely. Make it dark enough that you can't see your hand in front of your face with your eyes open, if possible. If that is not achievable, use an eye shade. I use a silky one called Alaska Bear that costs about $8 and works better than the fancier versions I've tried.
8. Give yourself enough time to sleep-what sleep scientists call a sleep opportunity. This means going to bed at least eight hours before you need to wake up, preferably nine. If you don't even give yourself a chance to get adequate sleep, then the rest of this chapter is moot.
9. Fix your wake-up time-and don't deviate from it, even on weekends. If you need flexibility, you can vary your bedtime, but make it a priority to budget for at least eight hours in bed each night.
10. Don't obsess over your sleep, especially if you're having problems. If you need an alarm clock, make sure it's turned away from you so you can't see the numbers. Clock-watching makes it harder to fall asleep. And if you find yourself worrying about poor sleep scores, give yourself a break from your sleep tracker.

But what if we *still* can't sleep? This brings us to the last and most vexing sleep problem, true insomnia. We have probably all experienced the inability to fall asleep at some point, but for many people it is a chronic problem. So the first question to ask is: Is it really insomnia? Or are you simply not prepared to sleep properly?

If you find yourself lying awake in bed, unable to get back to sleep, my advice is to stop fighting it. Get up, go into another room and do something relaxing. Fix a cup of tea (noncaffeinated, obviously), and read a (preferably boring) book until you feel sleepy again. The key, says Vikas Jain, is to find something that is relaxing and enjoyable but that serves no function; you never want to give your insomnia a purpose, such as doing work or paying bills, because if you do, your brain will make sure to wake you up for it on a regular basis. Keep in mind, too, that you might not actually have insomnia; you might simply be a night-owl chronotype, thinking you "should" go to bed much earlier than your brain or your body is ready for. So adjust your bedtime and waking time, if possible.

If the sleeplessness persists, even after following the advice outlined above, the most effective treatment is a form of psychotherapy called Cognitive Behavioral Therapy for Insomnia, or CBT-I. The goal of CBT-I is to help restore confidence in one's ability to sleep, by helping the patient break bad sleep habits and eliminate any anxieties that may be preventing them from getting to sleep. Therapists will also use sleep restriction, again, as a way of increasing sleep pressure. That, in turn, helps restore confidence in their ability to sleep. Studies of CBT-I techniques have found that they are more effective than sleeping medications.

After ignoring sleep for decades, I'm now a fan. I consider it a kind of performance-enhancing substance, not only physically but cognitively. Long term, this thing called sleep also has the power to improve our healthspan in remarkable ways. Just like exercise, sleep is its own kind of wonder drug, with both global and localized benefits to the brain, to the heart, and especially to our metabolism.

So if evolution has made sleep non-negotiable, I'm no longer going to argue the point. Rather, I've embraced it.

CHAPTER 17

Work in Progress

The High Price of Ignoring Emotional Health

Every man is a bridge, spanning the legacy he
inherited and the legacy he passes on.

—Terrence Real

New patients arrive every Monday, and I was the first to show up. It was a few weeks before Christmas, and I had flown from San Diego to Nashville and then gotten into a beat-up minivan taxi that reeked of nicotine for a two-hour ride to a place that I'd never heard of called Bowling Green, Kentucky. It was a cold morning, and the driver would not stop looking at his phone as he drove. Strangely, I was not upset by this. I *wanted* us to crash. At least then I would be spared what was to come.

By late morning I was sitting in the common area of a facility called The Bridge to Recovery, an isolated place set deep in the woods. It smelled musty. Waiting for the others to arrive, I wandered through the kitchen and saw a sign that read, "Religion is for people who are afraid of Hell. Spirituality is for people who have been there."

Where the hell am I? I wondered.

The first of the other newcomers was a woman who looked to be about fifty. We stared at each other without saying anything. She looked so sad, as if she had been crying for a year straight. I wondered if I looked the same way to her. By that evening, all the "newbies" were there. They were exhausted, pale, totally depleted. Several were addicts—to drugs, alcohol, sex, or some combination thereof. I looked at them in dismay, thinking I wasn't like them.

After some introductory remarks, we did something called a check-in, where we all took turns describing our emotional state. How we felt at that exact moment. I had no words for how I felt. I was angry beyond words. A simmering rage. I just couldn't do it; I lacked the emotional awareness to even understand my own feelings, let alone articulate them. I was furious that I had needed to come to this place. I was furious that I had failed. I believed that I did not belong here, with these broken people. Every cell in my body wanted to call the Texting Death Cab Company and get out of there.

Then one of the veterans, a woman my age named Sarah who was in her third week there (and who always had a way of saying the right thing, I would learn), must have seen the look on my face. Without even knowing my name, she turned to me and said, "Hey, it's okay—nobody shows up here on a winning streak."

I may not have felt like I was at rock bottom, but I was headed in that direction fast. A few weeks earlier, I had nearly gotten into a fistfight with a random guy in a parking lot. I was standing right in his face, begging him to throw the first punch so I could rip his larynx out, a procedure I described in surgical detail, with a few choice epithets to boot. I'm pretty certain that I would have won that fight, but I also could have lost everything: my house, my medical license, my freedom, probably what was left of my marriage. Outwardly, I was a successful-seeming guy with a thriving medical practice, a beautiful wife and kids, wonderful friends, robust health, and a contract to write this book. But in reality, I was out of control.

I wasn't just some garden-variety road-raging maniac either. It was much worse than that. A few months earlier—on Tuesday, July 11, 2017, at 5:45 p.m., to be exact—I had received a call from Jill, my wife. She was in an ambulance with our infant son, Ayrton, on the way to the hospital. For some reason, he had suddenly stopped breathing and fallen unconscious. His eyes were completely rolled back in their sockets and he was lifeless and blue, with no heartbeat. Only the quick reaction of our nanny had saved him. She rushed him to Jill, who is a nurse. Her instincts took over and she immediately put him on the floor and began performing CPR, rhythmically but carefully pressing her fingers on his tiny sternum as the nanny frantically dialed 911. He was barely a month old.

By the time the firefighters stormed into the house, about five minutes later, Ayrton was breathing again, and his skin was turning from blue back to pink as oxygen returned to his body. The firemen were stunned. We never see these kids come back, they told Jill. To this day, we still don't know how or why it happened, but this is likely what occurs when babies die suddenly in their sleep: they choke for a moment on their own saliva, or some other vasovagal insult occurs, and their very immature nervous system fails to restart their breathing.

When Jill called me from the ambulance, I was in New York, in a taxi on Fifty-Fourth Street, on my way to dinner. After she finished telling me the story, I just said, without a shred of emotion, "Okay, call me when you get to the hospital, so I can talk to the doctors in the ICU."

She got off the phone pretty quickly, and, of course, it's obvious why she was upset: our son had nearly died, and the right thing for me to say, the *only* thing to say, was that I was getting the next flight home.

Jill stayed in the hospital with Ayrton, alone, for four days. She pleaded with me to come home. I called in daily to talk to the doctors and discuss each day's test results, but I stayed in New York, busy with my "important" work. Ayrton's cardiac arrest happened on a Tuesday, but I did not come home to San Diego until Friday of the following week. *Ten days later.*

Even today, just thinking about what happened, I feel nauseous about my behavior. I can't believe I did that to my family. I can't believe what a blind, selfish, checked-out husband and father I was. And I know I may never fully forgive myself for it, for as long as I live.

I must have been giving off a very troubled vibe during this period, because around then my close friend Paul Conti, a medical school classmate who is now a brilliant and very intuitive psychiatrist, began urging me to go to this place in Kentucky. I looked it up, and it seemed to be a place for addicts. "This doesn't make sense," I told him. "I'm not an addict."

He explained to me, over several months of gentle discussion, that addiction can take many forms, not merely to drugs or alcohol. Often, he continued, it is an outgrowth of some trauma that has happened in a person's past. Paul is an expert in trauma, and he saw that I displayed all the behavioral signs: anger, detachment, obsessiveness, a need to achieve that was fueled by insecurity. "I don't know what it was [that happened], but you just have to trust me on this," he said. He was relentless.

I agreed to go to Kentucky, but I was still looking for any excuse to get out of it. In early November, a woman from the Bridge called to do my intake interview. It was a long, tedious conversation, and my patience finally expired when she asked, "Have you ever been subject to any kind of abuse?"

I got so angry I yelled, "Fuck you!" and hung up the phone. After this call I decided to cancel my planned stay. What was wrong with these people, asking such idiotic questions?

That Thanksgiving weekend is still a blur. It was the only Thanksgiving in our life together when we didn't go to a dinner with friends or family, or host one ourselves. We just stayed home alone. On Sunday night, Jill begged me again to go to Kentucky. I can't just go off the grid for that long, I said. My patients need me, and you need help with the kids. This was total bullshit, and we both knew it. She replied point blank, "You're of no help to me; in fact, you're hurting me, and your kids, very badly."

Confronted with the brutal truth, I knew I had to go.

As should be obvious by now, this chapter will be different from the rest of this book, because in it I am not the physician; I am the patient. And I am a patient who considers himself lucky to be alive. Up until this point, I have focused almost entirely on the physical aspects of healthspan and longevity, but here I will explore their emotional and mental sides, which in some ways are more important than everything else that I've laid out thus far.

My journey transformed not only my own life, and the life of my family, but also the way that I think about longevity. The process is ongoing, requiring daily work on my part—nearly as much time and effort as I devote to exercise (which is a lot, as you know by now). This is as it should be, I've come to realize. Emotional health and physical health are closely intertwined, in ways that mainstream medicine, Medicine 2.0, is still only beginning to grasp. On the most obvious level, an angry episode like my confrontation in that parking lot could have easily triggered a cardiac event, particularly given my own presumed genetic propensity for heart disease. I could have dropped dead that very afternoon.

Another very direct way in which mental health affects lifespan is via suicide, which ranks among the top ten causes of death across all age groups, from our teens into our eighties. When I think of suicide, I often think of a man named Ken Baldwin, who leaped off the Golden Gate Bridge in 1985, when he was twenty-eight. Unlike 99 percent of jumpers from that bridge, he survived. As he fell, he later told the author Tad Friend, "I instantly realized that everything in my life that I'd thought was unfixable, was totally fixable—except for having just jumped."

Not all suicides jump from bridges. Many more people sort of slow-roll into misery and early death via various roundabout routes, letting stress and anger erode their health, or falling into self-medicating addictions to alcohol and drugs, or engaging in other reckless, life-endangering behaviors that mental health professionals call parasuicide. It's not a surprise that deaths related to alcohol and drug abuse have surged over the last two decades, especially among people ages thirty to sixty-five; the CDC estimates that more than one hundred thousand Americans died from drug overdoses between April 2020 and April 2021, about as many as died from diabetes.

These "accidental" overdoses account for almost 40 percent of all accidental deaths, a category that also includes automobile accidents and deaths from falls. Some of these overdoses were no doubt truly accidental, but I'd wager that the vast majority were ultimately attributable to the victims' mental health issues, on some level. They were slow-motion suicides, deaths of despair—an agonizing but often invisible form of the "slow death" we talked about earlier.

This category of death has grown so much over the last two decades or so, fueled by the prevalence of addictive opioids in our society, that it has actually helped to diminish life expectancy for some segments of the American population—the first time that this has happened in more than a century. Middle-aged white men and women, in particular, are succumbing to drug and alcohol overdoses, liver disease, and suicide at unprecedented rates, as Anne Case and Angus Deaton first observed in 2015. The substance-abuse crisis has created a longevity crisis, because it is really *a mental health crisis in disguise.*

This type of suffering is far more prevalent than suicide rates would suggest. It simply robs you of the joy that enables you to focus on your health, life, and relationships with others, so that instead of living, you are merely waiting to die. This is why I've come to believe that emotional health may represent the most important component of healthspan. Nothing else about longevity is really worth much without some degree of happiness, fulfillment, and connection to others. And misery and unhappiness can also destroy your physical health, just as surely as cancer, heart disease, neurodegenerative disease, and orthopedic injury.

Even just living alone, or feeling lonely, is linked to a much higher risk of mortality. While most issues around emotional health are not age dependent, this is the one emotional health "risk factor" that does seem to grow worse with increasing age. Surveys show that older Americans report spending more time alone every day—an average of about seven hours daily, for those age seventy-five—and are far more likely to live alone than people in middle age and younger. And the way things were going for me, I was looking at a sad, lonely, miserable old age.

It took me a while to recognize this, but feeling connected and having healthy relationships with others, *and with oneself,* is as imperative as maintaining efficient glucose metabolism or an optimal lipoprotein profile. It is just as important to get your emotional house in order as it is to have a colonoscopy or an Lp(a) test, if not more so. It's just a lot more complicated.

It's a two-way street between emotional and physical health. In my own practice, I witness firsthand how many of my patients' physical and longevity issues are rooted in, or exacerbated by, their emotional health. I see it on a daily basis. It is harder to motivate a patient who is feeling depressed to go and start an exercise program; someone who is overstressed at work and miserable in their personal life may not see the point of early cancer screening or monitoring their blood glucose levels. So they drift along, as their emotional misery drags their physical health down along with it.

My own situation was almost the opposite: I was doing *everything* to live longer, despite being completely miserable emotionally. I was as physically healthy as I'd ever been, circa 2017, but to what end? I was on a horrible path, both emotionally and in terms of my interpersonal relationships. The words of my therapist, Esther Perel, rang in my head practically every day: "Why would you want to live longer if you're so unhappy?"

The one thing that I had in common with some of my patients was that we all found it easier to just avoid dealing with problems that seemed so complex and overwhelming. I didn't even know where to begin—scratch that, I didn't even recognize that I needed help, until long after it was obvious to everyone around me. I had to reach pretty much the end of my rope before I could make myself face up to the truth and go to the Bridge, that godforsaken, difficult, ultimately wonderful place in the woods of Kentucky, and begin to do the work that needed to be done: to begin to acquire the tools that I needed to function better, emotionally.

My first few days at the Bridge felt like weeks, possibly months. The time just crept by. I had no phone, and they had even taken away my books. This was part of the plan, to force us to sit in our own misery. There was literally nothing else to do. I moved like a zombie through the daily activities, from our one cup of morning coffee to inner-child work to equine therapy. My only solace was my 4:30 a.m. morning workout, which also represented the one addiction in which I was still permitted to indulge. Otherwise, there was no relief, and no solitude.

Before I arrived, I had my assistant call to request a private room. The person on the phone had basically laughed at her. "Tell your Very Important Person that we don't do that. Everybody has a roommate." So I had a roommate, who seemed like a nice enough guy, and he had some pretty cool tattoos, but in my rush to judge him (and everyone else) all I could see were the differences. He hadn't gone to college. He worked in a machine shop. He liked strippers and cocaine. His wife hated him, which is actually something that we might have had in common at that point in time.

At first, I clammed up. The part of the day that I dreaded most was the twice-daily emotional check-ins, where we were supposed to describe exactly what we were feeling at that moment. I couldn't do it. I just sat there seething. By Wednesday or Thursday, it had almost become a joke. We had all heard at least bits and pieces of everyone else's story, but nobody knew anything about mine. At one point someone said, "C'mon, dude, are you like a serial killer or something? Like, what's up?"

I said nothing. I don't think my roommate slept well that night.

Finally, after four or five days, I could no longer remain silent. They had set aside almost an entire day when we were all supposed to tell our life stories from the beginning. We had an hour each, and we were supposed to prepare. So I was finally telling my life story for the first time to this group of perfect strangers—not even Jill had heard the whole thing—but I was telling it in a way that was very matter-of-fact: this happened when I was five, that happened when I was seven, and so on. Some of it was sexual; some of it was physical. But it was not all bad, I explained. These events, terrible as they were, had led me to take up boxing and martial arts at age thirteen. I got to punch bags, and people, and that channeled my anger. I learned how to protect myself, but I also gained discipline and focus, qualities that proved invaluable when, at around age nineteen, I pivoted from pugilism to mathematics.

Terrible as it was, my past was also what had set me on the path to becoming a doctor, I continued, growing somewhat defensive. Throughout college, I volunteered at a shelter for sexually abused teenagers, and I became close to many of them over four years, including one young woman who had been abused by her father. When she attempted suicide—one of many attempts—I went to visit her in the hospital. I was a senior by then, and I had already applied to the top PhD programs in aerospace engineering. But I wasn't really sure it was my calling. Spending so much time in the hospital with her helped lead to the epiphany that I was meant to care for people, not solve equations.

So do you see? I concluded. Parts of my past may have been bad, but in a way they also ended up setting me on a course toward a better life. Some of the kids I grew up with and boxed with, meanwhile, were getting arrested for armed robbery, and getting girls pregnant in high school, and all kinds of other stuff. That could easily have been me. So in a way, I said, my abuse may have actually saved my life—I don't really even need to be here!

Right then, one of our therapists, Julie Vincent, cut me off. There are many rules at the Bridge, and one of the most important ones is *no minimizing*. You are not allowed to minimize anything that someone else is saying, and you are especially not allowed to minimize your own experiences. But she didn't flag me for that. Instead, she asked a simple question: "You were five years old when this first happened to you, right?"

"That's right," I replied.

"And your son Reese is almost five years old now, right?"

I nodded.

"So you're saying it's okay that this happened to you when you were his age—but would you be okay with people doing that to Reese now?"

Another rule at the Bridge is that you're not supposed to hand anyone a Kleenex when they're crying. They're supposed to get up and fetch it themselves. Now it was my turn to stand up and walk over to the Kleenex box. It all came pouring out of me and, finally, I was able to embrace why I was there and begin the hard work of unpacking the last forty years of my life.

One framework that the therapists at the Bridge work with, and that I found helpful, is called the Trauma Tree. The idea behind it is that certain undesirable behaviors that we manifest as adults, such as addiction and uncontrolled anger, are actually adaptations to the various types of trauma we suffered in childhood. So while we only see the manifestation of the tree above the ground, the trunk and branches, we need to look underground, at the roots, to understand the tree completely. But the roots are often very well hidden, as they were with me.

Trauma generally falls into five categories: (1) abuse (physical or sexual, but also emotional or spiritual); (2) neglect; (3) abandonment; (4) enmeshment (the blurring of boundaries between adults and children); and (5) witnessing tragic events. Most of the things that wound children fit into these five categories.

Trauma is a pretty loaded word, and the therapists at the Bridge were careful to explain that there can be "big-T" trauma or "little-t" traumas. Being a victim of rape would qualify as a big-T trauma, while having an alcoholic parent might subject a child to a host of little-t traumas. But in large enough doses over a long enough time, little-t traumas can shape a person's life just as much as one major terrible event.

Both types can do tremendous damage, but little-t trauma is more challenging to address—in part, I suspect, because we are more inclined to dismiss it. Jeff English, one of the therapists I was working with, offered a useful blanket definition: Trauma, big T or little t, means having experienced moments of perceived helplessness. The situations in question may or may not have been life-or-death, he explained, "but to a child with an undeveloped brain, it may have seemed that way."

This perfectly described how I had felt at certain times in my childhood. The feeling of powerlessness was a large source of my pain (and in later life, my anger). But I also want to make an important distinction between trauma and adversity. They are not the same. I am not suggesting that it is ideal for children to grow up without experiencing any adversity at all, which sometimes seems to be a primary goal of modern parenting. Many stressors can be beneficial, while others are not. There is no bright line between trauma and adversity; terrible as it was, my own experience had made me stronger in some ways. Julie's question is a pretty good litmus test: Would I want my child to experience it? If my daughter finished dead last in a cross-country race (for example), and didn't get a medal, that would be okay. Sure, she might feel upset in the moment, but it could also motivate her to train harder and give her a better appreciation for the joy of placing in the top three one day. What would not be okay is if I had then screamed at her, in front of the other runners, for getting beaten by the shortest kid on the team.

Just as an aside, a 2019 study provides an elegant demonstration of the principle that setbacks can be net positive. The researchers looked at junior scientists who had applied for NIH grants and separated them into two groups: One group had scored just above the threshold for funding, while the other had scored just below the funding line, meaning their grants were not funded. While the near-miss group were more likely to drop out of science in the immediate aftermath, those who stuck with it eventually outperformed their peers who had received funding on their first try. The early setback had not impaired their careers but may have had an opposite effect.

The most important thing about childhood trauma is not the event itself but the way the child adapts to it. Children are remarkably resilient, and wounded children become adaptive children. The problems begin when these adaptive children grow up to become maladaptive, dysfunctional adults. This dysfunction is represented by the four branches of the trauma tree: (1) addiction, not only to vices such as drugs, alcohol, and gambling, but also to socially acceptable things such as work, exercise, and perfectionism (check); (2) codependency, or excessive psychological reliance on another person; (3) habituated survival strategies, such as a propensity to anger and rage (check); (4) attachment disorders, difficulty forming and maintaining connections or meaningful relationships with others (check). These branches are often fairly obvious and easy to spot; the tricky part is digging down to the roots and beginning to disentangle them. All of this is highly individual; everyone responds and adapts to trauma in a unique way. And it's not as if there is some sort of pill that can make someone's trauma, or their adaptations to it, simply go away. It requires hard work—and, as I would come to understand, it can also take a very long time.

This is yet another realm where Medicine 2.0 too often falls short. Most therapists diagnose patients based on the bible of mental health, the *Diagnostic and Statistical Manual of Mental Disorders,* 5th edition (DSM-5), a 991-page-long compendium of every conceivable psychological condition. The DSM is a valiant attempt to organize and codify all of the myriad forms of mental disorders—to scientize it, in effect, and also to facilitate insurance reimbursement. But in reality, as Paul Conti observes, our stories and our conditions are really unique to each of us. Not all of them fall into tidy diagnostic categories. Everyone is different; everyone's story is different. No person is a "code." Therefore, he believes, such rigorous codification "presents an obstacle to *actually understanding the person.*"

This is also what makes it difficult to offer blanket advice to everyone about this topic; every reader will have their own emotional makeup, their own history, and their own issues to address. Yet one difficulty that we all share is that Medicine 2.0 is set up to treat mental and emotional health in pretty much the same way that it treats everything else: diagnose, prescribe, and, of course, bill. While antidepressants and other psychoactive medications have helped many patients, including me, finding a complete solution is rarely simple. For one thing, this is primarily a disease-based model, which is how Medicine 2.0 addresses and solves other problems, such as infections and acute illnesses: treat the symptoms and send the patient home. Or if the situation is more serious, as it was with me, send the patient off for a couple of weeks at a place like the Bridge, and then send them home—*voilà*, problem solved.

One reason this approach has proved less effective in the psychological realm is that mental health and emotional health are not the same thing. Mental health encompasses disease-like states such as clinical depression and schizophrenia, which are complex and difficult to treat but do present with recognizable symptoms. Here, we are more interested in *emotional health,* which incorporates mental health but is also much broader—and less easy to codify and categorize. Emotional health has more to do with the way we regulate our emotions and manage our interpersonal relationships. I did not have a mental *illness,* per se, but I did have serious issues with my emotional health that impaired my ability to live a happy, well-adjusted life—and potentially did put my life in danger. Medicine 2.0 has a harder time dealing with situations such as this.

Taking care of our emotional health requires a paradigm shift similar to the shift from Medicine 2.0 to Medicine 3.0. It's about long-term prevention, just like our approach to preventing cardiovascular disease. We have to be able to recognize potential problems early and be willing to put in hard work to address these problems over a long period of time. And our approach must be tailored to each individual, with their unique history and set of issues.

Our Medicine 3.0 thesis is that if we address our emotional health, and do so early on, we will have a better chance of avoiding clinical mental health issues such as depression and chronic anxiety—and our overall health will benefit as well. But there is rarely a simple cure or a quick fix, any more than we have a quick fix for cancer or metabolic disease.

Addressing emotional health takes just as much constant effort and daily practice as maintaining other aspects of our physical health by creating an exercise routine, following a nutritional program, adhering to sleep rituals, and so on. The key is to be as proactive as possible, so that we can continue to thrive in all domains of healthspan, throughout the later decades of our lives.

What makes dealing with emotional health harder than physical health, I suspect, is that we are often less able to recognize the need to make changes. Few people who are overweight and out of shape fail to realize they need to make a change. Making the change might be another story. But countless people are in desperate need of help with their emotional health, yet fail to recognize the signs and symptoms of their condition. I was the poster child for this group.

After two weeks, I left the Bridge. My therapists there were uneasy about letting me go so soon; they wanted me to stay for another month, but I felt that I had made tremendous progress in that relatively short time. Acknowledging my past felt like a huge deal to me. I felt hopeful, and they finally agreed that I could leave. So I flew home the day before Christmas.

This was probably a mistake.

I wish I could say that this marked the end of the story, the point where Old Peter said goodbye, with his selfishness and his anger, and New Peter took his place, and we all lived happily ever after. Alas, that was not the case; it was, at best, only the end of the beginning.

I had a lot of work to do when I came home, to process what had been unearthed at the Bridge and to begin to try to heal my relationships with my wife and my children. With the help of two wonderful therapists, Esther Perel (alone) and Lorie Teagno (with my wife), I made slow progress as the weeks and months went by. Lorie and Esther both felt I needed a male therapist, one who could model healthy male emotions. I tried out several good male therapists, but I did not feel a connection to any of them the way I had felt connected to Jeff English, my primary therapist at the Bridge.

I was ready to give up when Esther suggested that I read Terrence Real's book *I Don't Want to Talk About It,* a groundbreaking treatise on the roots of male depression. Once I started, I could not put it down. It was almost creepy that this guy seemed to be writing *about me,* despite never having met me. His main thesis is that with women, depression is generally overt, or obvious, but men are socialized to conceal their depression, channeling it inward or into other emotions, such as anger, without ever wanting to discuss it. (Hence the title.) I could relate to the stories that he shared about his patients. So I began to work with Terry as well. After having gone far too long without any therapy at all, I was now seeing three therapists.

Terry had grown up working-class in Camden, New Jersey, with a father whom he describes as a "loving, smart, and brutally violent man." It turned out that the driving force was his father's hidden depression, which he had adeptly handed on to Terry. "My father beat his depression into me with a strap," he told me. Trying to cope with his father's anger and violence was what had pushed him in the direction of studying psychotherapy. "I needed to make sense of my father and his violence, so I would not repeat it," he said.

Terry helped me continue to connect the dots between my own childhood and the kinds of dysfunction that had marked my adolescence and my life as an adult. Looking back at my teenage self, and the way I was in college, I realize now that I was morbidly depressed—clinically, off-my-rocker depressed. I just didn't know it at the time. I had the classical symptoms of covert male depression, which were a tendency to isolate myself and, above all, a propensity to anger, perhaps my most potent addiction. One of the first things I wrote in my journal, after an early discussion with Terry, still resonates today: "90% of male rage is helplessness masquerading as frustration."

Terry helped me make sense of the helplessness that I still felt. I came to understand that the crucial factor for me was the shame I felt about having been victimized. As is the case with many men, I had flipped that shame into a feeling of grandiosity. "Shame feels bad; grandiosity feels good," he told me. "It is central to masculinity and traditional manhood, this flip from the one-down victim to the one-up avenger. What's devilish about flipping from shame into grandiosity like this is that it works. It makes you feel better in the short run, but it just creates havoc in your life in the long run."

Even worse was the realization of what my behavior had been doing to my family, especially my kids. I was not so delusional as to think I was being a particularly good dad, at that point, but I took at least some modicum of pride in the fact that I could protect my kids from the trauma I had suffered. I was a great "provider" and "protector." They would never have to suffer my specific childhood shame. But I knew they saw my overflowing anger, even though it was rarely directed toward them or Jill.

At the Bridge, I learned that children don't respond to a parent's anger in a logical way. If they see me screaming at a driver who just cut me off, they internalize that rage as though it were directed to them. Second, trauma is generational, although not necessarily linear. Children of alcoholics are not inevitably destined to become alcoholics themselves, but one way or another, trauma finds its way down the line.

As Terry had written: "Family pathology rolls from generation to generation like a fire in the woods taking down everything in its path until one person, in one generation, has the courage to turn and face the flames. That person brings peace to his ancestors and spares the children that follow."

I wanted to be that person.

Slowly, with the help of Terry as well as Esther and Lorie, I began to pick up some tools to help me deal with my past and to guide my day-to-day behavior onto a better path. One helpful model that Terry had taught me was to think about my relationships as akin to a delicate ecosystem, a kind of emotional ecology. Why would I want to poison the environment in which I had to live?

This sounds so basic, but it took some thought and consideration, and even strategizing, to put into practice. It meant pulling back from the little things that used to make me mad at the people around me, on a daily or even hourly basis; that, I now recognized, was poisoning the drinking well. I had to learn new ways of dealing with day-to-day problems and frustrations. This is an important stage in Terry's framework, the stage of teaching: *This is how you do it right. This is how you listen to your partner's complaint and be compassionate.*

"These are all skills," Terry told me. "And like all the skills you have tried to master over your life, you can learn these, also."

Some of the changes I made seem like no-brainers. I made sure to spend time with my kids—one on one, no phones—every day that I was home. I would check in with Jill on her experience (not "events") each day. I limited my phone time and my work hours to a strict window. One day a week, typically Saturday or Sunday, I would refrain from doing any work at all, something that went against decades of ingrained habit. Even more amazing, Jill and I went on an actual vacation for the first time in years, just the two of us, no kids.

One skill I worked on that is a bit more complicated is called "reframing." Reframing is basically the ability to look at a given situation from someone else's point of view—literally reframing it. This is an incredibly difficult thing for most of us to do, as David Foster Wallace explained in his now famous

2005 commencement address to the graduating class at Kenyon College, "This Is Water":

> Everything in my own immediate experience supports my deep belief that I am the absolute center of the universe; the realest, most vivid and important person in existence. We rarely think about this sort of natural, basic self-centeredness because it's so socially repulsive. But it's pretty much the same for all of us. It is our default setting, hard-wired into our motherboards at birth.
>
> Think about it: there is no experience you have had that you are not the absolute center of. The world as you experience it is there in front of YOU or behind YOU, to the left or right of YOU, on YOUR TV or YOUR monitor. And so on. Other people's thoughts and feelings have to be communicated to you somehow, but your own are so immediate, urgent, real.

I could relate. This had certainly been my own default setting, for as long as I could remember. It's tempting to try to pin it on my own history of trauma, and my need to adapt to protect myself, but obviously it had stopped serving me so well. Easier described than accomplished, reframing entails taking a step back from a situation and then asking yourself, What does this situation look like through the other person's eyes? How do *they* see it? And why is your time, your convenience, or your agenda any more important than theirs?

This comes in handy almost every single day. For example, if my wife comes home and snips at me because I didn't help put away the groceries, my tendency might be to think, *Hey, I'm working really hard and I can't always pitch in.* And that sense of entitlement would sneak up inside me because, well, I am working very hard, and someone else can put away the groceries.

But then I ask myself, *Wait, what has Jill's day been like today?*

She had to pick up our boys from school and take them to the grocery store, where they probably fought like wild animals and made everyone in the store think Jill is the worst mother on the planet because she can't control her spoiled little brats, while she stood in line at the deli counter just to get me the perfectly sliced deli meat that can't be found with the prepackaged deli meat, and then on the way home she hit every single red light while the boys threw Lego bricks at each other.

And you know what? When I view it through her lens, I quickly get over myself and realize that I'm the one who's being selfish and that next time I have to do better. That's the power of reframing. You realize that you have to step back from a situation, temper your reflexive reaction, and try to see what is actually happening.

Somewhere along the line, in a random airport on a long work trip, I had picked up David Brooks's book *The Road to Character.* On the plane, I read the part where Brooks makes a key distinction between "résumé virtues," meaning the accomplishments that we list on our CV, our degrees and fellowships and jobs, versus "eulogy virtues," the things that our friends and family will say about us when we are gone. And it shook me.

For my entire life, I had been accumulating mostly résumé virtues. I had plenty of those. But I had also recently attended a funeral for a woman about my age who had died of cancer, and I was struck by how lovingly and movingly her family had spoken about her—with hardly a mention of her impressive professional or educational success. What mattered to them was the person she had been and the things she had done for others, most of all her children.

Would anyone be speaking that way about me when it was my turn in the casket?

I doubted it. And I decided that that had to change.

I began using these tools and strategies on a daily basis, forming an emotional health routine of sorts. I focused on eulogy virtues, not résumé virtues. I worked on being more relational, more present with my family. I tried to practice reframing. But something still felt off. Even as I worked on my relationships with those closest to me, I still had a major blind spot: my relationship with myself. I had become a much better husband and father, but inside,

I was just as hard on myself as ever. My deep self-hatred and loathing still contaminated most of my thoughts and emotions, and I didn't even realize it—nor did I understand why it was happening.

I know I was not alone in this feeling. I was speaking with a patient of mine once, an incredibly successful and well-known person, and he said something that stunned me. "I need to be great," he said, "in order to feel like I'm not worthless."

That stunned me. Even *he* feels this way?

Yet my own insecurity and self-hatred still gnawed at me. While I was getting better at dealing with other people—that constituted some progress—I was as hard on myself as ever. Anger still ruled me, even when I was supposedly having fun. Simply missing a shot at archery or spinning out of a turn in my driving simulator would send me into a seething, self-loathing rage. I would constantly lose my temper with myself and throw tantrums, yelling out loud and even snapping an arrow across my thigh if I missed a shot. That hurt a lot, but I kept doing it.

It was as if I had my own personal Bobby Knight, the Indiana University basketball coach who was famed for his red-faced sideline meltdowns (and who ultimately lost his job because of them), living inside my head. Whenever I made a mistake or felt I performed poorly, even in tiny ways, my own personal Coach Knight jumped up from the bench to scream at me. Make a mistake cooking dinner? *How do you not know how to grill a fucking steak?* Flub the intro recording to a podcast? *You are a worthless sack of shit who has no business being alive, let alone having a podcast!*

The crazy part is that I actually believed that voice served me well. This rage and self-doubt had fueled much of my personal drive and whatever success I had enjoyed, I told myself. It was simply the price I had to pay. But in reality, all it had produced was more résumé virtues. And I wasn't even all that proud of my résumé. It would never be good enough.

For the first time in my life, I had a radical thought: *Who cares how well you perform if you're so utterly miserable?*

During this time, Paul Conti, who continued to keep tabs on my declining emotional health as a friend, sensed another rising storm. He began suggesting that I go into another residential treatment facility. The Bridge had helped me greatly, and without it I would have lost my family. But Paul felt I had left the Bridge too soon, staying for only two weeks, and thus had not yet scratched the surface when it came to examining and healing my relationship with myself. But I stubbornly refused. *I'll be fine.*

Something had to give, and soon enough, it did.

I imagine that if 2020 had been like any other year, I could have kicked the can down the road for a few more years and just gotten by somehow. But there is nothing like a crisis to bring every other simmering issue right to a full boil.

When COVID hit, our practice was already maxed out. We bring on most of our new patients in the first two quarters of each year, so I had already committed my ancillary bandwidth to learning the ins and outs of the new patients. COVID instantly doubled or tripled our workload. There were daily calls with the research team to discuss everything we could find out about the disease, starting very early in the morning, as well as a new and daunting slate of COVID-related podcasts. I gave up my morning meditation practice in order to field the countless calls from patients, who were understandably panicked and looking for reassurance.

As March bled into April it became clear there was no end in sight. One day in late April 2020, I was on a routine morning call with my practice manager when I couldn't take it anymore and started venting. I've lost control, I told her. I can't keep my patients' stories straight anymore. Was it patient X or patient Y who just last week told me about his daughter's struggle at school? Was it patient A or patient B whom I needed to reach out to that evening about an issue she was having? She tried to soothe me, saying I was doing the best I could under the circumstances and that our patients were grateful. But the more she talked, the angrier I got.

And just like that, I spun into a radical, self-destructive episode, one like I've never experienced before or since. Even remembering it now is terrifying. I threw a table across our living room. I tore my T-shirt to pieces. I screamed, in rage and pain. My wife begged me to leave the house for fear I would harm her or the kids. I thought about driving myself into a bridge abutment or other structure fast enough that I'd be killed. I was convinced that I was broken, defective; when they autopsied my brain, they would discover just how screwed up I was. I was beyond fixing. Nothing could make it right.

I ended up holed up in a motel, on the phone with Paul, Esther, and Terry. They insisted that I needed to go back to a place like the Bridge. Now. True to form, I stubbornly disagreed, claiming that I could fix this with just a little more time and support, if only I could go home and get some rest. After pleading with them for forty-eight hours, I finally relented. In the middle of the night, I drove myself to Phoenix, Arizona, to be admitted to a place called Psychological Counseling Services, or PCS.

Terry had been telling me about PCS for nearly a year. He said it was a place that worked miracles, healing wounds that seemed beyond permanent. I asked how he could be so sure. He said I just needed to trust him.

Just as with my visit to the Bridge two and a half years earlier, it took a few days to get settled in. Because it was the beginning of the pandemic, I was alone, dealing with therapists remotely on Zoom for twelve hours each day while I sat in a tiny Airbnb a few miles from the facility.

It was not until the second week that I began to make progress. Slowly, I came to accept that I had built a structure of perfectionism and workaholism on the pillars of performance-based esteem. This structure rested on a foundation of my shame, some of which was brought on by trauma and some of which was inherited, as children take on the shame of those around them. But all of it was exacerbated by my own vicious cycle of self-loathing and guilt for my actions. It's not a coincidence that I have gravitated toward sports that demand perfection, like archery and driving race cars.

I ended up spending three weeks at PCS—twenty-one agonizing, uninterrupted days—finishing the work I had begun at the Bridge and going far beyond what I had imagined was possible. We covered an enormous amount of ground, but one task absolutely stymied me. On my second day, I was assigned to write out a list of forty-seven affirmations, representing one positive statement about myself for each year of my life. I made it to about five or six before I got completely stuck. For days and days, I couldn't come up with anything good to say about myself. My perfectionism and my shame did not permit me to believe anything nice about myself. I just couldn't do it.

Finally, on the nineteenth day—a blistering hot Wednesday morning—it happened. One of my therapists, Marcus, was pushing deeper and deeper into a story I had told him earlier about how I had stopped wanting to celebrate my birthdays when I was about seven; in fact, I revealed, I would keep my birthday a "secret" until well into my twenties. His questions made it clear that this was not something a healthy child would do, and it likely masked something more deeply wrong. He just kept digging and would not let it go.

That recognition pushed me into an emotional freefall. It had been two and a half years in the making, but I finally was able to let go and accept the truth about my past and how it had shaped me, without any excuses or rationalizations. All that I had become—good and bad—was in response to what I had experienced. It wasn't simply the big-T traumas, either; we uncovered many, many more little-t traumas, hidden in the cracks, that had affected me even more profoundly. I hadn't been protected. I hadn't felt safe. My trust had been broken by people who were close to me. I felt abandoned. All of that had manifested itself as my own self-loathing as an adult; I had become my own worst enemy. And I hadn't deserved any of it. This was the key insight. That little, sweet boy did not deserve any of it. And he was still with me.

Once I had accepted all this, it was easy to write out the forty-seven affirmations.

I am flawed, but not defective.
I am a good husband and father.
I am a good cook.
I am not my shame.
I will find a way to love myself.

They just poured out of me. It reminded me of this observation by Jacob Riis, the great Danish American journalist and social reformer: "When nothing seems to help, I go back and look at a stonecutter hammering away at his rock perhaps a hundred times without as much as a crack showing in it. Yet at the hundred-and-first blow it will split in two, and I know it was not the last blow that did it, but all that had gone before."

Looking back on all this, one of the most important lessons that I learned is that the type of change I describe in this chapter is not possible unless we are equipped with a set of effective tools and sensors with which to monitor, maintain, and restore our emotional equilibrium. These tools and sensors are not innate; for most of us, they must be learned, and refined, and practiced daily. And neither are they quick fixes.

Yes, medications such as antidepressants and mood stabilizers matter and can help. Yes, a mindfulness meditation practice can make all this easier. Yes, molecules such as MDMA and psilocybin, when used with skilled guidance and in the correct setting, can be powerful; I have used both at critical points in my recovery, with remarkable results. But too often I see people tethering their hopes of transformation *solely* to a ketamine trip or a journey to the jungles of Peru with a shaman to guide them through the mind-blowing experience of an ayahuasca journey, or some other singular experience (or even, as in my case, thinking that two weeks in a facility such as the Bridge is enough, after which we can continue as though nothing fundamental has changed).

All of these modalities are powerful and potentially useful, but we need to think of them as merely adjuncts to the deep and often very unpleasant, uncomfortable, at times very slow—at other times too fast—self-exploration that is required in real psychotherapy. True recovery requires probing the depths of what shaped you, how you adapted to it, and how those adaptations are now serving you (or not, as in my case). This also takes time, as I found out the hard way; the biggest mistake of all is to believe that you're "cured," by a few months on a drug or a handful of therapy sessions, when in fact you're not even halfway there.

My progress upon returning from PCS was rooted in daily action, much of it uncomfortable. My most pressing challenge was quite simply to avoid having another one of my meltdowns, like the one that had led to me going to PCS in the first place. I had had other, lesser episodes leading up to it, but this one had felt like the explosion of the space shuttle *Challenger*, which blew up over the Atlantic Ocean just after launch in 1986.

At the time, that disaster seemed completely unexpected, but a lengthy investigation revealed that was not the case at all. There had been warning signs and system failures building up inside the space shuttle program for years prior. These problems had been documented by the engineers, but they were ignored or covered up by management, because doing so seemed "easier" than delaying the launch. The result was a catastrophe that could have been prevented. My goal was to learn to understand the warning signs and the systems failures that could lead to a blow-up in my own life, to prevent it from ever happening again. The idea is somewhat similar to what we've been talking about with Medicine 3.0, only applied to emotional health: spotting potential problems early and taking preventive action as soon as possible.

The way in which I do this, the tools that I use, derive from a school of psychology known as dialectical behavior therapy, or DBT, developed in the 1990s by Marsha Linehan. Based on the principles of cognitive behavioral therapy, which seeks to teach patients new ways of thinking about or acting on their problems, DBT was developed to help individuals with more serious and potentially dangerous issues, such as an inability to regulate their emotions and a propensity to harm themselves or even attempt suicide. These people are lumped into something called borderline personality disorder, which is a bit of a catch-all diagnosis, but DBT has also been found to be helpful in patients with less dramatic and dangerous emotional health issues, a category that encompasses many more of us. I liken it, naturally, to Formula One: the race circuit is a high-stakes, high-risk laboratory where car manufacturers develop and test technologies that trickle down to our everyday street cars.

One thing I like about DBT is that it is backed up by evidence: clinical trials have found it to be effective in helping suicidal and self-harming patients stop their dangerous behavior. Another thing that draws me to DBT is that it is skills-based, not just theoretical. Practicing DBT means literally working through a workbook with a DBT therapist, doing exercises every day. I'm better at doing than thinking sometimes. The practice of DBT is predicated on learning to execute concrete skills, repetitively, under stress, that aim to break the chain reaction of **negative stimulus → negative emotion → negative thought → negative action.**

DBT consists of four pillars joined by one overarching theme. The overarching theme is mindfulness, which gives you the ability to work through the other four: *emotional regulation* (getting control over our emotions), *distress tolerance* (our ability to handle emotional stressors), *interpersonal effectiveness* (how well we make our needs and feelings known to others), and *self-management* (taking care of ourselves, beginning with basic tasks like getting up in time to go to work or school). The first two—emotion regulation and distress tolerance—are the ones I need to work on most, so that's where I've focused with my DBT therapist, Andy White.

I visualize my distress tolerance as a window that opens and closes vertically. The narrower this window becomes, the more likely I am to become dysregulated. My goals are to keep this window as wide as possible and to be very attentive to anything that might narrow it, even factors outside my control (see figure 15).

Many behaviors expand this window: exercise, sound sleep, good nutrition, time with my family, medications such as antidepressants or mood stabilizers, deep social connections, spending time in nature, and recreational activities that do not emphasize self-judgment. These are the things I have control over. I don't have as much control over the things that compress my window, but I still have some—for example, overcommitting to projects and saying yes to more than I should. Managing this window (in part by learning to say no) and trying to keep it as wide as possible is something I think about and work on almost every day.

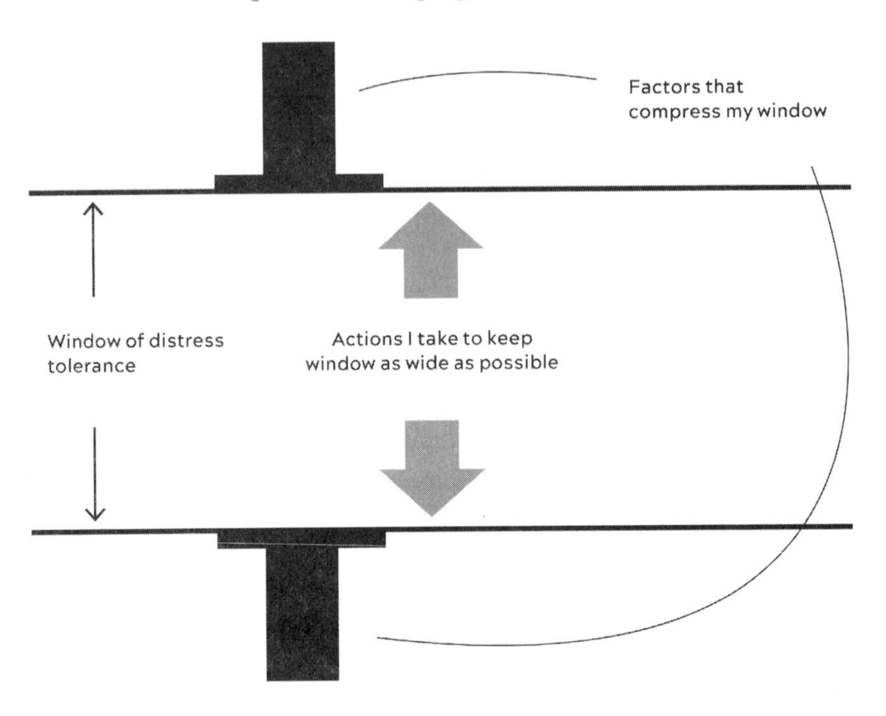

This is how I visualize my daily efforts to maintain and increase my distress tolerance, represented by the "window" or gap shown here. I try to focus on doing whatever I can to keep this window as wide open as possible.

They are linked: I needed to increase my distress tolerance in order to regain control over my emotions. And the better I regulate my emotions, the less I need to rely on that distress tolerance window. I found that as I worked on those two, my interpersonal effectiveness, which was obviously far from perfect, improved naturally. Self-management has never really been an issue for me, but someone else might have different needs; DBT is highly adaptable.

DBT is rooted in mindfulness, which is one of those mushy buzzwords that I'd always despised until I began to understand it was a really effective tool to create distance between my thoughts and myself, to wedge even a sliver of space between some stimulus and my knee-jerk response. I needed that.

I had been practicing mindfulness meditation since I left the Bridge, with obviously mixed results, but I did begin to develop occasional flashes of insight, moments when I was able to detach myself from my thoughts and emotions. It's not complete detachment in the sense that we're checking out, but we want to create enough of a gap between stimulus and response so that we are not simply reacting reflexively to things that happen, like a driver who cuts us off in traffic or angry or distressing thoughts that we might have. That gap, in turn, allows us to process the situation in a calmer and more rational way. Do we really need to honk and curse, and potentially make the situation worse (even if the guy deserves it)? Or is it better to simply accept what happened and move on? Mindfulness helps us reframe it: The other driver may be rushing to the hospital with a sick child, for all we know.

Another way in which mindfulness helps is by reminding us that when we are suffering, it is rarely because of some direct cause, like a rock that is crushing our leg at this very moment. Much more often, it is because we are thinking about some painful event that occurred in the past or worrying about something bad that may occur in the future. This, too, was an enormous revelation to me. Simply put, I experience less pain because I am able to recognize when the source of that pain is inside my own head. This was not an original insight, but it was nevertheless profound. I was about 2,500 years behind the Buddha, who said that "your worst enemy cannot harm you as much as your own unguarded thoughts." Seneca improved on that in the first century AD, observing that "we suffer more often in imagination than in reality." And later, in the sixteenth century, Shakespeare's Hamlet noted, "There is nothing either good or bad, but thinking makes it so."

One obvious way this applies is in how we think about ourselves. What does our inner dialogue sound like? Is it kind and forgiving and wise, or is it harsh and judgmental, like my inner Bobby Knight? One of the most powerful exercises I learned was to simply listen to my self-talk. I would record voice memos to myself on my phone, after I did anything that could produce self-judgment, such as archery or driving my race-car simulator, or even just cooking dinner, and send each one to my therapist. My instinct in these situations was typically to scream at myself for failing somehow. My therapist at PCS told me to imagine instead that my best friend had performed exactly as I had done. How would I speak to him? Would I berate him the way I often berated myself? Of course not.

This was a slightly different take on reframing, forcing me to step outside myself and really see the disconnect between my "mistakes" (minor) and the way I talked to myself about those mistakes (brutal). I did this multiple times a day, every single day, for about four months; you can imagine how much space it took up on my phone. Over time, my inner Bobby Knight became fainter and fainter, and today it's almost hard for me to remember what that voice used to sound like.

Another important goal of DBT is to help people learn to regulate their emotions. When I arrived at the Bridge, I had very little ability to recognize how I was feeling, let alone change or manage my emotional state. All I knew was overflowing anger. This came to a head with me at the beginning of COVID, where I became so overloaded and so overwhelmed that I just exploded. I lost the ability to regulate my emotions, up and down. My close friend Jim Kochalka, a clinical psychologist, calls this type of emotional dysregulation "the inflammation of the psyche," which feels about right to me.

This anger had long been an obstacle in my personal relationships, even with my family. As Terry Real had pointed out long ago, this anger was rooted in shame, but very often my anger would also *create* more shame. If I yell at my kids, for example, especially when I do it because I'm upset about something else, I feel shame. That shame then becomes an obstacle to my ability to reconcile with them, so I feel more shame. It's like I'm digging myself into a hole, and it's not only with my kids. Until I can reconcile and own my behavior, I can't move on. This used to be a much bigger problem, but at least now I can usually spot it in real time, before the hole gets too deep.

DBT teaches a variety of techniques to enable people to maintain and improve their distress tolerance, and to recognize and cope with their emotions—and not be controlled *by* them, as I had been for so long. One simple tactic that I use to cope with mounting emotional distress is inducing an abrupt sensory change—typically, by throwing ice water on my face or, if I'm really struggling, taking a cold shower or stepping into an ice bath. This simple intervention stimulates an important cranial nerve, the vagus nerve, which causes our heart rate and respiratory rate to slow and switches us into a calm, parasympathetic mode (and out of our fight-or-flight sympathetic mode). Interventions like these are often enough to help refocus and think about a situation more calmly and constructively. Another technique I have grown very fond of is slow, deep breathing: four seconds to inhale, six seconds to exhale. Repeat. As the breath goes, the nervous system follows.

It is also important to note that DBT is not a passive modality. It requires conscious thought and action on a daily basis. One tactic that I've found especially helpful is called *opposite action*—that is, if I feel like doing one thing (generally, not a helpful or positive thing), I'll force myself instead to do the exact opposite. By doing so, I also change the underlying emotions.

The first time I experienced this was a pleasant Sunday afternoon shortly after we moved to Austin. I had made a commitment to my wife that I would take one day off each week, presumably Sunday, to be with the family. Sunday rolled around and I was drowning in work. I was stressed out and grumpy, and I didn't want to see or hear anyone. I just wanted to grind through my work. All too conditioned to my selfish ways, Jill barely pushed back when I said I was too busy to take the kids to a nearby creek. But as I watched her piling the kids into the minivan, I spotted a perfect chance to put theory into practice. I ran out to the van, hopped in the front seat, and said, "Let's go." We got to Barton Creek and really didn't do anything special beyond walking around, skipping rocks across the water, and seeing who could hop from boulder to boulder without getting wet. Much to my surprise, my mood completely changed. I even insisted we stop for burgers and fries (!) on the way home.

This is an easy example, obviously. Who wouldn't want to play with their kids instead of working? But for Old Peter, it would have been impossible. This small lesson, which I have implemented countless times since, taught me something very important: changing the behavior can change the mood. You do not need to wait for your mood to improve to make a behavior change. This is also why cognitive therapies alone sometimes come up short; simply thinking about problems might not help if our thinking itself is disordered.

Exercise is another important component of my overall emotional health program, particularly my practice of rucking, discussed in chapter 12. I find that spending time moving in nature, simply enjoying the feeling of the wind in my face and the smell of the budding spring leaves (and a heavily loaded pack on my back) helps me cultivate what Ryan Holiday calls "stillness," the ability to remain calm and focused amid all the distractions that our world offers and that we create for ourselves. When my family comes along, it's important bonding time. When I'm alone, rucking serves as a mindfulness practice, a kind of walking meditation. No phone, no music, no podcasts. Just the sounds of nature, and of my heavy breathing. This is another example of how action can lead us into a better mental state. And as Michael Easter pointed out to me, there is actual research suggesting that exposing oneself to the fractal geometric patterns in nature can reduce physiological stress, and that these effects show up on an EEG.

The most important "tactic" by far is my regular weekly therapy session (down from three or four per week when I left PCS). This is not optional. Each session begins with a physical check-in: How am I feeling? How have I slept (a big one)? Am I in physical pain? Am I in conflict? Then we dissect and discuss the events and issues of the week in minute detail. No topic is too insignificant. If, for example, I found myself getting really upset at a TV show or movie, this might be worth exploring. But we also tackle big-picture issues, the ones that propelled me into crisis in the first place. I complement my therapy sessions by writing in my journal, a place where I can practice articulating my emotions and understanding them, holding nothing back. I feel strongly that there is no substitute for this kind of work with a trained therapist.

Most days, I try to stick to my "green-light" behaviors, even when I don't automatically want to or feel too busy, or whatever. Every day I make mistakes, and every day I try to forgive myself for them. Some days are better than others, but over time I've made tangible progress. It's important to note that my list of go-to activities and behaviors might not be the same as someone else's, and even mine are not the same today as they were in the six months after I left PCS; there's a line in the DBT literature about how it's important to seek pleasurable activities "consistent with your own values." Everyone has different problems and a different mental makeup, and everyone can find their own unique solutions. The techniques of DBT are adaptable and flexible, which is what makes them useful to a wide range of people.

If you take nothing else from my story, take this: *If I can change, you can change.* All of this has to begin with the simple belief that real change is possible. That's the most important step. I believed I was the most horrible, incorrigible, miserable son of a bitch that was ever shat into civilization. For as long as I could remember, I believed that I was defective and that my flaws were hard-wired. Unchangeable. Only when I at least entertained the notion that maybe I was not actually a monster was I able to start chipping away at the narrative that had nearly destroyed my life and everyone in my wake.

This is the key step. You have to believe you can change—and that you deserve better.

Yet it can be a very difficult step for many people to take, for a number of reasons—the social stigma that persists around mental and emotional health, to name just one. It's difficult for many people, myself included at one point, to recognize that they have a problem, admit that they need help, and then take action, particularly if it means talking about it openly with others, or taking time off work, or dealing with the expense of treatment.

This is part of the shift in our mindset that needs to happen if we are to begin to address the epidemic of emotional health disorders, along with the attendant drug use, alcohol abuse, eating disorders, suicide, and violence that goes along with it. We have to make it okay to be vulnerable, to ask for and receive help.

I resisted seeking help for the longest time. It was only when I was confronted with unbearable choices—losing my family, or even losing my life at my own hands—that I reluctantly agreed to do what I should have done much sooner, and to pay as much attention to my emotional health as I had always paid to my physical health.

As I settled into the next phase of my recovery, I began to notice something I had never experienced before: I found more joy in *being* than in *doing.*

For the first time in my life, I felt that I could *be* a good father. I could *be* a good husband. I could *be* a good person. After all, this is the whole point of living. And the whole point of *outliving*.

There's a quote from Paulo Coelho that I think about often: "Maybe the journey isn't so much about becoming anything," he writes. "Maybe it's about unbecoming everything that isn't really you, so you can be who you were meant to be in the first place."

EPILOGUE

It was only after much reflection on this whole experience that I really began to understand how emotional health relates to longevity, and how my journey helped redefine my perspective.

I had long subscribed to a kind of Silicon Valley approach to longevity and health, believing that it is possible to hack our biology, and hack it, and hack it, until we become these perfect little humanoids who can live to be 120 years old. I used to be all about that, constantly tinkering and experimenting with new fasting protocols or sleep gadgets to maximize my own longevity. Everything in my life needed to be optimized. And longevity was basically an engineering problem. Or so I thought.

It took five years, two stints in inpatient treatment centers, and the near loss of my marriage and my kids to change my mind. What I eventually realized, after this long and very painful journey, is that longevity is meaningless if your life sucks. Or if your relationships suck. None of it matters if your wife hates you. None of it matters if you are a shitty father, or if you are consumed by anger or addiction. Your résumé doesn't really matter, either, when it comes time for your eulogy.

All these need to be addressed if your life is to be worth prolonging—because the most important ingredient in the whole longevity equation is the *why.* Why do we want to live longer? For what? For whom?

My obsession with longevity was really about my fear of dying. And something about having children was making my obsession with longevity ever more frenetic. I was running away from dying as fast as I could. Yet at the same time, ironically, I was also avoiding actually living. My tactics might have succeeded in my living longer, with optimal glucose regulation and ideal lipoprotein levels, but my strategy was unquestionably accumulating more regrets. My physical and cognitive health were great, but my emotional health was tanking.

My biggest regret is that so much of the misery that I've experienced, and the pain that I have inflicted on other people, could have been avoided if I had reached a better understanding of this sooner in life, preferably much sooner. The saddest part is that I wasted so much time being so detached, so miserable, and so misguided. So much time pursuing an empty goal.

But as my recovery progressed, I noticed that my preoccupation with dying began to fade away. And my quest for longevity no longer felt like a grim, desperate task; now the things I did every day felt welcome, necessary. I was enhancing my life and looking forward to the future. My journey to outlive finally had clarity, purpose, and meaning.

It brought me back to something my dear friend Ric Elias had said to me. Ric had been one of 155 passengers on the US Airways flight that emergency-landed in the Hudson River in January of 2009. As the plane was coming down, Ric and most of the other passengers were certain that they were going to die. Only the pilot's skill and more than a little luck prevented disaster. If the plane had been going a little bit faster, it would have broken apart on impact; a few miles per hour slower, and the nose would have tipped forward and it would have sunk into the river. A handful of tiny factors like that made the difference between everyone on that plane living and many or most (or all) of them dying.

That day changed Ric's outlook on longevity in a way that really resonates with me. All that time, I had been obsessed about longevity for the wrong reason. I was not thinking about a long, healthy life ahead; instead, I was mourning the past. I was trapped by the pain that my past had caused and was continuing to cause. I wanted to live longer, I think, only because deep down I knew I needed more runway to try to make things right. But I was only looking backward, not forward.

"I think people get old when they stop thinking about the future," Ric told me. "If you want to find someone's *true* age, listen to them. If they talk about the past and they talk about all the things that happened that they did, they've gotten old. If they think about their dreams, their aspirations, what they're still looking forward to—they're young."

Here's to staying young, even as we grow older.

ACKNOWLEDGMENTS

This book came perilously close to never seeing the light of day. In early 2020, after my book agent and publisher fired me for failing to deliver a manuscript that was already a year late, I was in no mood to put any additional effort into it and decided to scrap the whole project. The draft sat untouched for about nine months until my friend Michael Ovitz asked if he could read it. A couple of weeks later, Michael called to tell me that he thought it had great potential and that it needed to be published. He suggested that my coauthor Bill Gifford and I send a cleaned-up version to his friend, Diana Baroni, at Penguin Random House. Had Michael not forced the issue, making the introduction to Diana and consummating the agreement with Penguin Random House, *Outlive* might still be a random Google doc seen only by Bill and me and a handful of others. I'm grateful for Diana's ability to see what that somewhat ragged manuscript could become, and, more importantly, for her guidance in helping us get it there.

Long before that, this book would have died on the vine without Bill's help. In mid 2017, after I had written about 30,000 words on my own, my then publisher said my draft was too technical and lacked any sense of me as a person and my own journey to understanding the importance of longevity. They suggested I find a coauthor, and so began a long search that led to Bill. I had read a story Bill wrote in 2015 about rapamycin, as well as his book *Spring Chicken,* and had a hunch that he was the right person to help me navigate a very delicate task: to convey this complex subject matter accurately and with attention to nuance and detail, while making it readable and accessible to a broader audience. As Bill put it, he is my translator. In the process, Bill also became a close friend and someone who, at times, saw the worst in me, but I hope also the best.

I cannot imagine having written this book without the help of Bob Kaplan. Bob was my head of research from 2015 through 2021, and he played an essential role in not only gathering and poring through all the studies that went into this book, but also pushing back on ideas and forcing me to be more rigorous in my thinking. If that wasn't enough, Bob came out of retirement in 2022 to take on the Herculean task of organizing the notes. Bob, along with Vin Miller, also did most of the fact-checking, while Rachel Harrus, Sam Lipman, and Kathryn Birkenbach helped with some of the research.

One thing that really surprised me about this process was how generous people were with their time and expertise. I sent many sections of the manuscript to experts for feedback. Without a single exception, every person that I asked said yes. My gratitude to the following people cannot be overstated: Kellyann Niotis and Richard Isaacson (neurodegenerative diseases), Matt Walker and Vik Jain (sleep), Lew Cantley and Keith Flaherty (cancer), Layne Norton, David Allison, and Kevin Bass (nutrition), Steve Austad (caloric restriction), Nir Barzilai (centenarians), Matt Kaeberlein and David Sabatini (rapamycin, mTOR), Tom Dayspring (atherosclerosis), and Beth Lewis, who was immensely helpful as I tried (and tried, and tried) to write about stability in a way that made sense.

So much of what I've written about in this book is rooted in my interactions with my patients and with my podcast guests. My patients' experiences comprise the substrate for this book, the raw material, and they remind me constantly of the need to be continually learning. This is why my podcast, *The Drive,* exists: It's a forcing function that requires me and my staff to learn at a breakneck pace. The knowledge I gain each week through interviewing experts has also informed much of what you have just read.

As indebted as I feel to the brilliant scientists and physicians who have mentored me throughout my career, I feel an equal if not greater debt to Paul Conti for forcing me to go to the Bridge, and to the therapists who saved my life: Esther Perel, Terry Real, Lorie Teagno, Katy Powell, Andy White, Jeff English, and entire team at PCS.

Several friends also read early sections of this book and provided great feedback: Rosie Kurmaniak, Deb and Hugh Jackman, David Buttaro, Jason Fried, and Judith Barker.

You might not know this about me (although maybe you do by now), but I'm kind of a particular guy, so getting the cover just "right" was no easy task. Thankfully, Rodrigo Corral and his team were able to come in and design a cover that Bill and I felt really represented the work inside. They remained incredibly patient with my micromanaging of every detail of this process without so much as a chirp.

One of the hardest things about writing this book was simply finding time to work on it. The clinical team at Early Medical worked overtime to enable me to spend large blocks of time uninterrupted. Lacey Stenson manages almost every facet of my personal and professional life and executed some very big lifts to make this book happen. Without Lacey, none of the trains run on time. Nick Stenson not only manages every aspect of our digital and podcast content, but he also oversaw the entire launch strategy and execution for this book, which turned out to be much more involved than he or I ever expected.

Lastly, and most importantly, I want to thank Jill. She lived through the highs and lows and never—not one single moment—stopped supporting me, even when any reasonable person would have been justified in kicking me to the curb. You never let go of the rope. Olivia, Reese, and Ayrton saw too much of their daddy in front of a computer screen on nights and weekends and repeatedly asked that I work less. Now that this book is done, I can finally give them more of what they rightly deserve.

Bill Gifford

I would like to thank Martha McGraw for her kindness, coaching, and support throughout this long and sometimes arduous project. I wouldn't have made it without you. Thanks also to Bob Kaplan for the massive research downloads and helping me make sense of many complex topics. And to my friend Stephen Dark for all the walks.

NOTES

Introduction

Page 2 **The reality was:** Yamamoto et al. (2015).

Chapter 1. The Long Game

Page 10 **In 1900, life expectancy:** Kinsella (1992).

Page 12 **we have cut mortality rates by two-thirds in the industrialized world:** Mensah et al. (2017).

Page 12 **Death rates from cancer:** Siegel et al. (2021).

Chapter 2. Medicine 3.0

Page 23 **"First, do no harm":** Sokol (2013).

Page 28 **as Steven Johnson points out:** S. Johnson (2021).

Page 28 **The Northwestern University economist:** Gordon (2016).

Page 32 **The study reported a 24 percent relative increase:** Manson et al. (2013).

Page 34 **The *Titanic*'s wireless operator:** *New York Times* (1985).

Chapter 3. Objective, Strategy, Tactics

Page 43 **"hallmarks of aging":** López-Otín et al. (2013).

Page 54 **low LDL cholesterol does not cause cancer:** Benn et al. (2011).
Page 54 **higher LDL cholesterol *is* causally linked:** Ference (2015).

Chapter 4. Centenarians

Page 59 **"cigarettes, whiskey, and wild, wild women":** Taylor (2009).
Page 59 **"I've only ever had one wrinkle":** Spencer (2005).
Page 60 **Mildred Bowers:** Picard (2018).
Page 61 **centenarians are no more health-conscious:** Rajpathak et al. (2011).
Page 61 **according to the Census Bureau:** United States Census Bureau (2022).
Page 62 **Studies of Scandinavian twins:** Hjelmborg et al. (2006).
Page 62 **Being the sister of a centenarian:** Sebastiani, Nussbaum, et al. (2016).
Page 63 **The overall mortality rate for Americans:** Xu (2016).
Page 63 **According to research by Thomas Perls:** Evert et al. (2003).
Page 64 **"a double-edged sword":** Perls (2017).
Page 65 **As Perls and his colleagues put it:** Hitt et al. (1999).
Page 68 **having one or two copies:** Michaelson (2014).
Page 68 **a large 2019 meta-analysis:** Sebastiani, Gurinovich, et al. (2019).
Page 69 **three SNPs (or variants) in *FOXO3*:** Willcox et al. (2008).
Page 69 **Since then, several other studies:** Revelas et al. (2018).
Page 70 **A genetic analysis of Spanish centenarians:** Serna et al. (2012).
Page 70 **For example, a 2007 study:** Melov et al. (2007).

Chapter 5. Eat Less, Live Longer?

Page 77 **Ultimately, he and others discovered:** E.J. Brown et al. (1994), Sabatini et al. (1994).
Page 77 **it is highly "conserved":** Tatebe and Shiozaki (2017).
Page 77 **The job of mTOR:** G.Y. Liu and Sabatini (2020).
Page 77 **"mTOR is like the general contractor":** Attia (2018a).
Page 78 **"a finger in every major process":** Attia (2018a).
Page 78 **"by postponing death from cancer":** D.E. Harrison, Strong, Sharp, et al. (2009).
Page 78 **Even better, other labs:** Selvarani, Mohammed, and Richardson (2021).
Page 78 **the well-publicized finding:** Baur et al. (2006).
Page 79 **When resveratrol was subjected:** Miller et al. (2011); Strong et al. (2013).
Page 79 **nicotinamide riboside:** D.E. Harrison, Strong, Reifsnyder, et al. (2021).
Page 79 **rapamycin can extend mouse lifespans:** Selvarani, Mohammed, and Richardson (2021).
Page 79 **more modern experiments have demonstrated:** Fontana and Partridge (2015).
Page 81 **Studies dating back to the 1930s:** McDonald and Ramsey (2010).
Page 82 **when it senses low levels of nutrients:** Hardie (2011).
Page 83 **Autophagy is essential to life:** Kourtis and Tavernarakis (2009).

Page 83 **Mice who lack one specific autophagy gene:** Karsli-Uzunbas et al. (2014).
Page 84 **the rapamycin analog everolimus:** Mannick et al. (2014).
Page 85 **Kaeberlein is doing a large clinical trial:** Creevy et al. (2022).
Page 85 **actually seemed to improve cardiac function:** Urfer et al. (2017).
Page 85 **"One thing that's been surprising to me":** Attia (2018b).
Page 87 **One large 2014 analysis:** Bannister et al. (2014).

Chapter 6. One Disease to Rule Them All

Page 90 **His curiosity piqued, Zelman recruited:** Zelman (1952).
Page 91 **nonalcoholic steatohepatitis, or NASH:** Ludwig et al. (1980).
Page 91 **More than one in four people:** S.A. Harrison et al. (2021).
Page 91 **the average American adult male weighed:** Fryar et al. (2018); Ogden et al. (2004).
Page 92 **With regard to ALT liver values:** Kwo, Cohen, and Lim (2017).
Page 92 **Even that may not be low enough:** Prati et al. (2002).
Page 93 **by 2025, NASH and cirrhosis:** Fayek et al. (2016).
Page 93 **According to the Centers for Disease Control (CDC):** CDC (2022b).
Page 94 **If you meet three:** Hirode and Wong (2020).
Page 94 **About 90 percent of the US population:** Araújo, Cai, and Stevens (2019).
Page 95 **A large meta-analysis of studies:** Stefan, Schick, and Häring (2017).
Page 97 **when he surgically implanted fat tissue:** Gavrilova et al. (2000).
Page 98 **NAFLD is just one of many undesirable consequences:** Tchernof and Després (2013).
Page 98 **people of Asian descent:** Anand et al. (2011); Sniderman, Bhopal, et al. (2007).
Page 99 **research by Mitch Lazar:** Ahima and Lazar (2013).
Page 100 **This is where insulin resistance likely begins:** M.C. Petersen and Shulman (2018).
Page 101 **But insulin seems to be the most potent:** Frayn (2019).
Page 102 **the Greek physician Aretaeus of Cappadocia:** Tuchman (2009).
Page 102 **The composer Johann Sebastian Bach:** Diamond (2003).
Page 102 **the famed diabetologist Elliott Joslin:** Joslin (1940).
Page 102 **By 1970, around the time I was born:** NIDDK (2018).
Page 102 **according to a 2022 CDC report:** CDC (2022e).
Page 103 **In 2020:** CDC (2020).
Page 103 **deaths were attributed to type 2 diabetes:** CDC (2020).
Page 104 **we humans have a unique capacity:** R.J. Johnson, Stenvinkel, et al. (2020).
Page 104 **"We realized fructose was having effects":** Attia (2020c).
Page 104 **Johnson and his team began investigating:** R.J. Johnson and Andrews (2015).
Page 106 **does *not* put the brakes on this ATP "spending":** R.J. Johnson, Sánchez-Lozada, et al. (2017).
Page 109 **insulin resistance itself is associated with huge increases:** Igwe et al. (2015); Matsuzaki et al. (2010); Zethelius and Cederholm (2015).

Chapter 7. The Ticker

Page 112 **Globally, heart disease and stroke:** Heron (2021); WHO (2019).
Page 112 **estimated 2,300 people:** CDC (2022c).
Page 113 **It's not just men who are at risk:** ACS (2022a); Heron (2021).
Page 115 **Fewer Sardinian men:** Caselli and Lipsi (2006).
Page 115 **Our vascular network is equally miraculous:** Bautch and Caron (2015).
Page 118 **The humble egg:** McNamara (2015).
Page 118 **Eating lots of saturated fat:** Mensink and Katan (1992).
Page 118 **most of the actual cholesterol that we consume:** Lammert and Wang (2005).
Page 118 **"There's no connection whatsoever":** Jaret (1997).
Page 118 **"cholesterol is not a nutrient of concern":** Dietary Guidelines Advisory Committee (2015).
Page 119 **Fully *half* of all major adverse cardiovascular events:** Sniderman, Thanassoulis, et al. (2016).
Page 119 ***Atlas of Atherosclerosis Progression and Regression:*** Stary (2003).
Page 122 **Autopsy data from young people:** Lawson (2016).
Page 124 **Approximately 15 percent of people:** Nasir et al. (2022); Uretsky et al. (2011).
Page 126 **in *JAMA Cardiology* in 2021:** Marston et al. (2022).
Page 129 **20 to 30 percent of the US population:** Tsimikas et al. (2018).
Page 129 **A class of drug called PCSK9:** O'Donoghue et al. (2019).
Page 131 **"Atherosclerosis *probably would not occur*":** Libby (2021).
Page 131 **there have been only thirteen reported cases:** Orphanet (2022).
Page 131 **Instead, we have over eighteen million cases:** Ritchie and Roser (2018).
Page 132 **scores of studies showing no ill effects:** Dietschy, Turley, and Spady (1993); Ference et al. (2019); Forrester (2010); Jakubowski et al. (2021); Karagiannis et al. (2021); R. Le, Zhao, and Hegele (2022); Libby and Tokgözoğlu (2022); Masana et al. (2018); O'Keefe et al. (2004); Soran, Ho, and Durrington (2018); N. Wang et al. (2020).
Page 132 **Does low HDL-C *causally* increase the risk:** Haase et al. (2012).
Page 132 **Does raising HDL-C *causally* lower the risk:** Voight et al. (2012).
Page 134 **deal-breaking side effects:** du Souich, Roederer, and Dufour (2017); Stroes et al. (2015).
Page 134 **statins are associated with a small increase:** Mach et al. (2018); C.B. Newman et al. (2019).
Page 134 **an asymptomatic rise in liver enzymes:** Jose (2016).
Page 136 **a 2018 *JAMA Cardiology* paper:** Thanassoulis, Sniderman, and Pencina (2018).

Chapter 8. The Runaway Cell

Page 141 **"This man had had a virulent and untreatable cancer":** Rosenberg and Barr (1992).
Page 143 **ten-year survival rates nearly doubled:** NCI (2022b).
Page 143 **in 2017 there were more cancer deaths:** NCI (2021).

Page 143 **This year, if recent trends continue:** NCI (2021).
Page 145 **a gene called *PTEN*:** Jamaspishvili et al. (2018).
Page 146 **"These are the starting blocks":** Pollack (2005).
Page 147 **only about 5 to 8 percent of US cancer research funding:** Sleeman and Steeg (2010).
Page 148 **the late author Christopher Hitchens:** Hitchens (2014).
Page 149 **two key hallmarks of cancer:** Hanahan and Weinberg (2011).
Page 149 **a German physiologist named Otto Warburg:** Warburg (1924, 1956).
Page 150 **the Warburg effect:** Liberti and Locasale (2016).
Page 150 **By the time he died in 1970:** Christofferson (2017).
Page 150 **Watson recounted:** J.D. Watson (2009).
Page 151 **Cantley, Matthew Vander Heiden, and Craig Thompson argued:** Vander Heiden, Cantley, and Thompson (2009).
Page 152 **Globally, about 12 to 13 percent:** Avgerinos et al. (2019).
Page 152 **Type 2 diabetes also:** Lega et al. (2019).
Page 152 **a family of enzymes called PI3-kinases:** Bradley (2004); Fruman et al. (2017).
Page 154 **one study of caloric restriction in humans:** Mercken et al. (2013).
Page 155 ***The Emperor of All Maladies*:** Mukherjee (2011).
Page 156 **Published in *Nature* in 2018:** Hopkins et al. (2018).
Page 157 **A randomized trial in 131 cancer patients:** de Groot et al. (2020).
Page 157 **using ice cream "as a topping on cake":** ACS (2022c).
Page 159 **In 2010, Rosenberg and his team:** Kochenderfer et al. (2010).
Page 162 ***The New England Journal of Medicine* had recently reported:** D.T. Le et al. (2015).
Page 163 **an analysis by oncologists:** Gay and Prasad (2017).
Page 163 **One recent paper:** Cavazzoni et al. (2020).
Page 163 **immune system can recognize:** Attia (2021b); Rosenberg (2021).
Page 164 **Between 80 and 90 percent:** Atkins et al. (2000).
Page 166 **and 67 percent of:** Taieb et al. (2020).
Page 167 **93 percent chance:** Waks et al. (2019).
Page 169 **About 70 percent of people:** Hofseth et al. (2020).
Page 170 **In 2020, some 3,640 Americans died:** ACS (2022b).
Page 174 **Galleri has been validated:** X. Chen et al. (2021).

Chapter 9. Chasing Memory

Page 178 **the *e2* version of *APOE*:** Reiman, Arboleda-Velasquez, et al. (2020).
Page 180 **a certain variant of the gene *Klotho*:** Belloy et al. (2020).
Page 182 **Luckily, these mutations are very rare:** Cacace, Sleegers, and Van Broeckhoven (2016); Cruchaga et al. (2012); Cuyvers and Sleegers (2016).
Page 183 **Every single one of them failed:** Cummings et al. (2022).
Page 184 **"Amyloid and tau define the disease":** Kolata (2020).
Page 184 **"plaque formation and other changes":** Blessed, Tomlinson, and Roth (1968).
Page 184 **Researchers from the Memory and Aging Center:** Rabinovici et al. (2019).

Page 185 **A 2013 analysis of preserved tissue:** Müller, Winter, and Graeber (2013).
Page 185 **Lewy body dementia as well as:** Kaivola et al. (2022).
Page 186 **"She was expecting Oliver Sacks":** Attia (2018c).
Page 187 **"Currently, firm conclusions cannot be drawn":** Daviglus et al. (2010).
Page 187 **A two-year randomized controlled trial:** Ngandu et al. (2015).
Page 188 **Two other large European:** Rosenberg et al. (2020); Andrieu et al. (2017); van Charante et al. (2016).
Page 189 **a rapid drop in estradiol in women:** Mosconi et al. (2018); Rahman et al. (2020); Ratnakumar et al. (2019); Zhou et al. (2020).
Page 189 **new research suggests that women:** Yan et al. (2022).
Page 189 **Yet Parkinson's also appears:** Cerri et al. (2019).
Page 189 **dementia has an extremely long prologue:** Langa and Levine (2014).
Page 190 **over forty-six million people:** Brookmeyer et al. (2018).
Page 192 **according to Francisco Gonzalez-Lima:** Attia (2019).
Page 193 **also known as "healthy user bias":** Yasuno et al. (2020).
Page 194 **In their seminal 1968 paper:** Blessed, Tomlinson, and Roth (1968).
Page 194 **The brain is a greedy organ:** Raichle and Gusnard (2002).
Page 195 **His "barf bag theory":** de la Torre (2016).
Page 195 **"We believed, and still do":** de la Torre (2018).
Page 196 **vascular dementia is currently considered distinct:** Wolters and Ikram (2019).
Page 196 **Having type 2 diabetes doubles or triples your risk:** Cholerton et al. (2016).
Page 196 **insulin resistance alone is enough:** Neth and Craft (2017).
Page 197 **spraying insulin right into subjects' noses:** Freiherr et al. (2013).
Page 197 **One study found that intranasal insulin:** Chapman et al. (2018).
Page 197 **The signal event here:** Kerrouche et al. (2006).
Page 197 **Brain imaging studies reveal:** Reiman, Caselli, et al. (1996); Small et al. (2000); Sperling et al. (2011).
Page 197 **Intriguingly, this reduction appears:** Kerrouche et al. (2006).
Page 197 **A woman with one copy:** Neu et al. (2017).
Page 198 **There is also some evidence:** Montagne et al. (2020).
Page 198 **It was the original human allele:** Trumble and Finch (2019).
Page 198 **children carrying *APOE e4:*** Mitter et al. (2012); Oriá et al. (2007).
Page 198 **higher levels of neuroinflammation in *e4* carriers:** Kloske and Wilcock (2020).
Page 200 **DHA, found in fish oil:** Yassine et al. (2017).
Page 200 **ketogenic therapies improved general cognition:** Grammatikopoulou et al. (2020).
Page 201 **heavier drinking is itself a risk factor:** Slayday et al. (2021).
Page 201 **stress and anxiety-related risk:** Maeng and Milad (2015).
Page 201 **grip strength, an excellent proxy:** Esteban-Cornejo et al. (2022).
Page 202 **Sleep disruptions and poor sleep are potential drivers:** C. Wang and Holtzman (2020).
Page 203 **Studies have found that hearing loss:** Zheng et al. (2017).

Page 203 ***P. gingivalis* has also shown up:** Dominy et al. (2019).
Page 204 **reduce the risk of Alzheimer's by about 65 percent:** Laukkanen et al. (2017).
Page 204 **and the risk of ASCVD by 50 percent:** Laukkanen et al. (2015).
Page 204 **lowering homocysteine with B vitamins:** A. Smith et al. (2010).
Page 204 **while optimizing omega-3 fatty acids:** Oulhaj et al. (2016).
Page 204 **Higher vitamin D levels:** Maddock et al. (2015).

Chapter 10. Thinking Tactically

Page 209 **"Cancer, like insanity":** Proctor (1995).
Page 212 **one person every twelve minutes:** NHTSA (2022a).
Page 212 **a very high proportion of fatalities occur at intersections:** NHTSA (2022b); Attia (2020b).

Chapter 11. Exercise

Page 218 **77 percent of the US population is like you:** Blackwell and Clarke (2018).
Page 218 **Going from zero weekly exercise:** Wen et al. (2011).
Page 218 **regular exercisers live as much as a *decade* longer:** Reimers, Knapp, and Reimers (2012).
Page 218 **habitual runners and cyclists:** Booth and Zwetsloot (2010).
Page 218 **The benefits of exercise begin:** I.-M. Lee and Buchner (2008).
Page 219 **The US government's physical activity guidelines:** HHS (2018).
Page 220 **the single most powerful marker:** Mandsager et al. (2018).
Page 221 **A 2018 study in *JAMA*:** Mandsager et al. (2018).
Page 221 **greater relative risk of death than smoking:** Mandsager et al. (2018).
Page 221 **Someone in the bottom quartile of VO_2 max:** Mandsager et al. (2018).
Page 221 **a much larger and more recent study:** Kokkinos et al. (2022).
Page 223 **"Cardiorespiratory fitness is inversely associated":** Mandsager et al. (2018).
Page 223 **A ten-year observational study:** Li et al. (2018).
Page 224 **at least one study suggests:** Artero et al. (2011).
Page 224 **side-by-side comparison of exercise studies versus drug studies:** Naci and Ioannidis (2015).
Page 225 **Endurance exercise such as running or cycling:** Seifert et al. (2010).
Page 225 **Exercise helps keep the brain vasculature healthy:** Barnes and Corkery (2018).
Page 225 **Longitudinal and cross-sectional studies:** Westerterp et al. (2021).
Page 226 **One Chilean study:** Bunout et al. (2011).
Page 226 **Having more muscle mass on your exoskeleton:** Jones et al. (2017).
Page 226 **correlated with a lower risk of falling:** Van Ancum et al. (2018).
Page 227 **Eight hundred thousand older people are hospitalized:** CDC (2021).
Page 227 **muscle atrophy and sarcopenia:** H.-S. Lin et al. (2016).
Page 227 **A recent study of older British adults:** Veronese et al. (2022).
Page 228 **subjects who are obese:** Nicklas et al. (2015).

Page 228 **or recovering from cancer treatment:** K.L. Campbell et al. (2019).
Page 228 **even those who are already elderly and frail:** Zhang et al. (2020).
Page 232 **your muscle strength will decline:** Danneskiold-Samsøe et al. (2009); Hughes et al. (2001); Lindle et al. (1997).

Chapter 12. Training 101

Page 237 **Zone 2 is one of five levels of intensity:** Allen and Coggan (2010).
Page 238 **a fascinating study:** San-Millán and Brooks (2018).
Page 242 **a process called mitochondrial biogenesis:** Lemasters (2005).
Page 242 **why zone 2 is such a powerful mediator:** Kawada and Ishii (2005).
Page 242 **glucose uptake increases as much as one-hundred-fold:** Richter (2021).
Page 243 **exercise also activates other pathways:** McMillin et al. (2017).
Page 244 **a side benefit of zone 2:** Seifert et al. (2010).
Page 245 **this measure of peak aerobic capacity:** Mandsager et al. (2018).
Page 247 **Studies suggest that your VO_2 max will decline:** C.-H. Kim et al. (2016).
Page 248 **it begins to threaten your ability:** Shephard (2009).
Page 248 **well-trained octogenarian endurance athletes:** Trappe et al. (2013).
Page 248 **increasing your VO_2 max:** Shephard (2009).
Page 248 **One study found that:** Shephard et al. (2009).
Page 249 **fitness of someone decades younger than you:** Booth and Zwetsloot (2010); Mandsager et al. (2018).
Page 249 **Robert Marchand:** Billat et al. (2017).
Page 252 **An eighty-year-old man:** Lexell (1995).
Page 253 **A study of twelve healthy volunteers:** Kortebein et al. (2007).
Page 253 **this muscle loss is called sarcopenia:** T.N. Kim and Choi (2013).
Page 253 **a broader clinical condition called frailty:** Xue (2011).
Page 253 **One study looked at sixty-two frail seniors:** Tieland, Dirks, et al. (2012).
Page 257 **"Carrying shaped our species":** Easter (2021).
Page 258 **Many studies suggest that grip strength:** Bohannon (2019); Hamer and O'Donovan (2017); Y. Kim et al. (2018); A.B. Newman et al. (2006).
Page 258 **The consensus definition of sarcopenia:** Cruz-Jentoft et al. (2019).
Page 259 **In 1985, men ages twenty to twenty-four:** Fain and Weatherford (2016).

Chapter 13. The Gospel of Stability

Page 263 **Epidemiological studies tell us:** Lieberman et al. (2021).
Page 264 **According to the CDC:** Dahlhamer (2018).
Page 264 **Back pain, in particular:** Shmagel et al. (2018).
Page 264 **It is a leading cause of disability:** Gaskin and Richard (2012).
Page 264 **Studies of college-age athletes:** Boneti Moreira et al. (2014).
Page 270 **The theory behind DNS:** Frank, Kobesova, and Kolar (2013).
Page 271 **"DNS beautifully integrates":** Attia (2021a).
Page 280 **Ten seconds is a respectable time:** Araujo et al. (2022).
Page 280 **"tech neck":** Tanweer (2021).

Chapter 14. Nutrition 3.0

Page 292 **Richard Feynman being asked:** Dye (1988).
Page 292 **eating an ounce of tree nuts:** Naghshi et al. (2020).
Page 297 **eating twelve hazelnuts every day:** Bao et al. (2013).
Page 297 **drinking diet sodas and abdominal fat:** Azad et al. (2017).
Page 298 **Austin Bradford Hill:** Hill (1965).
Page 299 **a very well-publicized 2017 study:** Schwingshackl, Schwedhelm, et al. (2018).
Page 299 **someone who smokes cigarettes:** Pesch et al. (2012); Proctor (2001); Sasco, Secretan, and Straif (2004); Youlden, Cramb, and Baade (2008).
Page 300 **food is so complex:** Ioannidis (2018); Moco et al. (2006); Ninonuevo et al. (2006); Wishart et al. (2007).
Page 300 **"Nutritional epidemiology is a scandal":** Crowe (2018).
Page 300 **David Allison ran an elegant experiment:** Ejima et al. (2016).
Page 302 **tainted by healthy user bias:** Naimi et al. (2017).
Page 302 **a recent study in *JAMA*:** Biddinger et al. (2022).
Page 303 **Now imagine trying to ensure:** WHI (n.d.).
Page 303 **In the end, despite all this effort:** Howard et al. (2006).
Page 305 **By the numbers, the nuts-or-olive-oil "Mediterranean" diet:** Estruch et al. (2013).
Page 305 **A post hoc analysis of PREDIMED data:** Martínez-Lapiscina et al. (2013).

Chapter 15. Putting Nutritional Biochemistry into Practice

Page 312 **a study published in *Science*:** Colman et al. (2009).
Page 312 **"Dieting Monkeys Offer Hope for Living Longer":** Wade (2009).
Page 313 **in August 2012, another monkey study:** Mattison et al. (2012).
Page 313 **"Severe Diet Doesn't Prolong Life":** Kolata (2012).
Page 317 **lots of plants, lots of starch:** Cordain, Miller, et al. (2000).
Page 317 **very metabolically healthy:** Cordain, Eaton, et al. (2002); Pontzer et al. (2018).
Page 317 **One reason carbohydrate restriction is so effective:** Gibson et al. (2015); Nymo et al. (2017); Phinney and Volek (2018); Sumithran et al. (2013).
Page 319 **Being Latino meant:** Oliveira, Cotrim, and Arrese (2019).
Page 321 **chronic drinking has strong associations:** Peng et al. (2020).
Page 321 **mainly via its negative effect on sleep:** C. Wang and Holtzman (2020).
Page 321 **numerous well-publicized studies:** Hines and Rimm (2001); Suzuki et al. (2009).
Page 321 **the Mendelian randomization study in *JAMA*:** Biddinger et al. (2022).
Page 322 **But there is also evidence suggesting:** Hanefeld et al. (1999); Kawano et al. (1999); H.-J. Lin et al. (2009); Standl, Schnell, and Ceriello (2011); Watanabe et al. (2011).
Page 325 **A 2011 study looking at twenty thousand people:** Pfister et al. (2011).
Page 325 **Another study in 2019:** Echouffo-Tcheugui et al. (2019).
Page 329 **Foods high in protein *and fat*:** Franz (1997).

Page 331 **More than one study has found:** W. Campbell et al. (2001).
Page 331 **elderly people consuming that RDA of protein (0.8 g/kg/day):** Wu (2016).
Page 332 **one gram per *pound* of body weight per day:** Baum, Kim, and Wolfe (2016).
Page 332 **the ideal way to achieve this:** Schoenfeld and Aragon (2018).
Page 332 **older people might require more protein:** Baum, Kim, and Wolfe (2016).
Page 334 **the Healthy Aging and Body Composition Study:** Houston et al. (2008).
Page 334 **boosting protein intake even moderately:** Rozentryt et al. (2010).
Page 334 **Adding thirty grams of milk protein:** Tieland, van de Rest, et al. (2012).
Page 334 **giving elderly people supplements:** Børsheim et al. (2009).
Page 334 **Another study in men with type 2 diabetes:** Nuttall and Gannon (2006).
Page 335 **dietary fat also tends to leave one feeling more satiated:** Boden et al. (2005); Holt et al. (1995); Samra (2010).
Page 337 **Our per capita consumption of soybean oil:** Blasbalg et al. (2011).
Page 338 **the most comprehensive review of this topic:** Abdelhamid et al. (2018).
Page 338 **A more recent publication:** Hooper et al. (2020).
Page 339 **yet another recent review:** Schwingshackl, Zähringer, et al. (2021).
Page 341 **Fasting over long periods:** Vendelbo et al. (2014).
Page 341 **lack of nutrients accelerates autophagy:** Bagherniya et al. (2018).
Page 341 **and it activates *FOXO*:** Gross, van den Heuvel, and Birnbaum (2008).
Page 342 **The time-restricted mice gained less weight:** Hatori et al. (2012).
Page 342 **a sixteen-hour fast for a mouse:** Jensen et al. (2013).
Page 342 **A 2020 clinical trial by Ethan Weiss:** Lowe et al. (2020).
Page 342 **Two similar studies also found minimal benefit:** Jamshed et al. (2019); D. Liu et al. (2022).
Page 343 **subjects can indeed lose weight on alternate-day fasting diets:** Varady and Gabel (2019).
Page 343 **One small but revealing study:** Templeman et al. (2021).

Chapter 16. The Awakening

Page 351 **We now know that even one sleepless night:** Dawson and Reid (1997); Lamond and Dawson (1999).
Page 351 **sleep-deprived medical personnel:** Mansukhani et al. (2012); Tang et al. (2019).
Page 351 **Poor sleep dramatically increases one's propensity:** Iftikhar et al. (2015).
Page 351 **up to and including type 2 diabetes:** Shan et al. (2015).
Page 351 **wreak havoc with the body's hormonal balance:** Leproult and Van Cauter (2010); Reutrakul and Van Cauter (2018); de Zambotti, Colrain, and Baker (2015).
Page 351 **Good sleep, in terms of not only quantity:** Goldstein and Walker (2014); Killgore (2013); Krause et al. (2017); Kuna et al. (2012); Motomura et al. (2013); Prather, Bogdan, and Hariri (2013); Rupp, Wesensten, and Balkin (2012); Van Dongen, Maislin, et al. (2003); Van Dongen, Baynard, et al. (2004); Yoo et al. (2007).

Page 352 **It even cleans itself:** Reddy and van der Werf (2020).
Page 352 **sleeping well is essential to preserving our cognition:** C. Wang and Holtzman (2020).
Page 352 **"[The] decimation of sleep":** Walker (2017).
Page 353 **Every animal engages in some form of sleep:** Cirelli and Tononi (2008).
Page 354 **studies conducted in dark caves:** Zuccarelli et al. (2019).
Page 354 **Even a single night of bad sleep:** Cullen et al. (2019); Fullagar et al. (2015).
Page 354 **Even our ability to perspire is impaired:** Dewasmes et al. (1993); Kolka and Stephenson (1988); Sawka, Gonzalez, and Pandolf (1984).
Page 354 **A 2014 observational study:** Milewski et al. (2014).
Page 354 **In one study, Stanford basketball players:** Mah et al. (2011).
Page 355 **LeBron James makes sleep a key part:** Ferriss (2018).
Page 355 **professional drivers displayed far worse reaction time:** Jackson et al. (2013).
Page 355 **But a survey conducted by AAA:** AAA Foundation (2016).
Page 355 **people who are sleep deprived:** Hafner et al. (2017); Killgore (2013); Krause et al. (2017); J. Lim and Dinges (2008); Van Dongen, Maislin, et al. (2003).
Page 356 **people who sleep less chronically:** Oyetakin-White et al. (2015).
Page 356 **Sleep researcher Eve van Cauter:** Broussard, Ehrmann, et al. (2012).
Page 357 **No fewer than nine different studies:** Broussard, Ehrmann, et al. (2012); Broussard, Chapotot, et al. (2015); Buxton et al. (2010); Leproult, Holmbäck, and Van Cauter (2014); Nedeltcheva et al. (2009); Rao et al. (2015); Spiegel, Leproult, and Van Cauter (1999); Stamatakis and Punjabi (2010); Tasali et al. (2008).
Page 357 **Multiple large meta-analyses:** Iftikhar et al. (2015); Itani et al. (2017); Shan et al. (2015).
Page 357 **Similar risk associations have been found:** Itani et al. (2017).
Page 357 **it also causes glucose to be released:** Kuo et al. (2015).
Page 358 **Studies by Eve van Cauter's group:** Spiegel, Tasali, et al. (2004); Spiegel, Leproult, L'hermite-Balériaux, et al. (2004).
Page 358 **Follow-up studies by van Cauter's group:** Bosy-Westphal et al. (2008); Brondel et al. (2010); Broussard, Kilkus, et al. (2016); Calvin et al. (2013); Spaeth, Dinges, and Goel (2015).
Page 359 **Two large meta-analyses:** Itani et al. (2017); Yin et al. (2017).
Page 359 **But one particularly interesting study:** Dashti et al. (2019).
Page 359 **The MR data confirmed the observational findings:** Daghlas et al. (2019).
Page 360 **chronic bad sleep as a powerful potential *cause*:** C. Wang and Holtzman (2020).
Page 360 **we sleep in a series of well-defined stages:** Lendner et al. (2020).
Page 361 **the brain clears out its cache:** Diekelmann and Born (2010); Wilson and McNaughton (1994).
Page 361 **Researchers have observed:** Walker (2009).
Page 361 **When we are young:** A.K. Patel, Reddy, and Araujo (2022).

Page 361 **REM sleep is especially helpful:** C. Smith and Lapp (1991); Stickgold et al. (2000).

Page 361 **Another very important function:** van der Helm and Walker (2009); Hutchison and Rathore (2015).

Page 362 **studies of combat veterans:** Repantis et al. (2020).

Page 362 **Perhaps most intriguing:** Goldstein-Piekarski et al. (2015).

Page 362 **enough to enter REM:** Rasking et al. (2007).

Page 362 **as social animals:** Yamazaki et al. (2020).

Page 362 **A few years ago, researchers in Rochester:** Iliff et al. (2013).

Page 362 **amyloid-beta and tau, the two proteins:** Lucey, McCullough, et al. (2019).

Page 362 **Broader studies have found:** Branger et al. (2016); B. Brown et al. (2016); Ju et al. (2013); Spira et al. (2013); Sprecher et al. (2015).

Page 363 **This can become a vicious cycle:** C. Wang and Holtzman (2020).

Page 363 **Also, up to half of people:** Emamian et al. (2016).

Page 363 **Insomnia affects 30 to 50 percent:** Benito-León et al. (2009); Jack et al. (2013); A.S.P. Lim, Kowgier, et al. (2013); A.S.P. Lim, Yu, et al. (2013); Lobo et al. (2008); Osorio et al. (2011).

Page 363 **One study linked poor sleep quality:** Potvin et al. (2012).

Page 363 **Meanwhile, superior sleep quality:** A.S.P. Lim, Kowgier, et al. (2013); A.S.P. Lim, Yu, et al. (2013).

Page 363 **Successfully treating sleep disturbance:** Ancoli-Israel et al. (2008); Moraes et al. (2006).

Page 364 **More research points:** Winer et al. (2019).

Page 364 **The first real blockbuster sleeping medication:** Saul (2006).

Page 365 **Currently, the US sleep medication market:** Business Wire (2021).

Page 365 **One study found that Ambien:** Arbon, Knurowska, and Dijk (2015).

Page 365 **a new class of sleep drugs:** Herring et al. (2016).

Page 365 **Quviviq (daridorexant):** Ziemichód et al. (2022).

Page 365 **Then there are the older benzodiazepine drugs:** Picton, Marino, and Nealy (2018).

Page 366 **especially if it also improves sleep architecture:** Zheng et al. (2022).

Page 366 **Keep in mind, however:** Shahid et al. (2011).

Page 368 **Studies have found that some individuals:** Kalmbach et al. (2017).

Page 370 **Not only does non-natural lighting:** Hardeland (2013).

Page 370 **One large-scale survey:** Gradisar et al. (2013).

Page 371 **according to research by Michael Gradisar:** Gradisar et al. (2013).

Page 371 **One of the signal events:** Harding, Franks, and Wisden (2020).

Page 371 **It's counterintuitive:** Ebrahim et al. (2013).

Page 372 **More alarming is the finding:** C. Smith and Smith (2003).

Page 372 **Most people think of caffeine as a stimulant:** Urry and Landolt (2015).

Page 372 **But the half-life of caffeine:** IOM (2001).

Page 373 **This is why one of the primary techniques:** Maurer et al. (2021).

Page 373 **Another way to help cultivate:** Dworak et al. (2007); Youngstedt et al. (2000).
Page 373 **Another way to turn down:** D. Kim et al. (2022).

Chapter 17. Work in Progress

Page 381 **Another very direct way:** CDC (2022f).
Page 381 **a man named Ken Baldwin:** Friend (2003).
Page 381 **It's not a surprise:** Spillane et al. (2020).
Page 381 **the CDC estimates:** Strobe (2021).
Page 381 **These "accidental" overdoses:** CDC (2022a).
Page 382 **They were slow-motion suicides:** Case et al. (2015).
Page 382 **This category of death:** CDC (2022d).
Page 382 **Middle-aged white men and women, in particular:** Case and Deaton (2015).
Page 382 **Surveys show that older Americans:** Livingston (2019).
Page 390 **Terrence Real's book:** Real (1998).
Page 392 **as David Foster Wallace explained:** Wallace (2009).
Page 394 **"résumé virtues":** Brooks (2016).
Page 399 **"When nothing seems to help":** Riis (1901).
Page 401 **Clinical trials have found it to be effective:** Asarnow et al. (2021); Linehan et al. (2006).
Page 406 **exposing oneself to the fractal geometric patterns:** Hagerhall (2008).

Disclosures

For an up-to-date list of all my disclosures, please see https://peterattiamd.com/about/ under the heading "Disclosures."

REFERENCES

AAA Foundation. (2016). 2015 Traffic Safety Culture Index. https://aaafoundation.org/2015-traffic-safety-culture-index/.

Abbasi, F., Chu, J.W., Lamendola, C., McLaughlin, T., Hayden, J., Reaven, G.M., and Reaven, P.D. (2004). Discrimination between obesity and insulin resistance in the relationship with adiponectin. *Diabetes 53*, 585–590. https://doi.org/10.2337/diabetes.53.3.585.

Abdelhamid, A.S., Martin, N., Bridges, C., Brainard, J.S., Wang, X., Brown, T.J., Hanson, S., Jimoh O.F., Ajabnoor S.M., Deane K.H.O., et al. (2018). Polyunsaturated fatty acids for the primary and secondary prevention of cardiovascular disease. *Cochrane Database Syst. Rev. 11*, CD012345. https://doi.org/10.1002/14651858.CD012345.pub3.

ACS (American Cancer Society). (2022a). Breast Cancer Statistics | How common is breast cancer? Last revised January 12. https://www.cancer.org/cancer/breast-cancer/about/how-common-is-breast-cancer.html.

———. (2022b). Colorectal cancer facts and figures, 2022–2022. https://www.cancer.org/content/dam/cancer-org/research/cancer-facts-and-statistics/colorectal-cancer-facts-and-figures/colorectal-cancer-facts-and-figures-2020-2022.pdf.

———. (2022c). Eating well during treatment. March 16. https://www.cancer.org/treatment/survivorship-during-and-after-treatment/coping/nutrition/once-treatment-starts.html.

ACSM (2017). *ACSM's guidelines for exercise testing and prescription.* Philadelphia: Lippincott Williams and Wilkins.

Ahima, R.S., and Lazar, M.A. (2013). The health risk of obesity—Better metrics imperative. *Science 341*, 856–858. https://doi.org/10.1126/science.1241244.

Alghamdi, B.S. (2018). The neuroprotective role of melatonin in neurological disorders. *J. Neurosci. Res. 96*, 1136–1149. https://doi.org/10.1002/jnr.24220.

Allen, H., and Coggan, A. (2010). *Training and racing with a power meter.* Boulder, CO: VeloPress.

Anand, S.S., Tarnopolsky, M.A., Rashid, S., Schulze, K.M., Desai, D., Mente, A., Rao, S., Yusuf, S., Gerstein, H.C., and Sharma, A.M. (2011). Adipocyte hypertrophy, fatty liver and metabolic risk factors in South Asians: The Molecular Study of Health and Risk in Ethnic Groups (mol-SHARE). *PLOS ONE 6*, e22112. https://doi.org/10.1371/journal.pone.0022112.

Ancoli-Israel, S., Palmer, B.W., Cooke, J.R., Corey-Bloom, J., Fiorentino, L., Natarajan, L., Liu, L., Ayalon, L., He, F., and Loredo, J.S. (2008). Cognitive effects of treating obstructive sleep apnea in Alzheimer's disease: A randomized controlled study. *J. Am. Geriatr. Soc. 56*, 2076–2081. https://doi.org/10.1111/j.1532-5415.2008.01934.x.

Andersson, C., Blennow, K., Almkvist, O., Andreasen, N., Engfeldt, P., Johansson, S.-E., Lindau, M., and Eriksdotter-Jönhagen, M. (2008). Increasing CSF phospho-tau levels during cognitive decline and progression to dementia. *Neurobiol. Aging 29*, 1466–1473. https://doi.org/10.1016/j.neurobiolaging.2007.03.027.

Andreasen, N., Hesse, C., Davidsson, P., Minthon, L., Wallin, A., Winblad, B., Vanderstichele, H., Vanmechelen, E., and Blennow, K. (1999). Cerebrospinal fluid beta-amyloid(1-42) in Alzheimer disease: Differences between early- and late-onset Alzheimer disease and stability during the course of disease. *Arch. Neurol. 56*, 673–680. https://doi.org/10.1001/archneur.56.6.673.

Andreasen, N., Vanmechelen, E., Van de Voorde, A., Davidsson, P., Hesse, C., Tarvonen, S., Räihä, I., Sourander, L., Winblad, B., and Blennow, K. (1998). Cerebrospinal fluid tau protein as a biochemical marker for Alzheimer's disease: A community-based follow up study. *J. Neurol. Neurosurg. Psychiatry 64*, 298–305. https://doi.org/10.1136/jnnp.64.3.298.

Andrieu, S., Guyonnet, S., Coley, N., Cantet, C., Bonnefoy, M., Bordes, S. (2017). Effect of long-term omega 3 polyunsaturated fatty acid supplementation with or without multidomain intervention on cognitive function in elderly adults with memory complaints (MAPT): A randomized placebo-controlled trial. *Lancet 16*, 377–389. https://doi.org/10.1016/S1474-4422(17)30040-6.

Araujo, C.G., de Souza e Silva, C.G., Laukkanen, J.A., Singh, M.F., Kunutsor, S.K., Myers, J., Franca, J.F., and Castro, C.L. (2022). Successful 10-second one-legged stance performance predicts survival in middle-aged and older individuals. *Br. J. Sports Med. 56*, 975–980. https://doi.org/10.1136/bjsports-2021-105360.

Araújo, J., Cai, J., and Stevens, J. (2019). Prevalence of optimal metabolic health in American adults: National Health and Nutrition Examination Survey 2009–2016. *Metab. Syndr. Relat. Disord. 17*, 46–52. https://doi.org/10.1089/met.2018.0105.

Arbon, E.L., Knurowska, M., and Dijk, D.-J. (2015). Randomised clinical trial of the effects of prolonged-release melatonin, temazepam and zolpidem on slow-wave activity during sleep in healthy people. *J. Psychopharmacol. 29*, 764–776. https://doi.org/10.1177/0269881115581963.

Artero, E.G., Lee, D.C., Ruiz, J.R. (2011). A prospective study of muscular strength and all-cause mortality in men with hypertension. *J. Am. Coll. Cardiol. 57*(18), 1831–1837. https:// doi:10.1016/j.jacc.2010.12.025.

Asarnow, J.R., Berk, M.S., Bedics, J., Adrian, M., Gallop, R., Cohen, J., Korslund, K., Hughes, J., Avina, C., Linehan, M.M., et al. (2021). Dialectical Behavior Therapy for suicidal self-harming youth: Emotion regulation, mechanisms, and mediators. *J. Am. Acad. Child Adolesc. Psychiatry 60*, 1105–1115.e4. https://doi.org/10.1016/j.jaac.2021.01.016.

Atkins, M.B., Kunkel, L., Sznol, M., and Rosenberg, S.A. (2000). High-dose recombinant interleukin-2 therapy in patients with metastatic melanoma: Long-term survival update. *Cancer J. Sci. Am. 6*, Suppl 1, S11–14.

Attia, P. (2018a). #09—David Sabatini, M.D., Ph.D.: Rapamycin and the discovery of mTOR—The nexus of aging and longevity? *The Drive* (podcast), episode 9, August 13. https://peterattiamd.com/davidsabatini/.

———. (2018b). #10—Matt Kaeberlein, Ph.D.: Rapamycin and dogs—man's best friends? Living longer, healthier lives and turning back the clock on aging and age-related diseases. *The Drive* (podcast), episode 10, August 20. https://peterattiamd.com/matt kaeberlein/.

———. (2018c). #18—Richard Isaacson, M.D.: Alzheimer's prevention. *The Drive* (podcast), episode 18, October 1. https://peterattiamd.com/richardisaacson/.

———. (2019). #38—Francisco Gonzalez-Lima, Ph.D.: Advancing Alzheimer's disease treatment and prevention: Is AD actually a vascular and metabolic disease? *The Drive* (podcast), episode 38, January 28. https://peterattiamd.com/franciscogonzalezlima/.

———. (2020a). Colorectal cancer screening. *peterattiamd.com* (blog), September 27. https:// peterattiamd.com/colorectal-cancer-screening/.

———. (2020b). The killer(s) on the road: Reducing your risk of automotive death. *peterattiamd.com* (blog), February 9. https://peterattiamd.com/the-killers-on-the-road -reducing-your-risk-of-automotive-death/.

———. (2020c). Rick Johnson, M.D.: Metabolic effects of fructose. *The Drive* (podcast), episode 87, January 6. https://peterattiamd.com/rickjohnson/.

———. (2021a). Michael Rintala, D.C.: Principles of Dynamic Neuromuscular Stabilization (DNS). *The Drive* (podcast), episode 152, March 8. https://peterattiamd.com /michaelrintala/.

———. (2021b). Steven Rosenberg, M.D., Ph.D.: The development of cancer immunotherapy and its promise for treating advanced cancers. *The Drive* (podcast), episode 177, September 27.

Avgerinos, K.I., Spyrou, N., Mantzoros, C.S., and Dalamaga, M. (2019). Obesity and cancer risk: Emerging biological mechanisms and perspectives. *Metabolism 91*, 121–135. https:// doi.org/10.1016/j.metabol.2018.11.001.

Azad, M.B., Abou-Setta, A.M., Chauhan, B.F., Rabbani, R., Lys, J., Copstein, L., Mann, A., Jeyaraman, M.M., Reid, A.E., Fiander, M., et al. (2017). Nonnutritive sweeteners and cardiometabolic health: A systematic review and meta-analysis of randomized controlled trials and prospective cohort studies. *CMAJ 189*, E929–E939. https://doi.org/10.1503 /cmaj.161390.

Bagherniya, M., Butler, A.E., Barreto, G.E., and Sahebkar, A. (2018). The effect of fasting or calorie restriction on autophagy induction: A review of the literature. *Ageing Res. Rev. 47*, 183–197. https://doi.org/10.1016/j.arr.2018.08.004.

Bannister, C.A., Holden, S.E., Jenkins-Jones, S., Morgan, C.L., Halcox, J.P., Schernthaner, G., Mukherjee, J., and Currie, C.J. (2014). Can people with type 2 diabetes live longer than those without? A comparison of mortality in people initiated with metformin or sulphonylurea monotherapy and matched, non-diabetic controls. *Diabetes Obes. Metab. 16*, 1165–1173. https://doi.org/10.1111/dom.12354.

Bao, Y., Han, J., Hu, F.B., Giovannucci, E.L., Stampfer, M.J., Willett, W.C., and Fuchs, C.S. (2013). Association of nut consumption with total and cause-specific mortality. *N. Engl. J. Med. 369*, 2001–2011. https://doi.org/10.1056/NEJMoa1307352.

Barnes, J.N., and Corkery, A.T. (2018). Exercise improves vascular function, but does this translate to the brain? *Brain Plast. 4*, 65–79. https://doi.org/10.3233/BPL-180075.

Baum, J.I., Kim, I.-Y., and Wolfe, R.R. (2016). Protein consumption and the elderly: What is the optimal level of intake? *Nutrients 8*, 359. https://doi.org/10.3390/nu8060359.

Baur, J.A., Pearson, K.J., Price, N.L., Jamieson, H.A., Lerin, C., Kalra, A., Prabhu, V.V., Allard, J.S., Lopez-Lluch, G., Lewis, K., et al. (2006). Resveratrol improves health and survival of mice on a high-calorie diet. *Nature 444*, 337–342. https://doi.org/10.1038/nature05354.

Bautch, V.L., and Caron, K.M. (2015). Blood and lymphatic vessel formation. *Cold Spring Harb. Perspect. Biol. 7*, a008268. https://doi.org/10.1101/cshperspect.a008268.

Beckett, L.A., Harvey, D.J., Gamst, A., Donohue, M., Kornak, J., Zhang, H., Kuo, J.H., and Alzheimer's Disease Neuroimaging Initiative (2010). The Alzheimer's Disease Neuroimaging Initiative: Annual change in biomarkers and clinical outcomes. *Alzheimers Dement. 6*, 257–264. https://doi.org/10.1016/j.jalz.2010.03.002.

Belloy, M.E., Napolioni, V., Han, S.S., Le Guen, Y., and Greicius, M.D. (2020). Association of *Klotho*-VS heterozygosity with risk of Alzheimer disease in individuals who carry *APOE4*. *JAMA Neurol. 77*, 849–862. https://doi.org/10.1001/jamaneurol.2020.0414.

Benito-León, J., Bermejo-Pareja, F., Vega, S., and Louis, E.D. (2009). Total daily sleep duration and the risk of dementia: A prospective population-based study. *Eur. J. Neurol. 16*, 990–997. https://doi.org/10.1111/j.1468-1331.2009.02618.x.

Benn, M., Tybjærg-Hansen, A., Stender, S., Frikke-Schmidt, R., and Nordestgaard, B.G. (2011). Low-density lipoprotein cholesterol and the risk of cancer: A Mendelian randomization study. *J. Natl. Cancer Inst. 103*, 508–519. https://doi.org/10.1093/jnci/djr008.

Biddinger, K.J., Emdin, C.A., Haas, M.E., Wang, M., Hindy, G., Ellinor, P.T., Kathiresan, S., Khera, A.V., and Aragam, K.G. (2022). Association of habitual alcohol intake with risk of cardiovascular disease. *JAMA Netw. Open 5*, e223849. https://doi.org/10.1001/jamanetworkopen.2022.3849.

Billat, V., Dhonneur, G., Mille-Hamard, L., Le Moyec, L., Momken, I., Launay, T., Koralsztein, J.P., and Besse, S. (2017). Case studies in physiology: Maximal oxygen consumption and performance in a centenarian cyclist. *J. Appl. Physiol. 122*, 430–434. https://doi.org/10.1152/japplphysiol.00569.2016.

Blackwell, D.L., and Clarke, T.C. (2018). State variation in meeting the 2008 federal guidelines for both aerobic and muscle-strengthening activities through leisure-time

physical activity among adults aged 18–64: United States, 2010–2015. *Natl. Health Stat. Rep. 112* (June), 1–22.
Blasbalg, T.L., Hibbeln, J.R., Ramsden, C.E., Majchrzak, S.F., and Rawlings, R.R. (2011). Changes in consumption of omega-3 and omega-6 fatty acids in the United States during the 20th century. *Am. J. Clin. Nutr. 93*, 950–962. https://doi.org/10.3945/ajcn.110.006643.
Blessed, G., Tomlinson, B.E., and Roth, M. (1968). The association between quantitative measures of dementia and of senile change in the cerebral grey matter of elderly subjects. *Br. J. Psychiatry J. Ment. Sci. 114*, 797–811. https://doi.org/10.1192/bjp.114.512.797.
Boden, G., Sargrad, K., Homko, C., Mozzoli, M., and Stein, T.P. (2005). Effect of a low-carbohydrate diet on appetite, blood glucose levels, and insulin resistance in obese patients with type 2 diabetes. *Ann. Intern. Med. 142*, 403–411. https://doi.org/10.7326/0003-4819-142-6-200503150-00006.
Bohannon, R.W. (2019). Grip strength: An indispensable biomarker for older adults. *Clin. Interv. Aging 14*, 1681–1691. https://doi.org/10.2147/CIA.S194543.
Boneti Moreira, N., Vagetti, G.C., de Oliveira, V., and de Campos, W. (2014). Association between injury and quality of life in athletes: A systematic review, 1980–2013. *Apunts Sports Med. 49*, 123–138.
Booth, F.W., and Zwetsloot, K.A. (2010). Basic concepts about genes, inactivity and aging. *Scand. J. Med. Sci. Sports 20*, 1–4. https://doi.org/10.1111/j.1600-0838.2009.00972.x.
Børsheim, E., Bui, Q.-U.T., Tissier, S., Cree, M.G., Rønsen, O., Morio, B., Ferrando, A.A., Kobayashi, H., Newcomer, B.R., and Wolfe, R.R. (2009). Amino acid supplementation decreases plasma and liver triglycerides in elderly. *Nutr. Burbank Los Angel. Cty. Calif. 25*, 281–288. https://doi.org/10.1016/j.nut.2008.09.001.
Bosy-Westphal, A., Hinrichs, S., Jauch-Chara, K., Hitze, B., Later, W., Wilms, B., Settler, U., Peters, A., Kiosz, D., and Müller, M.J. (2008). Influence of partial sleep deprivation on energy balance and insulin sensitivity in healthy women. *Obes. Facts 1*, 266–273. https://doi.org/10.1159/000158874.
Bouwman, F.H., van der Flier, W.M., Schoonenboom, N.S.M., van Elk, E.J., Kok, A., Rijmen, F., Blankenstein, M.A., and Scheltens, P. (2007). Longitudinal changes of CSF biomarkers in memory clinic patients. *Neurology 69*, 1006–1011. https://doi.org/10.1212/01.wnl.0000271375.37131.04.
Bradley, D. (2004). Biography of Lewis C. Cantley. *Proc. Natl. Acad. Sci. 101*, 3327–3328. https://doi.org/10.1073/pnas.0400872101.
Branger, P., Arenaza-Urquijo, E.M., Tomadesso, C., Mézenge, F., André, C., de Flores, R., Mutlu, J., de La Sayette, V., Eustache, F., Chételat, G., et al. (2016). Relationships between sleep quality and brain volume, metabolism, and amyloid deposition in late adulthood. *Neurobiol. Aging 41*, 107–114. https://doi.org/10.1016/j.neurobiolaging.2016.02.009.
Brondel, L., Romer, M.A., Nougues, P.M., Touyarou, P., and Davenne, D. (2010). Acute partial sleep deprivation increases food intake in healthy men. *Am. J. Clin. Nutr. 91*, 1550–1559. https://doi.org/10.3945/ajcn.2009.28523.
Brookmeyer, R., Abdalla, N., Kawas, C.H., and Corrada, M.M. (2018). Forecasting the prevalence of preclinical and clinical Alzheimer's disease in the United States. *Alzheimers Dement. 14*, 121–129. https://doi.org/10.1016/j.jalz.2017.10.009.

Brooks, D. (2016). *The road to character.* Farmington Hills, MI: Large Print Press.
Broussard, J.L., Chapotot, F., Abraham, V., Day, A., Delebecque, F., Whitmore, H.R., and Tasali, E. (2015). Sleep restriction increases free fatty acids in healthy men. *Diabetologia 58*, 791–798. https://doi.org/10.1007/s00125-015-3500-4.
Broussard, J.L., Ehrmann, D.A., Van Cauter, E., Tasali, E., and Brady, M.J. (2012). Impaired insulin signaling in human adipocytes after experimental sleep restriction. *Ann. Intern. Med. 157*, 549–557. https://doi.org/10.7326/0003-4819-157-8-201210160-00005.
Broussard, J.L., Kilkus, J.M., Delebecque, F., Abraham, V., Day, A., Whitmore, H.R., and Tasali, E. (2016). Elevated ghrelin predicts food intake during experimental sleep restriction. *Obesity 24*, 132–138. https://doi.org/10.1002/oby.21321.
Brown, B.M., Rainey-Smith, S.R., Villemagne, V.L., Weinborn, M., Bucks, R.S., Sohrabi, H.R., Laws, S.M., Taddei, K., Macaulay, S.L., Ames, D., et al. (2016). The relationship between sleep quality and brain amyloid burden. *Sleep 39*, 1063–1068. https://doi.org/10.5665/sleep.5756.
Brown, E.J., Albers, M.W., Shin, T.B., Ichikawa, K., Keith, C.T., Lane, W.S., and Schreiber, S.L. (1994). A mammalian protein targeted by G1-arresting rapamycin-receptor complex. *Nature 369*, 756–758. https://doi.org/10.1038/369756a0.
Brys, M., Pirraglia, E., Rich, K., Rolstad, S., Mosconi, L., Switalski, R., Glodzik-Sobanska, L., De Santi, S., Zinkowski, R., Mehta, P., et al. (2009). Prediction and longitudinal study of CSF biomarkers in mild cognitive impairment. *Neurobiol. Aging 30*, 682–690. https://doi.org/10.1016/j.neurobiolaging.2007.08.010.
Bunout, D., de la Maza, M.P., Barrera, G., Leiva, L., and Hirsch, S. (2011). Association between sarcopenia and mortality in healthy older people. *Australas. J. Ageing 30*, 89–92. https://doi.org/10.1111/j.1741-6612.2010.00448.x.
Business Wire (2021). U.S. sleep aids market worth $30 billion as Americans battle insomnia, sleep disorders—ResearchAndMarkets.com, June 30. https://www.businesswire.com/news/home/20210630005428/en/U.S.-Sleep-Aids-Market-Worth-30-Billion-as-Americans-Battle-Insomnia-Sleep-Disorders---ResearchAndMarkets.com.
Buxton, O.M., Pavlova, M., Reid, E.W., Wang, W., Simonson, D.C., and Adler, G.K. (2010). Sleep restriction for 1 week reduces insulin sensitivity in healthy men. *Diabetes 59*, 2126–2133. https://doi.org/10.2337/db09-0699.
Buysse, D.J., Reynolds, C.F., Charles, F., Monk, T.H., Berman, S.R., and Kupfer, D.J. (1989). The Pittsburgh Sleep Quality Index: A new instrument for psychiatric practice and research. *Psychiat. Res. 28*(2), 193–213.
Cacace, R., Sleegers, K., and Van Broeckhoven, C. (2016). Molecular genetics of early-onset Alzheimer's disease revisited. *Alzheimers Dement. 12*, 733–748. https://doi.org/10.1016/j.jalz.2016.01.012.
Calle, E.E., Rodriguez, C., Walker-Thurmond, K., and Thun, M.J. (2003). Overweight, obesity, and mortality from cancer in a prospectively studied cohort of U.S. adults. *N. Engl. J. Med. 348*, 1625. https://doi.org/10.1056/NEJMoa021423.
Calvin, A.D., Carter, R.E., Adachi, T., Macedo, P.G., Albuquerque, F.N., van der Walt, C., Bukartyk, J., Davison, D.E., Levine, J.A., and Somers, V.K. (2013). Effects of experimental

sleep restriction on caloric intake and activity energy expenditure. *Chest 144*, 79–86. https://doi.org/10.1378/chest.12-2829.

Campbell, K.L., Winters-Stone, K., Wiskemann, J., May, A.M., Schwartz, A.L., Courneya, K.S., Zucker, D., Matthews, C., Ligibel, J., Gerber, L., et al. (2019). Exercise guidelines for cancer survivors: Consensus statement from International Multidisciplinary Roundtable. *Med. Sci. Sports Exerc. 51*, 2375–2390. https://doi.org/10.1249/MSS.0000000000002116.

Campbell, W.W., Trappe, T.A., Wolfe, R.R., and Evans, W.J. (2001). The recommended dietary allowance for protein may not be adequate for older people to maintain skeletal muscle. *J. Gerontol. A. Biol. Sci. Med. Sci. 56*, M373–380. https://doi.org/10.1093/gerona/56.6.m373.

Case, A., and Deaton, A. (2015). Rising morbidity and mortality in midlife among white non-Hispanic Americans in the 21st century. *Proc. Natl. Acad. Sci. 112*(49), 15078–15083. https://doi.org/10.1073/pnas.151839311.

Caselli, G., and Lipsi, R.M. (2006). Survival differences among the oldest old in Sardinia: Who, what, where, and why? *Demogr. Res. 14*, 267–294.

Cavazzoni, A., Digiacomo, G., Alfieri, R., La Monica, S., Fumarola, C., Galetti, M., Bonelli, M., Cretella, D., Barili, V., Zecca, A., et al. (2020). Pemetrexed enhances membrane PD-L1 expression and potentiates T cell-mediated cytotoxicity by anti-PD-L1 antibody therapy in non-small-cell lung cancer. *Cancers 12*, E666. https://doi.org/10.3390/cancers12030666.

Cerri, S., Mus, L., and Blandini, F. (2019). Parkinson's disease in women and men: What's the difference? *J. Parkinson's Dis. 9*(3), 501–515. https://doi.org/10.3233/JPD-191683.

CDC (Centers for Disease Control). (2020a). The influence of metabolic syndrome in predicting mortality risk among US adults: Importance of metabolic syndrome even in adults with normal weight. https://www.cdc.gov/pcd/issues/2020/20_0020.htm.

———. (2020b). Diabetes. FastStats. https://www.cdc.gov/nchs/fastats/diabetes.htm.

———. (2021). Facts about falls. Injury Center. https://www.cdc.gov/falls/facts.html.

———. (2022a). Accidents or unintentional injuries. FastStats. https://www.cdc.gov/nchs/fastats/accidental-injury.htm.

———. (2022b) Adult obesity facts. https://www.cdc.gov/obesity/data/adult.html.

———. (2022c). Heart disease facts. https://www.cdc.gov/heartdisease/facts.htm.

———. (2022d). Life expectancy in the U.S. dropped for the second year in a row in 2021. Press release, August 31. https://www.cdc.gov/nchs/pressroom/nchs_press_releases/2022/20220831.htm.

———. (2022e). National diabetes statistics report. https://www.cdc.gov/diabetes/data/statistics-report/index.html?ACSTrackingID=DM72996&ACSTrackingLabel=New%20Report%20Shares%20Latest%20Diabetes%20Stats%20&deliveryName=DM72996.

———. (2022f). Ten leading causes of death and injury. https://www.cdc.gov/injury/wisqars/LeadingCauses_images.html.

Chan, J.M., Rimm, E.B., Colditz, G.A., Stampfer, M.J., and Willett, W.C. (1994). Obesity, fat distribution, and weight gain as risk factors for clinical diabetes in men. *Diabetes Care 17*, 961–969. https://doi.org/10.2337/diacare.17.9.961.

Chapman, C.D., Schiöth, H.B., Grillo, C.A., and Benedict, C. (2018). Intranasal insulin in Alzheimer's disease: Food for thought. *Neuropharmacology 136*, 196–201. https://doi.org/10.1016/j.neuropharm.2017.11.037.

Chen, D.L., Liess, C., Poljak, A., Xu, A., Zhang, J., Thoma, C., Trenell, M., Milner, B., Jenkins, A.B., Chisholm, D.J., et al. (2015). Phenotypic characterization of insulin-resistant and insulin-sensitive obesity. *J. Clin. Endocrinol. Metab. 100*, 4082–4091. https://doi.org/10.1210/jc.2015-2712.

Chen, X., Dong, Z., Hubbell, E., Kurtzman, K.N., Oxnard, G.R., Venn, O., Melton, C., Clarke, C.A., Shaknovich, R., Ma, T., et al. (2021). Prognostic significance of blood-based multi-cancer detection in plasma cell-free DNA. *Clin. Cancer Res. 27*, 4221–4229. https://doi.org/10.1158/1078-0432.CCR-21-0417.

Cholerton, B., Baker, L.D., Montine, T.J., and Craft, S. (2016). Type 2 diabetes, cognition, and dementia in older adults: Toward a precision health approach. *Diabetes Spectr. 29*, 210–219. https://doi.org/10.2337/ds16-0041.

Christofferson, Travis. *Tripping Over the Truth: How the Metabolic Theory of Cancer Is Overturning One of Medicine's Most Entrenched Paradigms.* Chelsea Green Publishing, 2017.

Cirelli, C., and Tononi, G. (2008). Is sleep essential? *PLOS Biol. 6*, e216. https://doi.org/10.1371/journal.pbio.0060216.

Colman, R.J., Anderson, R.M., Johnson, S.C., Kastman, E.K., Kosmatka, K.J., Beasley, T.M., Allison, D.B., Cruzen, C., Simmons, H.A., Kemnitz, J.W., et al. (2009). Caloric restriction delays disease onset and mortality in rhesus monkeys. *Science 325*, 201–204. https://doi.org/10.1126/science.1173635.

Copinschi, G., and Caufriez, A. (2013). Sleep and hormonal changes in aging. *Endocrinol. Metab. Clin. North Am. 42*, 371–389. https://doi.org/10.1016/j.ecl.2013.02.009.

Cordain, L., Eaton, S.B., Miller, J.B., Mann, N., and Hill, K. (2002). The paradoxical nature of hunter-gatherer diets: Meat-based, yet non-atherogenic. *Eur. J. Clin. Nutr. 56*, S42–S52. https://doi.org/10.1038/sj.ejcn.1601353.

Cordain, L., Miller, J.B., Eaton, S.B., Mann, N., Holt, S.H., and Speth, J.D. (2000). Plant-animal subsistence ratios and macronutrient energy estimations in worldwide hunter-gatherer diets. *Am. J. Clin. Nutr. 71*, 682–692. https://doi.org/10.1093/ajcn/71.3.682.

Creevy, K.E., Akey, J.M., Kaeberlein, M., and Promislow, D.E.L. (2022). An open science study of ageing in companion dogs. *Nature 602*, 51–57. https://doi.org/10.1038/s41586-021-04282-9.

Crispim, C.A., Zimberg, I.Z., dos Reis, B.G., Diniz, R.M., Tufik, S., and de Mello, M.T. (2011). Relationship between food intake and sleep pattern in healthy individuals. *J. Clin. Sleep Med. 7*, 659–664. https://doi.org/10.5664/jcsm.1476.

Crowe, K. (2018). University of Twitter? Scientists give impromptu lecture critiquing nutrition research. CBC Health, May 5. https://www.cbc.ca/news/health/second-opinion-alcohol180505-1.4648331.

Cruchaga, C., Haller, G., Chakraverty, S., Mayo, K., Vallania, F.L.M., Mitra, R.D., Faber, K., Williamson, J., Bird, T., Diaz-Arrastia, R., et al. (2012). Rare variants in *APP, PSEN1* and *PSEN2* increase risk for AD in late-onset Alzheimer's disease families. *PLOS ONE 7*, e31039. https://doi.org/10.1371/journal.pone.0031039.

Cruz-Jentoft, A.J., Bahat, G., Bauer, J., Boirie, Y., Bruyère, O., Cederholm, T., Cooper, C., Landi, F., Rolland, Y., Sayer, A.A., et al. (2019). Sarcopenia: Revised European consensus on definition and diagnosis. *Age Ageing 48*, 16–31. https://doi.org/10.1093/ageing/afy169.

Cullen, T., Thomas, G., Wadley, A.J., and Myers, T. (2019). The effects of a single night of complete and partial sleep deprivation on physical and cognitive performance: A Bayesian analysis. *J. Sports Sci. 37*, 2726–2734. https://doi.org/10.1080/02640414.2019.1662539.

Cummings, J.L., Goldman, D.P., Simmons-Stern, N.R., and Ponton, E. (2022). The costs of developing treatments for Alzheimer's disease: A retrospective exploration. *Alzheimers Dement. 18*, 469–477. https://doi.org/10.1002/alz.12450.

Cuyvers, E., and Sleegers, K. (2016). Genetic variations underlying Alzheimer's disease: Evidence from genome-wide association studies and beyond. *Lancet Neurol. 15*, 857–868. https://doi.org/10.1016/S1474-4422(16)00127-7.

Daghlas, I., Dashti, H.S., Lane, J., Aragam, K.G., Rutter, M.K., Saxena, R., and Vetter, C. (2019). Sleep duration and myocardial infarction. *J. Am. Coll. Cardiol. 74*, 1304–1314. https://doi.org/10.1016/j.jacc.2019.07.022.

Dahlhamer, J. (2018). Prevalence of chronic pain and high-impact chronic pain among adults—United States, 2016. *MMWR 67*. https://doi.org/10.15585/mmwr.mm6736a2.

Danneskiold-Samsøe, B., Bartels, E.M., Bülow, P.M., Lund, H., Stockmarr, A., Holm, C.C., Wätjen, I., Appleyard, M., and Bliddal, H. (2009). Isokinetic and isometric muscle strength in a healthy population with special reference to age and gender. *Acta Physiol. 197*, 1–68. https://doi.org/10.1111/j.1748-1716.2009.02022.x.

Dashti, H.S., Jones, S.E., Wood, A.R., Lane, J.M., van Hees, V.T., Wang, H., Rhodes, J.A., Song, Y., Patel, K., Anderson, S.G., et al. (2019). Genome-wide association study identifies genetic loci for self-reported habitual sleep duration supported by accelerometer-derived estimates. *Nat. Commun. 10*, 1100. https://doi.org/10.1038/s41467-019-08917-4.

Daviglus, M.L., Bell, C.C., Berrettini, W., Bowen, P.E., Connolly, E.S., Cox, N.J., Dunbar-Jacob, J.M., Granieri, E.C., Hunt, G., McGarry, K., et al. (2010). NIH state-of-the-science conference statement: Preventing Alzheimer's disease and cognitive decline. *NIH Consens. State Sci. Statements 27*, 1–30.

Dawson, D., and Reid, K. (1997). Fatigue, alcohol and performance impairment. *Nature 388*, 235–235. https://doi.org/10.1038/40775.

de Groot, S., Lugtenberg, R.T., Cohen, D., Welters, M.J.P., Ehsan, I., Vreeswijk, M.P.G., Smit, V.T.H.B.M., de Graaf, H., Heijns, J.B., Portielje, J.E.A., et al. (2020). Fasting mimicking diet as an adjunct to neoadjuvant chemotherapy for breast cancer in the multicentre randomized phase 2 DIRECT trial. *Nat. Commun. 11*, 3083. https://doi.org/10.1038/s41467-020-16138-3.

de la Torre, J. (2016). *Alzheimer's turning point: A vascular approach to clinical prevention.* Cham, Switzerland: Springer International, 169–183.

———. (2018). The vascular hypothesis of Alzheimer's disease: A key to preclinical prediction of dementia using neuroimaging. *J. Alzheimers Dis. 63*, 35–52. https://doi.org/10.3233/JAD-180004.

de Leon, M.J., DeSanti, S., Zinkowski, R., Mehta, P.D., Pratico, D., Segal, S., Rusinek, H., Li,

J., Tsui, W., Saint Louis, L.A., et al. (2006). Longitudinal CSF and MRI biomarkers improve the diagnosis of mild cognitive impairment. *Neurobiol. Aging 27*, 394–401. https://doi.org/10.1016/j.neurobiolaging.2005.07.003.

Dewasmes, G., Bothorel, B., Hoeft, A., and Candas, V. (1993). Regulation of local sweating in sleep-deprived exercising humans. *Eur. J. Appl. Physiol. 66*, 542–546. https://doi.org/10.1007/BF00634307.

de Zambotti, M., Colrain, I.M., and Baker, F.C. (2015). Interaction between reproductive hormones and physiological sleep in women. *J. Clin. Endocrinol. Metab. 100*, 1426–1433. https://doi.org/10.1210/jc.2014-3892.

Diamond, J. (2003). The double puzzle of diabetes. *Nature 423*, 599–602. https://doi.org/10.1038/423599a.

Diekelmann, S., and Born, J. (2010). The memory function of sleep. *Nat. Rev. Neurosci. 11*, 114–126. https://doi.org/10.1038/nrn2762.

Dietary Guidelines Advisory Committee. (2015). Scientific report of the 2015 Dietary Guidelines Advisory Committee: Advisory report to the Secretary of Health and Human Services and the Secretary of Agriculture. Washington, D.C.: U.S. Department of Agriculture, Agricultural Research Service. https://health.gov/sites/default/files/2019-09/Scientific-Report-of-the-2015-Dietary-Guidelines-Advisory-Committee.pdf.

Dietschy, J.M., Turley, S.D., and Spady, D.K. (1993). Role of liver in the maintenance of cholesterol and low density lipoprotein homeostasis in different animal species, including humans. *J. Lipid Res. 34*, 1637–1659.

Dominy, S.S., Lynch, C., Ermini, F., Benedyk, M., Marczyk, A., Konradi, A., Nguyen, M., Haditsch, U., Raha, D., Griffin, C., et al. (2019). Porphyromonas gingivalis in Alzheimer's disease brains: Evidence for disease causation and treatment with small-molecule inhibitors. *Sci. Adv. 5*, eaau3333. https://doi.org/10.1126/sciadv.aau3333.

du Souich, P., Roederer, G., and Dufour, R. (2017). Myotoxicity of statins: Mechanism of action. *Pharmacol. Ther. 175*, 1–16. https://doi.org/10.1016/j.pharmthera.2017.02.029.

Dworak, M., Diel, P., Voss, S., Hollmann, W., and Strüder, H.K. (2007). Intense exercise increases adenosine concentrations in rat brain: Implications for a homeostatic sleep drive. *Neuroscience 150*, 789–795. https://doi.org/10.1016/j.neuroscience.2007.09.062.

Dye, L. (1988). Nobel physicist R. P. Feynman of Caltech dies. *Los Angeles Times*, February 16. https://www.latimes.com/archives/la-xpm-1988-02-16-mn-42968-story.html.

Easter, M. (2021). *The comfort crisis: Embrace discomfort to reclaim your wild, happy, healthy self.* New York: Rodale Books.

Ebrahim, I.O., Shapiro, C.M., Williams, A.J., and Fenwick, P.B. (2013). Alcohol and sleep I: Effects on normal sleep. *Alcohol. Clin. Exp. Res. 37*, 539–549. https://doi.org/10.1111/acer.12006.

Echouffo-Tcheugui, J.B., Zhao, S., Brock, G., Matsouaka, R.A., Kline, D., and Joseph, J.J. (2019). Visit-to-visit glycemic variability and risks of cardiovascular events and all-cause mortality: The ALLHAT study. *Diabetes Care 42*, 486–493. https://doi.org/10.2337/dc18-1430.

Ejima, K., Li, P., Smith, D.L., Nagy, T.R., Kadish, I., van Groen, T., Dawson, J.A., Yang, Y.,

Patki, A., and Allison, D.B. (2016). Observational research rigor alone does not justify causal inference. *Eur. J. Clin. Invest. 46*, 985–993. https://doi.org/10.1111/eci.12681.

Emamian, F., Khazaie, H., Tahmasian, M., Leschziner, G.D., Morrell, M.J., Hsiung, G.-Y.R., Rosenzweig, I., and Sepehry, A.A. (2016). The association between obstructive sleep apnea and Alzheimer's disease: A meta-analysis perspective. *Front. Aging Neurosci. 8*, 78. https://doi.org/10.3389/fnagi.2016.00078.

Esteban-Cornejo, I., Ho, F.K., Petermann-Rocha, F., Lyall, D.M., Martinez-Gomez, D., Cabanas-Sánchez, V., Ortega, F.B., Hillman, C.H., Gill, J.M.R., Quinn, T.J., et al. (2022). Handgrip strength and all-cause dementia incidence and mortality: Findings from the UK Biobank prospective cohort study. *J. Cachexia Sarcopenia Muscle 13*, 1514–1525. https://doi.org/10.1002/jcsm.12857.

Estruch, R., Ros, E., Salas-Salvadó, J., Covas, M.-I., Corella, D., Arós, F., Gómez-Gracia, E., Ruiz-Gutiérrez, V., Fiol, M., Lapetra, J., et al. (2013). Primary prevention of cardiovascular disease with a Mediterranean diet. *N. Engl. J. Med. 368*, 1279–1290. https://doi.org/10.1056/NEJMoa1200303.

Evert, J., Lawler, E., Bogan, H., and Perls, T. (2003). Morbidity profiles of centenarians: Survivors, delayers, and escapers. *J. Gerontol. Ser. A 58*, M232–M237. https://doi.org/10.1093/gerona/58.3.M232.

Fagan, A.M., Mintun, M.A., Mach, R.H., Lee, S.-Y., Dence, C.S., Shah, A.R., LaRossa, G.N., Spinner, M.L., Klunk, W.E., Mathis, C.A., et al. (2006). Inverse relation between in vivo amyloid imaging load and cerebrospinal fluid Abeta42 in humans. *Ann. Neurol. 59*, 512–519. https://doi.org/10.1002/ana.20730.

Fain, E., and Weatherford, C. (2016). Comparative study of millennials' (age 20–34 years) grip and lateral pinch with the norms. *J. Hand Ther. 29*, 483–488. https://doi.org/10.1016/j.jht.2015.12.006.

Fayek, S.A., Quintini, C., Chavin, K.D., and Marsh, C.L. (2016). The current state of liver transplantation in the United States. *Am. J. Transplant. 16*, 3093–3104. https://doi.org/10.1111/ajt.14017.

Ference, B.A. (2015). Mendelian randomization studies: Using naturally randomized genetic data to fill evidence gaps. *Curr. Opin. Lipidol. 26*, 566–571. https://doi.org/10.1097/MOL.0000000000000247.

Ference, B.A., Bhatt, D.L., Catapano, A.L., Packard, C.J., Graham, I., Kaptoge, S., Ference, T.B., Guo, Q., Laufs, U., Ruff, C.T., et al. (2019). Association of genetic variants related to combined exposure to lower low-density lipoproteins and lower systolic blood pressure with lifetime risk of cardiovascular disease. *JAMA 322*, 1381–1391. https://doi.org/10.1001/jama.2019.14120.

Ferriss, T. (2018). LeBron James and his top-secret trainer, Mike Mancias (#349). *Tim Ferriss Show* (podcast), episode 349, November 27.

Fontana, L., and Partridge, L. (2015). Promoting health and longevity through diet: From model organisms to humans. *Cell 161*, 106–118. https://doi.org/10.1016/j.cell.2015.02.020.

Forrester, J.S. (2010). Redefining normal low-density lipoprotein cholesterol: A strategy to unseat coronary disease as the nation's leading killer. *J. Am. Coll. Cardiol. 56*, 630–636. https://doi.org/10.1016/j.jacc.2009.11.090.

Frank, C., Kobesova, A., and Kolar, P. (2013). Dynamic neuromuscular stabilization and sports rehabilitation. *Int. J. Sports Phys. Ther. 8*, 62–73.

Franz, M.J. (1997). Protein: Metabolism and effect on blood glucose levels. *Diabetes Educ. 23*, 643–646, 648, 650–651. https://doi.org/10.1177/014572179702300603.

Frayn, K. (2019). *Human metabolism: A regulatory perspective.* 4th ed. New York: Wiley.

Freiherr, J., Hallschmid, M., Frey, W.H., Brünner, Y.F., Chapman, C.D., Hölscher, C., Craft, S., De Felice, F.G., and Benedict, C. (2013). Intranasal insulin as a treatment for Alzheimer's disease: A review of basic research and clinical evidence. *CNS Drugs 27*, 505–514. https://doi.org/10.1007/s40263-013-0076-8.

Friend, T. (2003). Jumpers. *New Yorker,* October 13. https://www.newyorker.com/magazine/2003/10/13/jumpers.

Fruman, D.A., Chiu, H., Hopkins, B.D., Bagrodia, S., Cantley, L.C., and Abraham, R.T. (2017). The PI3K pathway in human disease. *Cell 170*, 605–635. https://doi.org/10.1016/j.cell.2017.07.029.

Fryar, C.D., Kruszon-Moran, D., Gu, Q., and Ogden, C.L. (2018). Mean body weight, height, waist circumference, and body mass index among adults: United States, 1999–2000 through 2015–2016. *Natl. Health Stat. Rep.* 1–16.

Fullagar, H.H.K., Skorski, S., Duffield, R., Hammes, D., Coutts, A.J., and Meyer, T. (2015). Sleep and athletic performance: The effects of sleep loss on exercise performance, and physiological and cognitive responses to exercise. *Sports Med. Auckl. NZ 45*, 161–186. https://doi.org/10.1007/s40279-014-0260-0.

Gaskin, D.J., and Richard, P. (2012). The economic costs of pain in the United States. *J. Pain 13*, 715–724. https://doi.org/10.1016/j.jpain.2012.03.009.

Gavrilova, O., Marcus-Samuels, B., Graham, D., Kim, J.K., Shulman, G.I., Castle, A.L., Vinson, C., Eckhaus, M., and Reitman, M.L. (2000). Surgical implantation of adipose tissue reverses diabetes in lipoatrophic mice. *J. Clin. Invest. 105*, 271–278.

Gay, N., and Prasad, V. (2017). Few people actually benefit from "breakthrough" cancer immunotherapy. Stat News, March 8. https://www.statnews.com/2017/03/08/immunotherapy-cancer-breakthrough/.

Gibala, M.J., Little, J.P., van Essen, M., Wilkin, G.P., Burgomaster, K.A., Safdar, A., Raha, S., and Tarnopolsky, M.A. (2006). Short-term sprint interval versus traditional endurance training: Similar initial adaptations in human skeletal muscle and exercise performance. *J. Physiol. 575*, 901–911. https://doi.org/10.1113/jphysiol.2006.112094.

Gibson, A.A., Seimon, R.V., Lee, C.M.Y., Ayre, J., Franklin, J., Markovic, T.P., Caterson, I.D., and Sainsbury, A. (2015). Do ketogenic diets really suppress appetite? A systematic review and meta-analysis. *Obes. Rev. 16*, 64–76. https://doi.org/10.1111/obr.12230.

Gillen, J.B., Percival, M.E., Skelly, L.E., Martin, B.J., Tan, R.B., Tarnopolsky, M.A., and Gibala, M.J. (2014). Three minutes of all-out intermittent exercise per week increases skeletal muscle oxidative capacity and improves cardiometabolic health. *PLOS ONE 9*, e111489. https://doi.org/10.1371/journal.pone.0111489.

Goldin, A., Beckman, J.A., Schmidt, A.M., and Creager, M.A. (2006). Advanced glycation end products. *Circulation 114*, 597–605. https://doi.org/10.1161/CIRCULATIONAHA.106.621854.

Goldstein, A.N., and Walker, M.P. (2014). The role of sleep in emotional brain function. *Annu. Rev. Clin. Psychol. 10*, 679–708. https://doi.org/10.1146/annurev-clinpsy-032813-153716.

Goldstein-Piekarski, A.N., Greer, S.M., Saletin, J.M., and Walker, M.P. (2015). Sleep deprivation impairs the human central and peripheral nervous system discrimination of social threat. *J. Neurosci. 35*, 10135–10145. https://doi.org/10.1523/JNEUROSCI.5254-14.2015.

Gordon, R.J. (2016). *The rise and fall of American growth: The U.S. standard of living since the Civil War.* Princeton, NJ: Princeton University Press.

Gradisar, M., Wolfson, A.R., Harvey, A.G., Hale, L., Rosenberg, R., and Czeisler, C.A. (2013). The sleep and technology use of Americans: Findings from the National Sleep Foundation's 2011 Sleep in America Poll. *J. Clin. Sleep Med. 9*, 1291–1299. https://doi.org/10.5664/jcsm.3272.

Graeber, C. (2018). *The breakthrough: Immunotherapy and the race to cure cancer.* New York: Twelve.

Grammatikopoulou, M.G., Goulis, D.G., Gkiouras, K., Theodoridis, X., Gkouskou, K.K., Evangeliou, A., Dardiotis, E., and Bogdanos, D.P. (2020). To keto or not to keto? A systematic review of randomized controlled trials assessing the effects of ketogenic therapy on Alzheimer disease. *Adv. Nutr. 11*, 1583–1602. https://doi.org/10.1093/advances/nmaa073.

Grandner, M.A., Sean, P.A., Drummond. (2007). Who are the long sleepers? Towards an understanding of the mortality relationship. *Sleep Medicine Reviews,* 11: 5, 341–360. https://doi.org/10.1016/j.smrv.2007.03.010.

Grimmer, T., Riemenschneider, M., Förstl, H., Henriksen, G., Klunk, W.E., Mathis, C.A., Shiga, T., Wester, H.-J., Kurz, A., and Drzezga, A. (2009). Beta amyloid in Alzheimer's disease: Increased deposition in brain is reflected in reduced concentration in cerebrospinal fluid. *Biol. Psychiatry 65*, 927–934. https://doi.org/10.1016/j.biopsych.2009.01.027.

Gross, D.N., van den Heuvel, A.P.J., and Birnbaum, M.J. (2008). The role of FoxO in the regulation of metabolism. *Oncogene 27*, 2320–2336. https://doi.org/10.1038/onc.2008.25.

Guyenet, S.J., and Carlson, S.E. (2015). Increase in adipose tissue linoleic acid of US adults in the last half century. *Adv. Nutr. 6*, 660–664. https://doi.org/10.3945/an.115.009944.

Haase, C.L., Tybjærg-Hansen, A., Ali Qayyum, A., Schou, J., Nordestgaard, B.G., and Frikke-Schmidt, R. (2012). LCAT, HDL cholesterol and ischemic cardiovascular disease: A Mendelian randomization study of HDL cholesterol in 54,500 Individuals. *J. Clin. Endocrinol. Metab. 97*, E248–E256. https://doi.org/10.1210/jc.2011-1846.

Hafner, M., Stepanek, M., Taylor, J., Troxel, W.M., and van Stolk, C. (2017). Why sleep matters: The economic costs of insufficient sleep. *Rand Health Q. 6*, 11.

Hagerhall, C.M., et al. 2008. Investigations of human EEG response to viewing fractal patterns. *Perception 37*, 1488–1494. https://doi.org/10.1068/p5918.

Hamer, M., and O'Donovan, G. (2017). Sarcopenic obesity, weight loss, and mortality: The English Longitudinal Study of Ageing. *Am. J. Clin. Nutr. 106*, 125–129. https://doi.org/10.3945/ajcn.117.152488.

Hanahan, D., and Weinberg, R.A. (2011). Hallmarks of cancer: The next generation. *Cell 144*, 646–674. https://doi.org/10.1016/j.cell.2011.02.013.

Hanefeld, M., Koehler, C., Schaper, F., Fuecker, K., Henkel, E., and Temelkova-Kurktschiev, T. (1999). Postprandial plasma glucose is an independent risk factor for increased carotid intima-media thickness in non-diabetic individuals. *Atherosclerosis 144*, 229–235. https://doi.org/10.1016/S0021-9150(99)00059-3.

Hardeland, R. (2013). Chronobiology of melatonin beyond the feedback to the suprachiasmatic nucleus: Consequences to melatonin dysfunction. *Int. J. Mol. Sci. 14*, 5817–5841. https://doi.org/10.3390/ijms14035817.

Hardie, D.G. (2011). AMP-activated protein kinase: An energy sensor that regulates all aspects of cell function. *Genes Dev. 25*, 1895–1908. https://doi.org/10.1101/gad.17420111.

Harding, E.C., Franks, N.P., and Wisden, W. (2020). Sleep and thermoregulation. *Curr. Opin. Physiol. 15*, 7–13. https://doi.org/10.1016/j.cophys.2019.11.008.

Harrison, D.E., Strong, R., Reifsnyder, P., Kumar, N., Fernandez, E., Flurkey, K., Javors, M.A., Lopez-Cruzan, M., Macchiarini, F., Nelson, J.F., et al. (2021). 17-a-estradiol late in life extends lifespan in aging UM-HET3 male mice; nicotinamide riboside and three other drugs do not affect lifespan in either sex. *Aging Cell 20*, e13328. https://doi.org/10.1111/acel.13328.

Harrison, D.E., Strong, R., Sharp, Z.D., Nelson, J.F., Astle, C.M., Flurkey, K., Nadon, N.L., Wilkinson, J.E., Frenkel, K., Carter, C.S., et al. (2009). Rapamycin fed late in life extends lifespan in genetically heterogeneous mice. *Nature 460*, 392–395. https://doi.org/10.1038/nature08221.

Harrison, S.A., Gawrieh, S., Roberts, K., Lisanti, C.J., Schwope, R.B., Cebe, K.M., Paradis, V., Bedossa, P., Aldridge Whitehead, J.M., Labourdette, A., et al. (2021). Prospective evaluation of the prevalence of non-alcoholic fatty liver disease and steatohepatitis in a large middle-aged US cohort. *J. Hepatol. 75*, 284–291. https://doi.org/10.1016/j.jhep.2021.02.034.

Hatori, M., Vollmers, C., Zarrinpar, A., DiTacchio, L., Bushong, E.A., Gill, S., Leblanc, M., Chaix, A., Joens, M., Fitzpatrick, J.A.J., et al. (2012). Time restricted feeding without reducing caloric intake prevents metabolic diseases in mice fed a high fat diet. *Cell Metab. 15*, 848–860. https://doi.org/10.1016/j.cmet.2012.04.019.

Heron, M. (2021). Deaths: Leading causes for 2018. *Natl. Vital Stat. Rep. 70*(4), 1–115.

Herring, W.J., Connor, K.M., Ivgy-May, N., Snyder, E., Liu, K., Snavely, D.B., Krystal, A.D., Walsh, J.K., Benca, R.M., Rosenberg, R., et al. (2016). Suvorexant in patients with insomnia: Results from two 3-month randomized controlled clinical trials. *Biol. Psychiatry 79*, 136–148. https://doi.org/10.1016/j.biopsych.2014.10.003.

HHS (US Department of Health and Human Services). (2018). *Physical activity guidelines for Americans.* 2nd ed. https://health.gov/sites/default/files/2019-09/Physical_Activity_Guidelines_2nd_edition.pdf.

Hill, A.B. (1965). The environment and disease: Association or causation? *Proc. R. Soc. Med. 58*, 295–300.

Hines, L., and Rimm, E. (2001). Moderate alcohol consumption and coronary heart disease: A review. *Postgrad. Med. J. 77*, 747–752. https://doi.org/10.1136/pmj.77.914.747.

Hirode, G., and Wong, R.J. (2020). Trends in the prevalence of metabolic syndrome in the United States, 2011-2016. *JAMA 323*, 2526–2528. https://doi.org/10.1001/jama.2020.4501.

Hitchens, C. (2014). *Mortality.* New York: Twelve.

Hitt, R., Young-Xu, Y., Silver, M., and Perls, T. (1999). Centenarians: The older you get, the healthier you have been. *Lancet 354*, 652.

Hjelmborg, J., Iachine, I., Skytthe, A., Vaupel, J.W., McGue, M., Koskenvuo, M., Kaprio, J., Pedersen, N.L., and Christensen, K. (2006). Genetic influence on human lifespan and longevity. *Hum. Genet. 119*, 312–321. https://doi.org/10.1007/s00439-006-0144-y.

Hofseth, L.J., Hebert, J.R., Chanda, A., Chen, H., Love, B.L., Pena, M.M., Murphy, E.A., Sajish, M., Sheth, A., Buckhaults, P.J., et al. (2020). Early-onset colorectal cancer: Initial clues and current views. *Nat. Rev. Gastroenterol. Hepatol. 17*, 352–364. https://doi.org/10.1038/s41575-019-0253-4.

Hoglund, K., Thelen, K.M., Syversen, S., Sjogren, M., von Bergmann, K., Wallin, A., Vanmechelen, E., Vanderstichele, H., Lutjohann, D., and Blennow, K. (2005). The effect of simvastatin treatment on the amyloid precursor protein and brain cholesterol metabolism in patients with Alzheimer's disease. *Dement. Geriatr. Cogn. Disord. 19*, 256–265. https://doi.org/10.1159/000084550.

Holt, S.H., Miller, J.C., Petocz, P., and Farmakalidis, E. (1995). A satiety index of common foods. *Eur. J. Clin. Nutr. 49*, 675–690.

Hooper, L., Martin, N., Jimoh, O.F., Kirk, C., Foster, E., and Abdelhamid, A.S. (2020). Reduction in saturated fat intake for cardiovascular disease. *Cochrane Database Syst. Rev.* https://doi.org/10.1002/14651858.CD011737.pub3.

Hopkins, B.D., Pauli, C., Du, X., Wang, D.G., Li, X., Wu, D., Amadiume, S.C., Goncalves, M.D., Hodakoski, C., Lundquist, M.R., et al. (2018). Suppression of insulin feedback enhances the efficacy of PI3K inhibitors. *Nature 560*, 499–503. https://doi.org/10.1038/s41586-018-0343-4.

Houston, D.K., Nicklas, B.J., Ding, J., Harris, T.B., Tylavsky, F.A., Newman, A.B., Lee, J.S., Sahyoun, N.R., Visser, M., Kritchevsky, S.B., et al. (2008). Dietary protein intake is associated with lean mass change in older, community-dwelling adults: The Health, Aging, and Body Composition (Health ABC) Study. *Am. J. Clin. Nutr. 87*, 150–155. https://doi.org/10.1093/ajcn/87.1.150.

Howard, B.V., Van Horn, L., Hsia, J., Manson, J.E., Stefanick, M.L., Wassertheil-Smoller, S., Kuller, L.H., LaCroix, A.Z., Langer, R.D., Lasser, N.L., et al. (2006). Low-fat dietary pattern and risk of cardiovascular disease: The Women's Health Initiative Randomized Controlled Dietary Modification Trial. *JAMA 295*, 655–666. https://doi.org/10.1001/jama.295.6.655.

Hughes, V.A., Frontera, W.R., Wood, M., Evans, W.J., Dallal, G.E., Roubenoff, R., and Singh, M.A.F. (2001). Longitudinal muscle strength changes in older adults: Influence of muscle mass, physical activity, and health. *J. Gerontol. Ser. A 56*, B209–B217. https://doi.org/10.1093/gerona/56.5.B209.

Hutchison, I.C., and Rathore, S. (2015). The role of REM sleep theta activity in emotional memory. *Front. Psychol. 6*, 1439. https://doi.org/10.3389/fpsyg.2015.01439.

Iftikhar, I.H., Donley, M.A., Mindel, J., Pleister, A., Soriano, S., and Magalang, U.J. (2015). Sleep duration and metabolic syndrome: An updated dose-risk metaanalysis. *Ann. Am. Thorac. Soc. 12*, 1364–1372. https://doi.org/10.1513/AnnalsATS.201504-190OC.

Igwe, E., Azman, A.Z.F., Nordin, A.J., and Mohtarrudin, N. (2015). Association between HOMA-IR and cancer. *Int. J. Public Health Clin. Sci. 2*, 21.

Iliff, J.J., Lee, H., Yu, M., Feng, T., Logan, J., Nedergaard, M., and Benveniste, H. (2013). Brain-wide pathway for waste clearance captured by contrast-enhanced MRI. *J. Clin. Invest. 123*, 1299–1309. https://doi.org/10.1172/JCI67677.

IOM (Institute of Medicine). Committee on Military Nutrition Research. (2001). *Caffeine for the sustainment of mental task performance: Formulations for military operations.* Washington, DC: National Academies Press.

Ioannidis, J.P.A. (2018). The challenge of reforming nutritional epidemiologic research. *JAMA 320*, 969–970. https://doi.org/10.1001/jama.2018.11025.

Itani, O., Jike, M., Watanabe, N., and Kaneita, Y. (2017). Short sleep duration and health outcomes: A systematic review, meta-analysis, and meta-regression. *Sleep Med. 32*, 246–256. https://doi.org/10.1016/j.sleep.2016.08.006.

Jack, C.R., Knopman, D.S., Jagust, W.J., Petersen, R.C., Weiner, M.W., Aisen, P.S., Shaw, L.M., Vemuri, P., Wiste, H.J., Weigand, S.D., et al. (2013). Update on hypothetical model of Alzheimer's disease biomarkers. *Lancet Neurol. 12*, 207–216. https://doi.org/10.1016/S1474-4422(12)70291-0.

Jackson, M.L., Croft, R.J., Kennedy, G.A., Owens, K., and Howard, M.E. (2013). Cognitive components of simulated driving performance: Sleep loss effects and predictors. *Accid. Anal. Prev. 50*, 438–444. https://doi.org/10.1016/j.aap.2012.05.020.

Jakubowski, B., Shao, Y., McNeal, C., Xing, C., and Ahmad, Z. (2021). Monogenic and polygenic causes of low and extremely low LDL-C levels in patients referred to specialty lipid clinics: Genetics of low LDL-C. *J. Clin. Lipidol. 15*, 658–664. https://doi.org/10.1016/j.jacl.2021.07.003.

Jamaspishvili, T., Berman, D.M., Ross, A.E., Scher, H.I., De Marzo, A.M., Squire, J.A., and Lotan, T.L.. (2018). Clinical implications of *PTEN* loss in prostate cancer. *Nat. Rev. Urol. 15*, 222–234. https://doi.org/10.1038/nrurol.2018.9.

Jamshed, H., Beyl, R.A., Della Manna, D.L., Yang, E.S., Ravussin, E., and Peterson, C.M. (2019). Early time-restricted feeding improves 24-hour glucose levels and affects markers of the circadian clock, aging, and autophagy in humans. *Nutrients 11*, 1234. https://doi.org/10.3390/nu11061234.

Jensen, T.L., Kiersgaard, M.K., Sørensen, D.B., and Mikkelsen, L.F. (2013). Fasting of mice: A review. *Lab. Anim. 47*, 225–240. https://doi.org/10.1177/0023677213501659.

Johnson, R.J., and Andrews, P. (2015). Ancient mutation in apes may explain human obesity and diabetes. *Scientific American*, October 1.

Johnson, R.J., Sánchez-Lozada, L.G., Andrews, P., and Lanaspa, M.A. (2017). Perspective: A historical and scientific perspective of sugar and its relation with obesity and diabetes. *Adv. Nutr. 8*, 412–422. https://doi.org/10.3945/an.116.014654.

Johnson, R.J., Stenvinkel, P., Andrews, P., Sánchez-Lozada, L.G., Nakagawa, T., Gaucher, E., Andres-Hernando, A., Rodriguez-Iturbe, B., Jimenez, C.R., Garcia, G., et al. (2020).

Fructose metabolism as a common evolutionary pathway of survival associated with climate change, food shortage and droughts. *J. Intern. Med. 287*, 252–262. https://doi.org/10.1111/joim.12993.

Johnson, S. (2021). *Extra life: A short history of living longer.* New York: Riverhead Books.

Jones, K., Gordon-Weeks, A., Coleman, C., and Silva, M. (2017). Radiologically determined sarcopenia predicts morbidity and mortality following abdominal surgery: A systematic review and meta-analysis. *World J. Surg. 41*, 2266–2279. https://doi.org/10.1007/s00268-017-3999-2.

Jose, J. (2016). Statins and its hepatic effects: Newer data, implications, and changing recommendations. *J. Pharm. Bioallied Sci. 8*, 23–28. https://doi.org/10.4103/0975-7406.171699.

Joslin, E.P. (1940). The universality of diabetes: A survey of diabetic morbidity in Arizona. The Frank Billings Lecture. *JAMA 115*, 2033–2038. https://doi.org/10.1001/jama.1940.02810500001001.

Ju, Y.-E.S., McLeland, J.S., Toedebusch, C.D., Xiong, C., Fagan, A.M., Duntley, S.P., Morris, J.C., and Holtzman, D.M. (2013). Sleep quality and preclinical Alzheimer disease. *JAMA Neurol. 70*, 587–593. https://doi.org/10.1001/jamaneurol.2013.2334.

Kalmbach, D.A., Schneider, L.D., Cheung, J., Bertrand, S.J., Kariharan, T., Pack, A.I., and Gehrman, P.R. (2017). Genetic basis of chronotype in humans: Insights from three landmark GWAS. *Sleep 40*, zsw048. https://doi.org/10.1093/sleep/zsw048.

Kanai, M., Matsubara, E., Isoe, K., Urakami, K., Nakashima, K., Arai, H., Sasaki, H., Abe, K., Iwatsubo, T., Kosaka, T., et al. (1998). Longitudinal study of cerebrospinal fluid levels of tau, A beta1-40, and A beta1-42(43) in Alzheimer's disease: A study in Japan. *Ann. Neurol. 44*, 17–26. https://doi.org/10.1002/ana.410440108.

Karagiannis, A.D., Mehta, A., Dhindsa, D.S., Virani, S.S., Orringer, C.E., Blumenthal, R.S., Stone, N.J., and Sperling, L.S. (2021). How low is safe? The frontier of very low (<30 mg/dL) LDL cholesterol. *Eur. Heart J. 42*, 2154–2169. https://doi.org/10.1093/eurheartj/ehaa1080.

Karsli-Uzunbas, G., Guo, J.Y., Price, S., Teng, X., Laddha, S.V., Khor, S., Kalaany, N.Y., Jacks, T., Chan, C.S., Rabinowitz, J.D., et al. (2014). Autophagy is required for glucose homeostasis and lung tumor maintenance. *Cancer Discov. 4*, 914–927. https://doi.org/10.1158/2159-8290.CD-14-0363.

Kaivola, K., Shah, Z., Chia, R., International LBD Genomics Consortium, and Scholz, S.W. (2022). Genetic evaluation of dementia with Lewy bodies implicates distinct disease subgroups. *Brain 145*(5), 1757–1762. https://doi.org/10.1093/brain/awab402.

Kawada, S., and Ishii, N. (2005). Skeletal muscle hypertrophy after chronic restriction of venous blood flow in rats. *Med. Sci. Sports Exerc. 37*, 1144–1150. https://doi.org/10.1249/01.mss.0000170097.59514.bb.

Kawano, H., Motoyama, T., Hirashima, O., Hirai, N., Miyao, Y., Sakamoto, T., Kugiyama, K., Ogawa, H., and Yasue, H. (1999). Hyperglycemia rapidly suppresses flow-mediated endothelium-dependent vasodilation of brachial artery. *J. Am. Coll. Cardiol. 34*, 146–154. https://doi.org/10.1016/S0735-1097(99)00168-0.

Keramidas, M.E., and Botonis, P.G. (2021). Short-term sleep deprivation and human thermoregulatory function during thermal challenges. *Exp. Physiol. 106*, 1139–1148. https://doi.org/10.1113/EP089467.

Kerrouche, N., Herholz, K., Mielke, R., Holthoff, V., and Baron, J.-C. (2006). 18FDG PET in vascular dementia: Differentiation from Alzheimer's disease using voxel-based multivariate analysis. *J. Cereb. Blood Flow Metab. 26*, 1213–1221. https://doi.org/10.1038/sj.jcbfm.9600296.

Killgore, W.D.S. (2013). Self-reported sleep correlates with prefrontal-amygdala functional connectivity and emotional functioning. *Sleep 36*, 1597–1608. https://doi.org/10.5665/sleep.3106.

Kim, C.-H., Wheatley, C.M., Behnia, M., and Johnson, B.D. (2016). The effect of aging on relationships between lean body mass and VO_2max in rowers. *PLOS ONE 11*, e0160275. https://doi.org/10.1371/journal.pone.0160275.

Kim, D.-Y., Hong, S.-H., Jang, S.-H., Park, S.-H., Noh, J.-H., Seok, J.-M., Jo, H.-J., Son, C.-G., and Lee, E.-J. (2022). Systematic review for the medical applications of meditation in randomized controlled trials. *Int. J. Environ. Res. Public. Health 19*, 1244. https://doi.org/10.3390/ijerph19031244.

Kim, T.N., and Choi, K.M. (2013). Sarcopenia: Definition, epidemiology, and pathophysiology. *J. Bone Metab. 20*, 1–10. https://doi.org/10.11005/jbm.2013.20.1.1.

Kim, Y., White, T., Wijndaele, K., Westgate, K., Sharp, S.J., Helge, J.W., Wareham, N.J., and Brage, S. (2018). The combination of cardiorespiratory fitness and muscle strength, and mortality risk. *Eur. J. Epidemiol. 33*, 953–964. https://doi.org/10.1007/s10654-018-0384-x.

Kinsella, K.G. (1992). Changes in life expectancy, 1900–1990. *Am. J. Clin. Nutr. 55*, 1196S–1202S. https://doi.org/10.1093/ajcn/55.6.1196S.

Kloske, C.M., and Wilcock, D.M. (2020). The important interface between apolipoprotein E and neuroinflammation in Alzheimer's disease. *Front. Immunol. 11*, 754. https://doi.org/10.3389/fimmu.2020.00754.

Kochenderfer, J.N., Wilson, W.H., Janik, J.E., Dudley, M.E., Stetler-Stevenson, M., Feldman, S.A., Maric, I., Raffeld, M., Nathan, D.-A.N., Lanier, B.J., et al. (2010). Eradication of B-lineage cells and regression of lymphoma in a patient treated with autologous T cells genetically engineered to recognize CD19. *Blood 116*, 4099–4102. https://doi.org/10.1182/blood-2010-04-281931.

Kokkinos, P., Faselis, C., Babu, H.S.I., Pittaras, A., Doumas, M., Murphy, R., Heimall, M.S., Sui, X., Zhang, J., and Myers, J. (2022). Cardiorespiratory fitness and mortality risk across the spectra of age, race, and sex. *J. Am. Coll. Cardiol.* 80, 598–609.

Kolata, G. (2012). Severe diet doesn't prolong life, at least in monkeys. *New York Times,* August 29, 2012. https://www.nytimes.com/2012/08/30/science/low-calorie-diet-doesnt-prolong-life-study-of-monkeys-finds.html?action=click&module=RelatedCoverage&pgtype=Article®ion=Footer.

———. (2020). An Alzheimer's treatment fails: "We don't have anything now." *New York Times,* February 10. https://www.nytimes.com/2020/02/10/health/alzheimers-amyloid-drug.html.

Kolka, M.A., an.d Stephenson, L.A. (1988). Exercise thermoregulation after prolonged wakefulness. *J. Appl. Physiol. 64*, 1575–1579. https://doi.org/10.1152/jappl.1988.64.4.1575.

Konstantinos, I., Avgerinos, N.S., Mantzoros, C.S., Dalamaga, M. (2019). Obesity and cancer risk: Emerging biological mechanisms and perspectives, *Metabolism 92,* 121–135. https://doi.org/10.1016/j.metabol.2018.11.001.

Kortebein, P., Ferrando, A., Lombeida, J., Wolfe, R., and Evans, W.J. (2007). Effect of 10 days of bed rest on skeletal muscle in healthy older adults. *JAMA 297,* 1769–1774. https://doi.org/10.1001/jama.297.16.1772-b.

Kourtis, N., and Tavernarakis, N. (2009). Autophagy and cell death in model organisms. *Cell Death Differ. 16,* 21–30. https://doi.org/10.1038/cdd.2008.120.

Krause, A.J., Simon, E.B., Mander, B.A., Greer, S.M., Saletin, J.M., Goldstein-Piekarski, A.N., and Walker, M.P. (2017). The sleep-deprived human brain. *Nat. Rev. Neurosci. 18,* 404–418. https://doi.org/10.1038/nrn.2017.55.

Kuna, S.T., Maislin, G., Pack, F.M., Staley, B., Hachadoorian, R., Coccaro, E.F., and Pack, A.I. (2012). Heritability of performance deficit accumulation during acute sleep deprivation in twins. *Sleep 35,* 1223–1233. https://doi.org/10.5665/sleep.2074.

Kuo, T., McQueen, A., Chen, T.-C., and Wang, J.-C. (2015). Regulation of glucose homeostasis by glucocorticoids. *Adv. Exp. Med. Biol. 872,* 99–126. https://doi.org/10.1007/978-1-4939-2895-8_5.

Kwo, P.Y., Cohen, S.M., and Lim, J.K. (2017). ACG clinical guideline: Evaluation of abnormal liver chemistries. *Am. J. Gastroenterol. 112,* 18–35. https://doi.org/10.1038/ajg.2016.517.

Kwok, C.S., Kontopantelis, E., Kuligowski, G., Gray, M., Muhyaldeen, A., Gale, C.P., Peat, G.M., Cleator, J., Chew-Graham, C., Loke, Y.K., Mamas, M.A. (2018). Self-reported sleep duration and quality and cardiovascular disease and mortality. *JAHA,* 7:15. https://doi.org/10.1161/JAHA.118.008552.

Lammert, F., and Wang, D.Q.-H. (2005). New insights into the genetic regulation of intestinal cholesterol absorption. *Gastroenterology 129,* 718–734. https://doi.org/10.1053/j.gastro.2004.11.017.

Lamond, N., and Dawson, D. (1999). Quantifying the performance impairment associated with fatigue. *J. Sleep Res. 8,* 255–262. https://doi.org/10.1046/j.1365-2869.1999.00167.x.

Langa, K.M., and Levine, D.A. (2014). The diagnosis and management of mild cognitive impairment: A clinical review. *JAMA 312,* 2551–2561. https://doi.org/10.1001/jama.2014.13806.

Laukkanen, T., Khan, H., Zaccardi, F., and Laukkanen, J.A. (2015). Association between sauna bathing and fatal cardiovascular and all-cause mortality events. *JAMA Intern. Med. 175,* 542–548. https://doi.org/10.1001/jamainternmed.2014.8187.

Laukkanen, T., Kunutsor, S., Kauhanen, J., and Laukkanen, J.A. (2017). Sauna bathing is inversely associated with dementia and Alzheimer's disease in middle-aged Finnish men. *Age Ageing 46,* 245–249. https://doi.org/10.1093/ageing/afw212.

Lawson, J.S. (2016). Multiple infectious agents and the origins of atherosclerotic coronary artery disease. *Front. Cardiovasc. Med. 3,* 30. https://doi.org/10.3389/fcvm.2016.00030.

Le, D.T., Uram, J.N., Wang, H., Bartlett, B.R., Kemberling, H., Eyring, A.D., Skora, A.D., Luber, B.S., Azad, N.S., Laheru, D., et al. (2015). PD-1 blockade in tumors with mismatch-repair deficiency. *N. Engl. J. Med. 372,* 2509–2520. https://doi.org/10.1056/NEJMoa1500596.

Le, R., Zhao, L., and Hegele, R.A. (2022). Forty year follow-up of three patients with complete absence of apolipoprotein B-containing lipoproteins. *J. Clin. Lipidol. 16*, 155–159. https://doi.org/10.1016/j.jacl.2022.02.003.

Lee, I.-M., and Buchner, D.M. (2008). The importance of walking to public health. *Med. Sci. Sports Exerc. 40*, S512–518. https://doi.org/10.1249/MSS.0b013e31817c65d0.

Lee, J.C., Kim, S.J., Hong, S., and Kim, Y. (2019). Diagnosis of Alzheimer's disease utilizing amyloid and tau as fluid biomarkers. *Exp. Mol. Med. 51*, 1–10. https://doi.org/10.1038/s12276-019-0250-2.

Lega, I.C., and Lipscombe, L.L. (2019). Review: diabetes, obesity, and cancer—pathophysiology and clinical implications. *Endocr. Rev. 41*(1), 33–52. https://doi.org/10.1210/endrev/bnz014.

Lemasters, J.J. (2005). Selective mitochondrial autophagy, or mitophagy, as a targeted defense against oxidative stress, mitochondrial dysfunction, and aging. *Rejuvenation Res. 8*, 3–5. https://doi.org/10.1089/rej.2005.8.3.

Lendner, J.D., Helfrich, R.F., Mander, B.A., Romundstad, L., Lin, J.J., Walker, M.P., Larsson, P.G., and Knight, R.T. (2020). An electrophysiological marker of arousal level in humans. *ELife 9*, e55092. https://doi.org/10.7554/eLife.55092.

Leproult, R., Holmbäck, U., and Van Cauter, E. (2014). Circadian misalignment augments markers of insulin resistance and inflammation, independently of sleep loss. *Diabetes 63*, 1860–1869. https://doi.org/10.2337/db13-1546.

Leproult, R., and Van Cauter, E. (2010). Role of sleep and sleep loss in hormonal release and metabolism. *Endocr. Dev. 17*, 11–21. https://doi.org/10.1159/000262524.

Lexell, J. (1995). Human aging, muscle mass, and fiber type composition. *J. Gerontol. A. Biol. Sci. Med. Sci. 50 Spec No*, 11–16. https://doi.org/10.1093/gerona/50a.special_issue.11.

Li, R., Xia, J., Zhang, X., Gathirua-Mwangi, W.G., Guo, J., Li, Y., McKenzie, S., and Song, Y. (2018). Associations of muscle mass and strength with all-cause mortality among US older adults. *Med. Sci. Sports Exerc. 50*, 458–467. https://doi.org/10.1249/MSS.0000000000001448.

Libby, P. (2021). The changing landscape of atherosclerosis. *Nature 592*, 524–533. https://doi.org/10.1038/s41586-021-03392-8.

Libby, P., and Tokgözoğlu, L. (2022). Chasing LDL cholesterol to the bottom: PCSK9 in perspective. *Nat. Cardiovasc. Res. 1*, 554–561. https://doi.org/10.1038/s44161-022-00085-x.

Liberti, M.V., and Locasale, J.W. (2016). The Warburg effect: How does it benefit cancer cells? *Trends Biochem. Sci. 41*, 211–218. https://doi.org/10.1016/j.tibs.2015.12.001.

Lieberman, D.E., Kistner, T.M., Richard, D., Lee, I.-M., and Baggish, A.L. (2021). The active grandparent hypothesis: Physical activity and the evolution of extended human healthspans and lifespans. *Proc. Natl. Acad. Sci. 118*, e2107621118. https://doi.org/10.1073/pnas.2107621118.

Liguori, G., ed. (2020). *ACSM's guidelines for exercise testing and prescription.* 10th ed. Philadelphia: Wolters Kluwer Health.

Lim, A.S.P., Kowgier, M., Yu, L., Buchman, A.S., and Bennett, D.A. (2013a). Sleep fragmentation and the risk of incident Alzheimer's disease and cognitive decline in older persons. *Sleep 36*, 1027–1032. https://doi.org/10.5665/sleep.2802.

Lim, A.S.P., Yu, L., Kowgier, M., Schneider, J.A., Buchman, A.S., and Bennett, D.A. (2013b). Sleep modifies the relation of APOE to the risk of Alzheimer disease and neurofibrillary tangle pathology. *JAMA Neurol. 70*, 10.1001/jamaneurol.2013.4215. https://doi.org/10.1001/jamaneurol.2013.4215.

Lim, J., and Dinges, D.F. (2008). Sleep deprivation and vigilant attention. *Ann. N.Y. Acad. Sci. 1129*, 305–322. https://doi.org/10.1196/annals.1417.002.

Lin, H.-J., Lee, B.-C., Ho, Y.-L., Lin, Y.-H., Chen, C.-Y., Hsu, H.-C., Lin, M.-S., Chien, K.-L., and Chen, M.-F. (2009). Postprandial glucose improves the risk prediction of cardiovascular death beyond the metabolic syndrome in the nondiabetic population. *Diabetes Care 32*, 1721–1726. https://doi.org/10.2337/dc08-2337.

Lin, H.-S., Watts, J.N., Peel, N.M., and Hubbard, R.E. (2016). Frailty and post-operative outcomes in older surgical patients: A systematic review. *BMC Geriatr. 16*, 157. https://doi.org/10.1186/s12877-016-0329-8.

Lindle, R.S., Metter, E.J., Lynch, N.A., Fleg, J.L., Fozard, J.L., Tobin, J., Roy, T.A., and Hurley, B.F. (1997). Age and gender comparisons of muscle strength in 654 women and men aged 20–93 yr. *J. Appl. Physiol. 83*, 1581–1587. https://doi.org/10.1152/jappl.1997.83.5.1581.

Linehan, M.M., Comtois, K.A., Murray, A.M., Brown, M.Z., Gallop, R.J., Heard, H.L., Korslund, K.E., Tutek, D.A., Reynolds, S.K., and Lindenboim, N. (2006). Two-year randomized controlled trial and follow-up of dialectical behavior therapy vs therapy by experts for suicidal behaviors and borderline personality disorder. *Arch. Gen. Psychiatry 63*, 757–766. https://doi.org/10.1001/archpsyc.63.7.757.

Little, J.P., Gillen, J.B., Percival, M.E., Safdar, A., Tarnopolsky, M.A., Punthakee, Z., Jung, M.E., and Gibala, M.J. (2011). Low-volume high-intensity interval training reduces hyperglycemia and increases muscle mitochondrial capacity in patients with type 2 diabetes. *J. Appl. Physiol. 111*, 1554–1560. https://doi.org/10.1152/japplphysiol.00921.2011.

Liu, D., Huang, Y., Huang, C., Yang, S., Wei, X., Zhang, P., Guo, D., Lin, J., Xu, B., Li, C., et al. (2022). Calorie restriction with or without time-restricted eating in weight loss. *N. Engl. J. Med. 386*, 1495–1504. https://doi.org/10.1056/NEJMoa2114833.

Liu, G.Y., and Sabatini, D.M. (2020). mTOR at the nexus of nutrition, growth, ageing and disease. *Nat. Rev. Mol. Cell Biol. 21*, 183–203. https://doi.org/10.1038/s41580-019-0199-y.

Livingston, G. (2019). On average, older adults spend over half their waking hours alone. *Grius,* July 19. https://qrius.com/on-average-older-adults-spend-over-half-their-waking-hours-alone/.

Lobo, A., López-Antón, R., de-la-Cámara, C., Quintanilla, M.A., Campayo, A., Saz, P., and ZARADEMP Workgroup (2008). Non-cognitive psychopathological symptoms associated with incident mild cognitive impairment and dementia, Alzheimer's type. *Neurotox. Res. 14*, 263–272. https://doi.org/10.1007/BF03033815.

López-Otín, C., Blasco, M.A., Partridge, L., Serrano, M., and Kroemer, G. (2013). The hallmarks of aging. *Cell 153*, 1194–1217. https://doi.org/10.1016/j.cell.2013.05.039.

Lowe, D.A., Wu, N., Rohdin-Bibby, L., Moore, A.H., Kelly, N., Liu, Y.E., Philip, E., Vittinghoff, E., Heymsfield, S.B., Olgin, J.E., et al. (2020). Effects of time-restricted eating on weight loss and other metabolic parameters in women and men with overweight and obesity: The TREAT randomized clinical trial. *JAMA Intern. Med. 180*, 1491–1499. https://doi.org/10.1001/jamainternmed.2020.4153.

Lucey, B.P., McCullough, A., Landsness, E.C., Toedebusch, C.D., McLeland, J.S., Zaza, A.M., Fagan, A.M., McCue, L., Xiong, C., Morris, J.C., et al. (2019). Reduced non-rapid eye movement sleep is associated with tau pathology in early Alzheimer's disease. *Sci. Transl. Med. 11*, eaau6550. https://doi.org/10.1126/scitranslmed.aau6550.

Ludwig, J., Viggiano, T.R., McGill, D.B., and Oh, B.J. (1980). Nonalcoholic steatohepatitis: Mayo Clinic experiences with a hitherto unnamed disease. *Mayo Clinic proceedings*, 55(7), 434–438.

Lüth, H.-J., Ogunlade, V., Kuhla, B., Kientsch-Engel, R., Stahl, P., Webster, J., Arendt, T., and Münch, G. (2005). Age- and stage-dependent accumulation of advanced glycation end products in intracellular deposits in normal and Alzheimer's disease brains. *Cereb. Cortex 15*, 211–220. https://doi.org/10.1093/cercor/bhh123.

Mach, F., Ray, K.K., Wiklund, O., Corsini, A., Catapano, A.L., Bruckert, E., De Backer, G., Hegele, R.A., Hovingh, G.K., Jacobson, T.A., et al. (2018). Adverse effects of statin therapy: perception vs. the evidence: Focus on glucose homeostasis, cognitive, renal and hepatic function, haemorrhagic stroke and cataract. *Eur. Heart J. 39*, 2526–2539. https://doi.org/10.1093/eurheartj/ehy182.

Maddock, J., Cavadino, A., Power, C., and Hyppönen, E. (2015). 25-hydroxyvitamin D, APOE ε4 genotype and cognitive function: Findings from the 1958 British birth cohort. *Eur. J. Clin. Nutr. 69*, 505–508. https://doi.org/10.1038/ejcn.2014.201.

Maeng, L.Y., and Milad, M.R. (2015). Sex differences in anxiety disorders: Interactions between fear, stress, and gonadal hormones. *Horm. Behav. 76*, 106–117. https://doi.org/10.1016/j.yhbeh.2015.04.002.

Mah, C.D., Mah, K.E., Kezirian, E.J., and Dement, W.C. (2011). The effects of sleep extension on the athletic performance of collegiate basketball players. *Sleep 34*, 943–950. https://doi.org/10.5665/SLEEP.1132.

Mandsager, K., Harb, S., Cremer, P., Phelan, D., Nissen, S.E., and Jaber, W. (2018). Association of cardiorespiratory fitness with long-term mortality among adults undergoing exercise treadmill testing. *JAMA Netw. Open 1*, e183605. https://doi.org/10.1001/jamanetworkopen.2018.3605.

Mannick, J.B., Del Giudice, G., Lattanzi, M., Valiante, N.M., Praestgaard, J., Huang, B., Lonetto, M.A., Maecker, H.T., Kovarik, J., Carson, S., et al. (2014). mTOR inhibition improves immune function in the elderly. *Sci. Transl. Med. 6*, 268ra179. https://doi.org/10.1126/scitranslmed.3009892.

Manson, J.E., Chlebowski, R.T., Stefanick, M.L., Aragaki, A.K., Rossouw, J.E., Prentice, R.L., Anderson, G., Howard, B.V., Thomson, C.A., LaCroix, A.Z., et al. (2013). The Women's Health Initiative hormone therapy trials: Update and overview of health outcomes during the intervention and post-stopping phases. *JAMA 310*, 1353–1368.

Mansukhani, M.P., Kolla, B.P., Surani, S., Varon, J., and Ramar, K. (2012). Sleep deprivation in resident physicians, work hour limitations, and related outcomes: A systematic review of the literature. *Postgrad. Med. 124*, 241–249. https://doi.org/10.3810/pgm.2012.07.2583.

Marston, N.A., Giugliano, R.P., Melloni, G.E.M., Park, J.-G., Morrill, V., Blazing, M.A., Ference, B., Stein, E., Stroes, E.S., Braunwald, E., et al. (2022). Association of Apolipoprotein B–containing lipoproteins and risk of myocardial infarction in individuals with and without atherosclerosis: Distinguishing between particle concentration, type, and content. *JAMA Cardiol.* 7(3), 250–256. http://doi.org/10.1001/jamacardio.2021.5083.

Martínez-Lapiscina, E.H., Clavero, P., Toledo, E., Estruch, R., Salas-Salvadó, J., Julián, B.S., Sanchez-Tainta, A., Ros, E., Valls-Pedret, C., and Martinez-Gonzalez, M.Á. (2013). Mediterranean diet improves cognition: The PREDIMED-NAVARRA randomised trial. *J. Neurol. Neurosurg. Psychiatry 84*, 1318–1325. https://doi.org/10.1136/jnnp-2012-304792.

Masana, L., Girona, J., Ibarretxe, D., Rodríguez-Calvo, R., Rosales, R., Vallvé, J.-C., Rodríguez-Borjabad, C., Guardiola, M., Rodríguez, M., Guaita-Esteruelas, S., et al. (2018). Clinical and pathophysiological evidence supporting the safety of extremely low LDL levels: The zero-LDL hypothesis. *J. Clin. Lipidol. 12*, 292–299.e3. https://doi.org/10.1016/j.jacl.2017.12.018.

Masters, C.L., and Selkoe, D.J. (2012). Biochemistry of amyloid β-protein and amyloid deposits in Alzheimer disease. *Cold Spring Harb. Perspect. Med. 2*, a006262. https://doi.org/10.1101/cshperspect.a006262.

Matsuzaki, T., Sasaki, K., Tanizaki, Y., Hata, J., Fujimi, K., Matsui, Y., Sekita, A., Suzuki, S.O., Kanba, S., Kiyohara, Y., et al. (2010). Insulin resistance is associated with the pathology of Alzheimer disease: The Hisayama study. *Neurology 75*, 764–770. https://doi.org/10.1212/WNL.0b013e3181eee25f.

Mattison, J.A., Roth, G.S., Beasley, T.M., Tilmont, E.M., Handy, A.H., Herbert, R.L., Longo, D.L., Allison, D.B., Young, J.E., Bryant, M., et al. (2012). Impact of caloric restriction on health and survival in rhesus monkeys: The NIA study. *Nature 489*, https://doi.org/10.1038/nature11432.

Maurer, L.F., Schneider, J., Miller, C.B., Espie, C.A., and Kyle, S.D. (2021). The clinical effects of sleep restriction therapy for insomnia: A meta-analysis of randomised controlled trials. *Sleep Med. Rev. 58*, 101493. https://doi.org/10.1016/j.smrv.2021.101493.

McDonald, R.B., and Ramsey, J.J. (2010). Honoring Clive McCay and 75 years of calorie restriction research. *J. Nutr. 140*, 1205–1210. https://doi.org/10.3945/jn.110.122804.

McLaughlin, T., Abbasi, F., Cheal, K., Chu, J., Lamendola, C., and Reaven, G. (2003). Use of metabolic markers to identify overweight individuals who are insulin resistant. *Ann. Intern. Med. 139*, 802–809. https://doi.org/10.7326/0003-4819-139-10-200311180-00007.

McMillin, S.L., Schmidt, D.L., Kahn, B.B., and Witczak, C.A. (2017). GLUT4 is not necessary for overload-induced glucose uptake or hypertrophic growth in mouse skeletal muscle. *Diabetes 66*, 1491–1500. https://doi.org/10.2337/db16-1075.

McNamara, D.J. (2015). The fifty year rehabilitation of the egg. *Nutrients 7*, 8716–8722. https://doi.org/10.3390/nu7105429.

Melov, S., Tarnopolsky, M.A., Beckman, K., Felkey, K., and Hubbard, A. (2007). Resistance exercise reverses aging in human skeletal muscle. *PLOS ONE 2*, e465. https://doi.org/10.1371/journal.pone.0000465.

Mensah, G. A., Wei, G. S., Sorlie, P. D., Fine, L. J., Rosenberg, Y., Kaufmann, P. G., Mussolino, M. E., Hsu, L. L., Addou, E., Engelgau, M. M., & Gordon, D. (2017). Decline in Cardiovascular Mortality: Possible Causes and Implications. *Circulation research, 120*(2), 366–380. https://doi.org/10.1161/CIRCRESAHA.116.309115.

Mensink, R.P., and Katan, M.B. (1992). Effect of dietary fatty acids on serum lipids and lipoproteins. A meta-analysis of 27 trials. *Arterioscler. Thromb. J. Vasc. Biol. 12*, 911–919. https://doi.org/10.1161/01.atv.12.8.911.

Mercken, E.M., Crosby, S.D., Lamming, D.W., JeBailey, L., Krzysik-Walker, S., Villareal, D., Capri, M., Franceschi, C., Zhang, Y., Becker, K., et al. (2013). Calorie restriction in humans inhibits the PI3K/AKT pathway and induces a younger transcription profile. *Aging Cell 12*, 645–651. https://doi.org/10.1111/acel.12088.

Michaelson, D.M. (2014). APOE ε4: The most prevalent yet understudied risk factor for Alzheimer's disease. *Alzheimers Dement. 10*, 861–868. https://doi.org/10.1016/j.jalz.2014.06.015.

Milewski, M.D., Skaggs, D.L., Bishop, G.A., Pace, J.L., Ibrahim, D.A., Wren, T.A.L., and Barzdukas, A. (2014). Chronic lack of sleep is associated with increased sports injuries in adolescent athletes. *J. Pediatr. Orthop. 34*, 129–133. https://doi.org/10.1097/BPO.0000000000000151.

Miller, R.A., Harrison, D.E., Astle, C.M., Baur, J.A., Boyd, A.R., de Cabo, R., Fernandez, E., Flurkey, K., Javors, M.A., Nelson, J.F., et al. (2011). Rapamycin, but not resveratrol or simvastatin, extends life span of genetically heterogeneous mice. *J. Gerontol. Ser. A 66A*, 191–201. https://doi.org/10.1093/gerona/glq178.

Mitter, S.S., Oriá, R.B., Kvalsund, M.P., Pamplona, P., Joventino, E.S., Mota, R.M.S., Gonçalves, D.C., Patrick, P.D., Guerrant, R.L., and Lima, A.A.M. (2012). Apolipoprotein E4 influences growth and cognitive responses to micronutrient supplementation in shantytown children from northeast Brazil. *Clinics 67*, 11–18. https://doi.org/10.6061/clinics/2012(01)03.

Moco, S., Bino, R.J., Vorst, O., Verhoeven, H.A., de Groot, J., van Beek, T.A., Vervoort, J., and de Vos, C.H.R. (2006). A liquid chromatography-mass spectrometry-based metabolome database for tomato. *Plant Physiol. 141*, 1205–1218. https://doi.org/10.1104/pp.106.078428.

Mollenhauer, B., Bibl, M., Trenkwalder, C., Stiens, G., Cepek, L., Steinacker, P., Ciesielczyk, B., Neubert, K., Wiltfang, J., Kretzschmar, H.A., et al. (2005). Follow-up investigations in cerebrospinal fluid of patients with dementia with Lewy bodies and Alzheimer's disease. *J. Neural Transm. 112*, 933–948. https://doi.org/10.1007/s00702-004-0235-7.

Montagne, A., Nation, D.A., Sagare, A.P., Barisano, G., Sweeney, M.D., Chakhoyan, A., Pachicano, M., Joe, E., Nelson, A.R., D'Orazio, L.M., et al. (2020). APOE4 leads to blood-brain barrier dysfunction predicting cognitive decline. *Nature 581*, 71–76. https://doi.org/10.1038/s41586-020-2247-3.

Moraes, W. dos S., Poyares, D.R., Guilleminault, C., Ramos, L.R., Bertolucci, P.H.F., and Tufik, S. (2006). The effect of donepezil on sleep and REM sleep EEG in patients with Alzheimer disease: A double-blind placebo-controlled study. *Sleep 29*, 199–205. https://doi.org/10.1093/sleep/29.2.199.

Mosconi, L., Rahman, A., Diaz, I., Wu, X., Scheyer, O., Hristov, H.W., Vallabhajosula, S., Isaacson, R.S., de Leon, M.J., and Brinton, R.D. (2018). Increased Alzheimer's risk during the menopause transition: A 3-year longitudinal brain imaging study. *PLOS ONE 13*, e0207885. https://doi.org/10.1371/journal.pone.0207885.

Motomura, Y., Kitamura, S., Oba, K., Terasawa, Y., Enomoto, M., Katayose, Y., Hida, A., Moriguchi, Y., Higuchi, S., and Mishima, K. (2013). Sleep debt elicits negative emotional reaction through diminished amygdala-anterior cingulate functional connectivity. *PLOS ONE 8*, e56578. https://doi.org/10.1371/journal.pone.0056578.

Mukherjee, S. (2011). *The emperor of all maladies: A biography of cancer.* New York: Scribner.

Mullane, K., and Williams, M. (2020). Alzheimer's disease beyond amyloid: Can the repetitive failures of amyloid-targeted therapeutics inform future approaches to dementia drug discovery? *Biochem. Pharmacol. 177*, 113945. https://doi.org/10.1016/j.bcp.2020.113945.

Müller, U., Winter, P., and Graeber, M.B. (2013). A presenilin 1 mutation in the first case of Alzheimer's disease. *Lancet Neurol. 12*, 129–130. https://doi.org/10.1016/S1474-4422(12)70307-1.

Naci, H., and Ioannidis, J.P.A. (2015). Comparative effectiveness of exercise and drug interventions on mortality outcomes: Metaepidemiological study. *Br. J. Sports Med. 49*, 1414–1422. https://doi.org/10.1136/bjsports-2015-f5577rep.

Naghshi, S., Sadeghian, M., Nasiri, M., Mobarak, S., Asadi, M., and Sadeghi, O. (2020). Association of total nut, tree nut, peanut, and peanut butter consumption with cancer incidence and mortality: A comprehensive systematic review and dose-response meta-analysis of observational studies. *Adv. Nutr. 12*, 793–808. https://doi.org/10.1093/advances/nmaa152.

Naimi, T.S., Stockwell, T., Zhao, J., Xuan, Z., Dangardt, F., Saitz, R., Liang, W., and Chikritzhs, T. (2017). Selection biases in observational studies affect associations between "moderate" alcohol consumption and mortality. *Addiction 112*, 207–214. https://doi.org/10.1111/add.13451.

Nakamura, T., Shoji, M., Harigaya, Y., Watanabe, M., Hosoda, K., Cheung, T.T., Shaffer, L.M., Golde, T.E., Younkin, L.H., and Younkin, S.G. (1994). Amyloid beta protein levels in cerebrospinal fluid are elevated in early-onset Alzheimer's disease. *Ann. Neurol. 36*, 903–911. https://doi.org/10.1002/ana.410360616.

Nasir, K., Cainzos-Achirica, M., Valero-Elizondo, J., Ali, S.S., Havistin, R., Lakshman, S., Blaha, M.J., Blankstein, R., Shapiro, M.D., Arias, L., et al. (2022). Coronary atherosclerosis in an asymptomatic U.S. population. *JACC Cardiovasc. Imaging 15*(9), 1619–1621. https://doi.org/10.1016/j.jcmg.2022.03.010.

NCI (National Cancer Institute). (2015). Risk factors: Age. https://www.cancer.gov/about-cancer/causes-prevention/risk/age.

———. (2021). Risk factors: Age. https://www.cancer.gov/about-cancer/causes-prevention/risk/age.

———. (2022a). Obesity and cancer. Fact sheet, April 5. https://www.cancer.gov/about-cancer/causes-prevention/risk/obesity/obesity-fact-sheet.

———. (2022b). SEER survival statistics—SEER Cancer Query Systems. https://seer.cancer.gov/canques/survival.html.

Nedeltcheva, A.V., Kessler, L., Imperial, J., and Penev, P.D. (2009). Exposure to recurrent sleep restriction in the setting of high caloric intake and physical inactivity results in increased insulin resistance and reduced glucose tolerance. *J. Clin. Endocrinol. Metab. 94*, 3242–3250. https://doi.org/10.1210/jc.2009-0483.

Neth, B.J., and Craft, S. (2017). Insulin resistance and Alzheimer's disease: Bioenergetic linkages. *Front. Aging Neurosci. 9*, 345. https://doi.org/10.3389/fnagi.2017.00345.

Neu, S.C., Pa, J., Kukull, W., Beekly, D., Kuzma, A., Gangadharan, P., Wang, L.-S., Romero, K., Arneric, S.P., Redolfi, A., et al. (2017). Apolipoprotein E genotype and sex risk factors for Alzheimer disease: A meta-analysis. *JAMA Neurol. 74*, 1178–1189. https://doi.org/10.1001/jamaneurol.2017.2188.

Newman, A.B., Kupelian, V., Visser, M., Simonsick, E.M., Goodpaster, B.H., Kritchevsky, S.B., Tylavsky, F.A., Rubin, S.M., and Harris, T.B. (2006). Strength, but not muscle mass, is associated with mortality in the Health, Aging and Body Composition Study cohort. *J. Gerontol. Ser. A 61*, 72–77. https://doi.org/10.1093/gerona/61.1.72.

Newman, C.B., Preiss, D., Tobert, J.A., Jacobson, T.A., Page, R.L., Goldstein, L.B., Chin, C., Tannock, L.R., Miller, M., Raghuveer, G., et al. (2019). Statin safety and associated adverse events: A scientific statement from the American Heart Association. *Arterioscler. Thromb. Vasc. Biol. 39*, e38–e81. https://doi.org/10.1161/ATV.0000000000000073.

New York Times. (1985). New evidence, old debate. September 12. https://www.nytimes.com/1985/09/12/us/new-evidence-old-debate.html.

Ngandu, T., Lehtisalo, J., Solomon, A., Levälahti, E., Ahtiluoto, S., Antikainen, R., Bäckman, L., Hänninen, T., Jula, A., Laatikainen, T., et al. (2015). A 2 year multidomain intervention of diet, exercise, cognitive training, and vascular risk monitoring versus control to prevent cognitive decline in at-risk elderly people (FINGER): A randomised controlled trial. *Lancet 385*, 2255–2263. https://doi.org/10.1016/S0140-6736(15)60461-5.

NHTSA (National Highway Traffic Safety Administration). (2022a). Early estimates of motor vehicle traffic fatalities and fatality rate by sub-categories in 2021. Traffic Safety Facts, May. https://crashstats.nhtsa.dot.gov/Api/Public/ViewPublication/813298.

———. (2022b). Fatality and Injury Reporting System Tool (FIRST). https://cdan.dot.gov/query.

Nicklas, B.J., Chmelo, E., Delbono, O., Carr, J.J., Lyles, M.F., and Marsh, A.P. (2015). Effects of resistance training with and without caloric restriction on physical function and mobility in overweight and obese older adults: A randomized controlled trial. *Am. J. Clin. Nutr. 101*, 991–999. https://doi.org/10.3945/ajcn.114.105270.

NIDDK (National Institute of Diabetes and Digestive and Kidney Diseases). (2018). *Diabetes in America.* 3rd ed. Bethesda, MD: NIDDK.

Ninonuevo, M.R., Park, Y., Yin, H., Zhang, J., Ward, R.E., Clowers, B.H., German, J.B., Freeman, S.L., Killeen, K., Grimm, R., et al. (2006). A strategy for annotating the human milk glycome. *J. Agric. Food Chem. 54*, 7471–7480. https://doi.org/10.1021/jf0615810.

Nuttall, F.Q., and Gannon, M.C. (2006). The metabolic response to a high-protein, low-carbohydrate diet in men with type 2 diabetes mellitus. *Metabolism 55*, 243–251. https://doi.org/10.1016/j.metabol.2005.08.027.

Nymo, S., Coutinho, S.R., Jørgensen, J., Rehfeld, J.F., Truby, H., Kulseng, B., and Martins, C.

(2017). Timeline of changes in appetite during weight loss with a ketogenic diet. *Int. J. Obes. 41*, 1224–1231. https://doi.org/10.1038/ijo.2017.96.

O'Donoghue, M.L., Fazio, S., Giugliano, R.P., et al. (2019). Lipoprotein(a), PCSK9 inhibition, and cardiovascular risk. *Circulation,139*(12):1483-1492. doi:10.1161/CIRCULATION AHA.118.037184.

Ogden, C.L., Fryar, C.D., Carroll, M.D., and Flegal, K.M. (2004). Mean body weight, height, and body mass index, United States 1960–2002. *Adv. Data* 1–17.

Ohayon, M.M., Carskadon, M.A., Guilleminault, C., and Vitiello, M.V. (2004). Meta-analysis of quantitative sleep parameters from childhood to old age in healthy individuals: Developing normative sleep values across the human lifespan. *Sleep 27*, 1255–1273. https://doi.org/10.1093/sleep/27.7.1255.

O'Keefe, J.H., Cordain, L., Harris, W.H., Moe, R.M., and Vogel, R. (2004). Optimal low-density lipoprotein is 50 to 70 mg/dl: Lower is better and physiologically normal. *J. Am. Coll. Cardiol. 43*, 2142–2146. https://doi.org/10.1016/j.jacc.2004.03.046.

Oliveira, C., Cotrim, H., and Arrese, M. (2019). Nonalcoholic fatty liver disease risk factors in Latin American populations: Current scenario and perspectives. *Clin. Liver Dis. 13*, 39–42. https://doi.org/10.1002/cld.759.

Oriá, R.B., Patrick, P.D., Blackman, J.A., Lima, A.A.M., and Guerrant, R.L. (2007). Role of apolipoprotein E4 in protecting children against early childhood diarrhea outcomes and implications for later development. *Med. Hypotheses 68*, 1099–1107. https://doi.org/10.1016/j.mehy.2006.09.036.

Orphanet (2022). Orphanet: 3 hydroxyisobutyric aciduria. https://www.orpha.net/consor/cgi-bin/OC_Exp.php?Lng=EN&Expert=939.

Osorio, R.S., Pirraglia, E., Agüera-Ortiz, L.F., During, E.H., Sacks, H., Ayappa, I., Walsleben, J., Mooney, A., Hussain, A., Glodzik, L., et al. (2011). Greater risk of Alzheimer's disease in older adults with insomnia. *J. Am. Geriatr.* Soc. *59*, 559–562. https://doi.org/10.1111/j.1532-5415.2010.03288.x.

Oulhaj, A., Jernerén, F., Refsum, H., Smith, A.D., and de Jager, C.A. (2016). Omega-3 fatty acid status enhances the prevention of cognitive decline by B vitamins in mild cognitive impairment. *J. Alzheimers Dis. 50*, 547–557. https://doi.org/10.3233/JAD-150777.

Oyetakin-White, P., Suggs, A., Koo, B., Matsui, M.S., Yarosh, D., Cooper, K.D., and Baron, E.D. (2015). Does poor sleep quality affect skin ageing? *Clin. Exp. Dermatol. 40*, 17–22. https://doi.org/10.1111/ced.12455.

Patel, A.K., Reddy, V., and Araujo, J.F. (2022). *Physiology, sleep stages.* Treasure Island, FL: StatPearls.

Patel, D., Steinberg, J., and Patel, P. (2018). Insomnia in the elderly: A review. *J. Clin. Sleep Med. 14*, 1017–1024. https://doi.org/10.5664/jcsm.7172.

Peng, B., Yang, Q., B Joshi, R., Liu, Y., Akbar, M., Song, B.-J., Zhou, S., and Wang, X. (2020). Role of alcohol drinking in Alzheimer's disease, Parkinson's disease, and amyotrophic lateral sclerosis. *Int. J. Mol. Sci. 21*, 2316. https://doi.org/10.3390/ijms21072316.

Perls, T.T. (2017). Male centenarians: How and why are they different from their female counterparts? *J. Am. Geriatr. Soc. 65*, 1904–1906. https://doi.org/10.1111/jgs.14978.

Pesch, B., Kendzia, B., Gustavsson, P., Jöckel, K.-H., Johnen, G., Pohlabeln, H., Olsson, A., Ahrens, W., Gross, I.M., Brüske, I., et al. (2012). Cigarette smoking and lung cancer—relative risk estimates for the major histological types from a pooled analysis of case-control studies. *Int. J. Cancer 131*, 1210–1219. https://doi.org/10.1002/ijc.27339.

Petersen, K.F., Dufour, S., Savage, D.B., Bilz, S., Solomon, G., Yonemitsu, S., Cline, G.W., Befroy, D., Zemany, L., Kahn, B.B., et al. (2007). The role of skeletal muscle insulin resistance in the pathogenesis of the metabolic syndrome. *Proc. Natl. Acad. Sci. 104*, 12587–12594. https://doi.org/10.1073/pnas.0705408104.

Petersen, M.C., and Shulman, G.I. (2018). Mechanisms of insulin action and insulin resistance. *Physiol. Rev. 98*, 2133–2223. https://doi.org/10.1152/physrev.00063.2017.

Pfister, R., Sharp, S.J., Luben, R., Khaw, K.-T., and Wareham, N.J. (2011). No evidence of an increased mortality risk associated with low levels of glycated haemoglobin in a non-diabetic UK population. *Diabetologia 54*, 2025–2032. https://doi.org/10.1007/s00125-011-2162-0.

Phinney, S., and Volek, J. (2018). The science of nutritional ketosis and appetite. *Virta* (blog), July 25. https://www.virtahealth.com/blog/ketosis-appetite-hunger.

Picard, C. (2018). The secrets to living to 100 (according to people who've done it). *Good Housekeeping.*

Picton, J.D., Marino, A.B., and Nealy, K.L. (2018). Benzodiazepine use and cognitive decline in the elderly. *Am. J. Health. Syst. Pharm. 75*, e6–e12. https://doi.org/10.2146/ajhp160381.

Pollack, A. (2005). Huge genome project is proposed to fight cancer. *New York Times*, March 28. https://www.nytimes.com/2005/03/28/health/huge-genome-project-is-proposed-to-fight-cancer.html.

Pontzer, H., Wood, B.M., and Raichlen, D.A. (2018). Hunter-gatherers as models in public health. *Obes. Rev. 19*, 24–35. https://doi.org/10.1111/obr.12785.

Potvin, O., Lorrain, D., Forget, H., Dubé, M., Grenier, S., Préville, M., and Hudon, C. (2012). Sleep quality and 1-year incident cognitive impairment in community-dwelling older adults. *Sleep 35*, 491–499. https://doi.org/10.5665/sleep.1732.

Powell-Wiley, T.M., Poirier, P., Burke, L.E., Després, J.-P., Gordon-Larsen, P., Lavie, C.J., Lear, S.A., Ndumele, C.E., Neeland, I.J., Sanders, P., et al. (2021). Obesity and cardiovascular disease: A scientific statement from the American Heart Association. *Circulation 143*, e984–e1010. https://doi.org/10.1161/CIR.0000000000000973.

Prather, A.A., Bogdan, R., and Hariri, A.R. (2013). Impact of sleep quality on amygdala reactivity, negative affect, and perceived stress. *Psychosom. Med. 75*, 350–358. https://doi.org/10.1097/PSY.0b013e31828ef15b.

Prati, D., Taioli, E., Zanella, A., del a Torre, E., Butelli, S., Del Vecchio, E., Vianello, L., Zanuso, F., Mozzi, F., Milani, S., et al. (2002). Updated definitions of healthy ranges for serum alanine aminotransferase levels. *Ann. Intern. Med. 137*, 1–10. https://doi.org/10.7326/0003-4819-137-1-200207020-00006.

Proctor, R.N. (1995). *Cancer wars: How politics shapes what we know and don't know about cancer.* New York: Basic Books.

———. (2001). Tobacco and the global lung cancer epidemic. *Nat. Rev. Cancer 1*, 82–86. https://doi.org/10.1038/35094091.

Rabinovici, G.D., Gatsonis, C., Apgar, C., Chaudhary, K., Gareen, I., Hanna, L., Hendrix, J., Hillner, B.E., Olson, C., Lesman-Segev, O.H., et al. (2019). Association of amyloid positron emission tomography with subsequent change in clinical management among medicare beneficiaries with mild cognitive impairment or dementia. *JAMA 321*, 1286–1294. https://doi.org/10.1001/jama.2019.2000.

Rahman, A., Schelbaum, E., Hoffman, K., Diaz, I., Hristov, H., Andrews, R., Jett, S., Jackson, H., Lee, A., Sarva, H., et al. (2020). Sex-driven modifiers of Alzheimer risk: A multimodality brain imaging study. *Neurology 95*, e166–e178. https://doi.org/10.1212/WNL.0000000000009781.

Raichle, M.E., and Gusnard, D.A. (2002). Appraising the brain's energy budget. *Proc. Natl. Acad. Sci. 99*, 10237–10239. https://doi.org/10.1073/pnas.172399499.

Rajpathak, S.N., Liu, Y., Ben-David, O., Reddy, S., Atzmon, G., Crandall, J., and Barzilai, N. (2011). Lifestyle factors of people with exceptional longevity. *J. Am. Geriatr. Soc. 59*, 1509–1512. https://doi.org/10.1111/j.1532-5415.2011.03498.x.

Rao, M.N., Neylan, T.C., Grunfeld, C., Mulligan, K., Schambelan, M., and Schwarz, J.-M. (2015). Subchronic sleep restriction causes tissue-specific insulin resistance. *J. Clin. Endocrinol. Metab. 100*, 1664–1671. https://doi.org/10.1210/jc.2014-3911.

Raskind, M.A., Peskind, E.R., Hoff, D.J., Hart, K.L., Holmes, H.A., Warren, D., Shofer, J., O'Connell, J., Taylor, F., Gross, C., et al. (2007). A parallel group placebo controlled study of prazosin for trauma nightmares and sleep disturbance in combat veterans with post-traumatic stress disorder. *Biol. Psychiatry 61*, 928–934. https://doi.org/10.1016/j.biopsych.2006.06.032.

Raskind, M.A., Peskind, E.R., Kanter, E.D., Petrie, E.C., Radant, A., Thompson, C.E., Dobie, D.J., Hoff, D., Rein, R.J., Straits-Tröster, K., et al. (2003). Reduction of nightmares and other PTSD symptoms in combat veterans by prazosin: A placebo-controlled study. *Am. J. Psychiatry 160*, 371–373. https://doi.org/10.1176/appi.ajp.160.2.371.

Ratnakumar, A., Zimmerman, S.E., Jordan, B.A., and Mar, J.C. (2019). Estrogen activates Alzheimer's disease genes. *Alzheimers Dement. 5*, 906–917. https://doi.org/10.1016/j.trci.2019.09.004.

Real, T. (1998). *I don't want to talk about it: Overcoming the secret legacy of male depression.* New York: Scribner.

Reddy, O.C., and van der Werf, Y.D. (2020). The sleeping brain: Harnessing the power of the glymphatic system through lifestyle choices. *Brain Sci. 10*, 868. https://doi.org/10.3390/brainsci10110868.

Reiman, E.M., Arboleda-Velasquez, J.F., Quiroz, Y.T., Huentelman, M.J., Beach, T.G., Caselli, R.J., Chen, Y., Su, Y., Myers, A.J., Hardy, J., et al. (2020). Exceptionally low likelihood of Alzheimer's dementia in APOE2 homozygotes from a 5,000-person neuropathological study. *Nat. Commun. 11*, 667. https://doi.org/10.1038/s41467-019-14279-8.

Reiman, E.M., Caselli, R.J., Yun, L.S., Chen, K., Bandy, D., Minoshima, S., Thibodeau, S.N., and Osborne, D. (1996). Preclinical evidence of Alzheimer's disease in persons homozygous for the epsilon 4 allele for apolipoprotein E. *N. Engl. J. Med. 334*, 752–758. https://doi.org/10.1056/NEJM199603213341202.

Reimers, C.D., Knapp, G., and Reimers, A.K. (2012). Does physical activity increase life expectancy? A review of the literature. *J. Aging Res. 2012*, 243958. https://doi.org/10.1155/2012/243958.

Repantis, D., Wermuth, K., Tsamitros, N., Danker-Hopfe, H., Bublitz, J.C., Kühn, S., and Dresler, M. (2020). REM sleep in acutely traumatized individuals and interventions for the secondary prevention of post-traumatic stress disorder. *Eur. J. Psychotraumatology 11*, 1740492. https://doi.org/10.1080/20008198.2020.1740492.

Reutrakul, S., and Van Cauter, E. (2018). Sleep influences on obesity, insulin resistance, and risk of type 2 diabetes. *Metabolism 84*, 56–66. https://doi.org/10.1016/j.metabol.2018.02.010.

Revelas, M., Thalamuthu, A., Oldmeadow, C., Evans, T.-J., Armstrong, N.J., Kwok, J.B., Brodaty, H., Schofield, P.R., Scott, R.J., Sachdev, P.S., et al. (2018). Review and meta-analysis of genetic polymorphisms associated with exceptional human longevity. *Mech. Ageing Dev. 175*, 24–34. https://doi.org/10.1016/j.mad.2018.06.002.

Richter, E.A. (2021). Is GLUT4 translocation the answer to exercise-stimulated muscle glucose uptake? *Am. J. Physiol.-Endocrinol. Metab. 320*, E240–E243. https://doi.org/10.1152/ajpendo.00503.2020.

Riis, J.A. (1901). *The making of an American.* United States: Aegypan.

Ritchie, H., and Roser, M. (2018). Causes of death. Our World in Data. https://ourworldindata.org/causes-of-death.

Rosenberg, A., Mangialasche, F., Ngandu, T., Solomon, A., Kivipelto, M. (2020). Multidomain interventions to prevent cognitive impairment, Alzheimer's disease, and dementia: From FINGER to world-wide FINGERS. *J. Prev. Alzheimers Dis. 7*(1): 29–36. https://doi.org/10.14283/jpad.2019.41.

Rosenberg, S.A., and Barr, J.M. (1992). *The transformed cell.* New York: Putnam.

Roy, J., and Forest, G. (2018). Greater circadian disadvantage during evening games for the National Basketball Association (NBA), National Hockey League (NHL) and National Football League (NFL) teams travelling westward. *J. Sleep Res. 27*, 86–89. https://doi.org/10.1111/jsr.12565.

Rozentryt, P., von Haehling, S., Lainscak, M., Nowak, J.U., Kalantar-Zadeh, K., Polonski, L., and Anker, S.D. (2010). The effects of a high-caloric protein-rich oral nutritional supplement in patients with chronic heart failure and cachexia on quality of life, body composition, and inflammation markers: A randomized, double-blind pilot study. *J. Cachexia Sarcopenia Muscle 1*, 35–42. https://doi.org/10.1007/s13539-010-0008-0.

Rupp, T.L., Wesensten, N.J., and Balkin, T.J. (2012). Trait-like vulnerability to total and partial sleep loss. *Sleep 35*, 1163–1172. https://doi.org/10.5665/sleep.2010.

Sabatini, D.M., Erdjument-Bromage, H., Lui, M., Tempst, P., and Snyder, S.H. (1994). RAFT1: A mammalian protein that binds to FKBP12 in a rapamycin-dependent fashion and is homologous to yeast TORs. *Cell 78*, 35–43. https://doi.org/10.1016/0092-8674(94)90570-3.

Samra, R.A. (2010). Fats and satiety. In *Fat detection: Taste, texture, and post ingestive effects*, ed. J.-P. Montmayeur and J. le Coutre. Boca Raton, FL: CRC Press/Taylor and Francis.

San-Millán, I., and Brooks, G.A. (2018). Assessment of metabolic flexibility by means of measuring blood lactate, fat, and carbohydrate oxidation responses to exercise in professional endurance athletes and less-fit individuals. *Sports Med. Auckl. NZ 48*, 467–479. https://doi.org/10.1007/s40279-017-0751-x.

Sasco, A.J., Secretan, M.B., and Straif, K. (2004). Tobacco smoking and cancer: A brief review of recent epidemiological evidence. *Lung Cancer Amst. Neth. 45,* Suppl 2, S3–9. https://doi.org/10.1016/j.lungcan.2004.07.998.

Saul, S. (2006). Record sales of sleeping pills are causing worries. *New York Times,* February 7. https://www.nytimes.com/2006/02/07/business/record-sales-of-sleeping-pills-are-causing-worries.html.

Sawka, M.N., Gonzalez, R.R., and Pandolf, K.B. (1984). Effects of sleep deprivation on thermoregulation during exercise. *Am. J. Physiol. 246,* R72–77. https://doi.org/10.1152/ajpregu.1984.246.1.R72.

Schoenfeld, B.J., and Aragon, A.A. (2018). How much protein can the body use in a single meal for muscle-building? Implications for daily protein distribution. *J. Int. Soc. Sports Nutr. 15,* 10. https://doi.org/10.1186/s12970-018-0215-1.

Schwingshackl, L., Schwedhelm, C., Hoffmann, G., Knüppel, S., Laure Preterre, A., Iqbal, K., Bechthold, A., De Henauw, S., Michels, N., Devleesschauwer, B., et al. (2018). Food groups and risk of colorectal cancer. *Int. J. Cancer 142,* 1748–1758. https://doi.org/10.1002/ijc.31198.

Schwingshackl, L., Zähringer, J., Beyerbach, J., Werner, S., Heseker, H., Koletzko, B., and Meerpoh, J. Total dietary fat intake, fat quality, and health outcomes: A scoping review of systematic reviews of prospective studies. *Ann. Nutr. Metab.* 77(1), 4–15. https://doi.org/10.1159/000515058.

Sebastiani, P., Gurinovich, A., Nygaard, M., Sasaki, T., Sweigart, B., Bae, H., Andersen, S.L., Villa, F., Atzmon, G., Christensen, K., et al. (2019). APOE alleles and extreme human longevity. *J. Gerontol. Ser. A 74,* 44–51. https://doi.org/10.1093/gerona/gly174.

Sebastiani, P., Nussbaum, L., Andersen, S.L., Black, M.J., and Perls, T.T. (2016). Increasing sibling relative risk of survival to older and older ages and the importance of precise definitions of "aging," "life span," and "longevity." *J. Gerontol. A. Biol. Sci. Med. Sci. 71,* 340–346. https://doi.org/10.1093/gerona/glv020.

Seidelin, K.N. (1995). Fatty acid composition of adipose tissue in humans: Implications for the dietary fat-serum cholesterol-CHD issue. *Prog. Lipid Res. 34,* 199–217. https://doi.org/10.1016/0163-7827(95)00004-J.

Seifert, T., Brassard, P., Wissenberg, M., Rasmussen, P., Nordby, P., Stallknecht, B., Adser, H., Jakobsen, A.H., Pilegaard, H., Nielsen, H.B., et al. (2010). Endurance training enhances BDNF release from the human brain. *Am. J. Physiol. Regul. Integr. Comp. Physiol. 298,* R372–377. https://doi.org/10.1152/ajpregu.00525.2009.

Selvarani, R., Mohammed, S., and Richardson, A. (2021). Effect of rapamycin on aging and age-related diseases—past and future. *GeroScience 43,* 1135–1158. https://doi.org/10.1007/s11357-020-00274-1.

Serna, E., Gambini, J., Borras, C., Abdelaziz, K.M., Mohammed, K., Belenguer, A., Sanchis, P., Avellana, J.A., Rodriguez-Mañas, L., and Viña, J. (2012). Centenarians, but not octogenarians, up-regulate the expression of microRNAs. *Sci. Rep. 2,* 961. https://doi.org/10.1038/srep00961.

Shahid, A., Wilkinson, K., Marcu, S., and Shapiro, C.M. (2011). Insomnia Severity Index (ISI). In *STOP, THAT and one hundred other sleep scales*, ed. A. Shahid, K. Wilkinson, S. Marcu, and C.M. Shapiro, 191–193. New York: Springer New York.

Shan, Z., Ma, H., Xie, M., Yan, P., Guo, Y., Bao, W., Rong, Y., Jackson, C.L., Hu, F.B., and Liu, L. (2015). Sleep duration and risk of type 2 diabetes: A meta-analysis of prospective studies. *Diabetes Care 38*, 529–537. https://doi.org/10.2337/dc14-2073.

Shephard, R.J. (2009). Maximal oxygen intake and independence in old age. *Br. J. Sports Med. 43*, 342–346. https://doi.org/10.1136/bjsm.2007.044800.

Shmagel, A., Ngo, L., Ensrud, K., and Foley, R. (2018). Prescription medication use among community-based US adults with chronic low back pain: A cross-sectional population based study. *J. Pain 19*, 1104–1112. https://doi.org/10.1016/j.jpain.2018.04.004.

Siegel, R.L., Miller, K.D., Fuchs, H.E., and Jemal, A. (2021). Cancer statistics, 2021. *CA. Cancer J. Clin. 71*, 7–33. https://doi.org/10.3322/caac.21654.

Slayday, R.E., Gustavson, D.E., Elman, J.A., Beck, A., McEvoy, L.K., Tu, X.M., Fang, B., Hauger, R.L., Lyons, M.J., McKenzie, R.E., et al. (2021). Interaction between alcohol consumption and apolipoprotein E (ApoE) genotype with cognition in middle-aged men. *J. Int. Neuropsychol. Soc. 27*, 56–68. https://doi.org/10.1017/S1355617720000570.

Sleeman, J., and Steeg, P.S. (2010). Cancer metastasis as a therapeutic target. *Eur. J. Cancer 46*, 1177–1180. https://doi.org/10.1016/j.ejca.2010.02.039.

Small, G.W., Ercoli, L.M., Silverman, D.H.S., Huang, S.-C., Komo, S., Bookheimer, S.Y., Lavretsky, H., Miller, K., Siddarth, P., Rasgon, N.L., et al. (2000). Cerebral metabolic and cognitive decline in persons at genetic risk for Alzheimer's disease. *Proc. Natl. Acad. Sci. 97*, 6037–6042.

Smith, A.D., Smith, S.M., de Jager, C.A., Whitbread, P., Johnston, C., Agacinski, G., Oulhaj, A., Bradley, K.M., Jacoby, R., and Refsum, H. (2010). Homocysteine-lowering by B Vitamins slows the rate of accelerated brain atrophy in mild cognitive impairment: A randomized controlled trial. *PLOS ONE 5*, e12244. https://doi.org/10.1371/journal.pone.0012244.

Smith, C., and Lapp, L. (1991). Increases in number of REMS and REM density in humans following an intensive learning period. *Sleep 14*, 325–330. https://doi.org/10.1093/sleep/14.4.325.

Smith, C., and Smith, D. (2003). Ingestion of ethanol just prior to sleep onset impairs memory for procedural but not declarative tasks. *Sleep 26*, 185–191.

Sniderman, A.D., Bhopal, R., Prabhakaran, D., Sarrafzadegan, N., and Tchernof, A. (2007). Why might South Asians be so susceptible to central obesity and its atherogenic consequences? The adipose tissue overflow hypothesis. *Int. J. Epidemiol. 36*, 220–225. https://doi.org/10.1093/ije/dyl245.

Sniderman, A.D., Thanassoulis, G., Williams, K., and Pencina, M. (2016). Risk of premature cardiovascular disease vs the number of premature cardiovascular events. *JAMA Cardiol. 1*, 492–494. https://doi.org/10.1001/jamacardio.2016.0991.

Sokol, D.K. (2013). "First do no harm" revisited. *BMJ 347*, f6426. https://doi.org/10.1136/bmj.f6426.

Soran, H., Ho, J.H., and Durrington, P.N. (2018). Acquired low cholesterol: Diagnosis and

relevance to safety of low LDL therapeutic targets. *Curr. Opin. Lipidol. 29*, 318–326. https://doi.org/10.1097/MOL.0000000000000526.
Spaeth, A.M., Dinges, D.F., and Goel, N. (2015). Resting metabolic rate varies by race and by sleep duration. *Obesity 23*, 2349–2356. https://doi.org/10.1002/oby.21198.
Spencer, C. (2005). *Genes, aging and immortality.* Upper Saddle River, NJ: Pearson.
Sperling, R.A., Aisen, P.S., Beckett, L.A., Bennett, D.A., Craft, S., Fagan, A.M., Iwatsubo, T., Jack, C.R., Kaye, J., Montine, T.J., et al. (2011). Toward defining the preclinical stages of Alzheimer's disease: Recommendations from the National Institute on Aging-Alzheimer's Association workgroups on diagnostic guidelines for Alzheimer's disease. *Alzheimers Dement. 7*, 280–292. https://doi.org/10.1016/j.jalz.2011.03.003.
Spiegel, K., Leproult, R., L'hermite-Balériaux, M., Copinschi, G., Penev, P.D., and Van Cauter, E. (2004b). Leptin levels are dependent on sleep duration: Relationships with sympathovagal balance, carbohydrate regulation, cortisol, and thyrotropin. *J. Clin. Endocrinol. Metab. 89*, 5762–5771. https://doi.org/10.1210/jc.2004-1003.
Spiegel, K., Leproult, R., and Van Cauter, E. (1999). Impact of sleep debt on metabolic and endocrine function. *Lancet 354*, 1435–1439. https://doi.org/10.1016/S0140-6736(99)01376-8.
Spiegel, K., Tasali, E., Penev, P., and Cauter, E.V. (2004a). Brief communication: Sleep curtailment in healthy young men is associated with decreased leptin levels, elevated ghrelin levels, and increased hunger and appetite. *Ann. Intern. Med. 141*, 846–850. https://doi.org/10.7326/0003-4819-141-11-200412070-00008.
Spillane, S., Shiels, M.S., Best, A.F., Haozous, E.A., Withrow, D.R., Chen, Y., Berrington de González, A., and Freedman, N.D. (2020). Trends in alcohol-induced deaths in the United States, 2000–2016. *JAMA Netw. Open 3*, e1921451. https://doi.org/10.1001/jamanetworkopen.2019.21451.
Spira, A.P., Gamaldo, A.A., An, Y., Wu, M.N., Simonsick, E.M., Bilgel, M., Zhou, Y., Wong, D.F., Ferrucci, L., and Resnick, S.M. (2013). Self-reported sleep and β-amyloid deposition in community-dwelling older adults. *JAMA Neurol. 70*, 1537–1543. https://doi.org/10.1001/jamaneurol.2013.4258.
Sprecher, K.E., Bendlin, B.B., Racine, A.M., Okonkwo, O.C., Christian, B.T., Koscik, R.L., Sager, M.A., Asthana, S., Johnson, S.C., and Benca, R.M. (2015). Amyloid burden is associated with self-reported sleep in non-demented late middle-aged adults. *Neurobiol. Aging 36*, 2568–2576. https://doi.org/10.1016/j.neurobiolaging.2015.05.004.
Stamatakis, K.A., and Punjabi, N.M. (2010). Effects of sleep fragmentation on glucose metabolism in normal subjects. *Chest 137*, 95–101. https://doi.org/10.1378/chest.09-0791.
Standl, E., Schnell, O., and Ceriello, A. (2011). Postprandial hyperglycemia and glycemic variability: Should we care? *Diabetes Care 34*, Suppl 2, S120–127. https://doi.org/10.2337/dc11-s206.
Stary, H.C. (2003). *Atlas of atherosclerosis progression and regression.* Boca Raton, FL: CRC Press.
Stefan, N., Schick, F., and Häring, H.-U. (2017). Causes, characteristics, and consequences of metabolically unhealthy normal weight in humans. *Cell Metab. 26*, 292–300. https://doi.org/10.1016/j.cmet.2017.07.008.

Stickgold, R., Whidbee, D., Schirmer, B., Patel, V., and Hobson, J.A. (2000). Visual discrimination task improvement: A multi-step process occurring during sleep. *J. Cogn. Neurosci. 12*, 246–254. https://doi.org/10.1162/089892900562075.

Stomrud, E., Hansson, O., Zetterberg, H., Blennow, K., Minthon, L., and Londos, E. (2010). Correlation of longitudinal cerebrospinal fluid biomarkers with cognitive decline in healthy older adults. *Arch. Neurol. 67*, 217–223. https://doi.org/10.1001/archneurol.2009.316.

Strobe, M. (2021). U.S. overdose deaths topped 100,000 in one year, officials say. AP News, November 17. https://apnews.com/article/overdodse-deaths-fentanayl-health-f34b022d75a1eb9776e27903ab40670f.

Stroes, E.S., Thompson, P.D., Corsini, A., Vladutiu, G.D., Raal, F.J., Ray, K.K., Roden, M., Stein, E., Tokgözoğlu, L., Nordestgaard, B.G., et al. (2015). Statin-associated muscle symptoms: impact on statin therapy—European Atherosclerosis Society Consensus Panel Statement on Assessment, Aetiology and Management. *Eur. Heart J. 36*, 1012–1022. https://doi.org/10.1093/eurheartj/ehv043.

Strong, R., Miller, R.A., Astle, C.M., Baur, J.A., de Cabo, R., Fernandez, E., Guo, W., Javors, M., Kirkland, J.L., Nelson, J.F., et al. (2013). Evaluation of resveratrol, green tea extract, curcumin, oxaloacetic acid, and medium-chain triglyceride oil on life span of genetically heterogeneous mice. *J. Gerontol. Ser. A 68*, 6–16. https://doi.org/10.1093/gerona/gls070.

Strozyk, D., Blennow, K., White, L.R., and Launer, L.J. (2003). CSF Abeta 42 levels correlate with amyloid-neuropathology in a population-based autopsy study. *Neurology 60*, 652–656. https://doi.org/10.1212/01.wnl.0000046581.81650.d0.

Sudimac, S., Sale, V., and Kühn, S. (2022). How nature nurtures: Amygdala activity decreases as the result of a one-hour walk in nature. *Mol Psychiatry.* https://doi.org/10.1038/s41380-022-01720-6.

Sumithran, P., Prendergast, L.A., Delbridge, E., Purcell, K., Shulkes, A., Kriketos, A., and Proietto, J. (2013). Ketosis and appetite-mediating nutrients and hormones after weight loss. *Eur. J. Clin. Nutr. 67*, 759–764. https://doi.org/10.1038/ejcn.2013.90.

Suzuki, K., Elkind, M.S., Boden-Albala, B., Jin, Z., Berry, G., Di Tullio, M.R., Sacco, R.L., and Homma, S. (2009). Moderate alcohol consumption is associated with better endothelial function: A cross sectional study. *BMC Cardiovasc. Disord. 9*, 8. https://doi.org/10.1186/1471-2261-9-8.

Tabata, I., Nishimura, K., Kouzaki, M., Hirai, Y., Ogita, F., Miyachi, M., and Yamamoto, K. (1996). Effects of moderate-intensity endurance and high-intensity intermittent training on anaerobic capacity and VO_{2max}. *Med. Sci. Sports Exerc. 28*, 1327–1330. https://doi.org/10.1097/00005768-199610000-00018.

Taieb, J., Gallois, C. (2020). Adjuvant chemotherapy for stage III colon cancer. *Cancers, 12*(9), 2679. https://doi.org/10.3390/cancers12092679.

Tang, C., Liu, C., Fang, P., Xiang, Y., and Min, R. (2019). Work-related accumulated fatigue among doctors in tertiary hospitals: A cross-sectional survey in six provinces of China. *Int. J. Environ. Res. Public. Health 16*, E3049. https://doi.org/10.3390/ijerph16173049.

Tanweer, S.A.W. (2021). How smart phones effects health. Tech neck: Causes and preventions. *Pak. J. Phys. Ther.* 02–02. https://doi.org/10.52229/pjpt.v2i04.1135.

Tapiola, T., Pirttilä, T., Mikkonen, M., Mehta, P.D., Alafuzoff, I., Koivisto, K., and Soininen, H. (2000). Three-year follow-up of cerebrospinal fluid tau, beta-amyloid 42 and 40 concentrations in Alzheimer's disease. *Neurosci. Lett. 280*, 119–122. https://doi.org/10.1016/s0304-3940(00)00767-9.

Tapiola, T., Alafuzoff, I., Herukka, S.-K., Parkkinen, L., Hartikainen, P., Soininen, H., and Pirttilä, T. (2009). Cerebrospinal fluid β-amyloid 42 and tau proteins as biomarkers of Alzheimer-type pathologic changes in the brain. *Arch. Neurol. 66*, 382–389. https://doi.org/10.1001/archneurol.2008.596.

Tasali, E., Leproult, R., Ehrmann, D.A., and Van Cauter, E. (2008). Slow-wave sleep and the risk of type 2 diabetes in humans. *Proc. Natl. Acad. Sci. 105*, 1044–1049. https://doi.org/10.1073/pnas.0706446105.

Tatebe, H., and Shiozaki, K. (2017). Evolutionary conservation of the components in the TOR signaling pathways. *Biomolecules 7*, 77. https://doi.org/10.3390/biom7040077.

Taylor, J. (2009). "Cigarettes, whisky, and wild, wild women." *Independent*, June 20. https://www.independent.co.uk/life-style/health-and-families/health-news/cigarettes-whisky-and-wild-wild-women-1710744.html.

Tchernof, A., and Després, J.-P. (2013). Pathophysiology of human visceral obesity: An update. *Physiol. Rev. 93*, 359–404. https://doi.org/10.1152/physrev.00033.2011.

Templeman, I., Smith, H.A., Chowdhury, E., Chen, Y.-C., Carroll, H., Johnson-Bonson, D., Hengist, A., Smith, R., Creighton, J., Clayton, D., et al. (2021). A randomized controlled trial to isolate the effects of fasting and energy restriction on weight loss and metabolic health in lean adults. *Sci. Transl. Med. 13*, eabd8034. https://doi.org/10.1126/scitranslmed.abd8034.

Thanassoulis, G., Sniderman, A.D., and Pencina, M.J. (2018). A long-term benefit approach vs standard risk-based approaches for statin eligibility in primary prevention. *JAMA Cardiol. 3*, 1090–1095. https://doi.org/10.1001/jamacardio.2018.3476.

Tieland, M., Dirks, M.L., van der Zwaluw, N., Verdijk, L.B., van de Rest, O., de Groot, L.C.P.G.M., and van Loon, L.J.C. (2012a). Protein supplementation increases muscle mass gain during prolonged resistance-type exercise training in frail elderly people: A randomized, double-blind, placebo-controlled trial. *J. Am. Med. Dir. Assoc. 13*, 713–719. https://doi.org/10.1016/j.jamda.2012.05.020.

Tieland, M., van de Rest, O., Dirks, M.L., van der Zwaluw, N., Mensink, M., van Loon, L.J.C., and de Groot, L.C.P.G.M. (2012b). Protein supplementation improves physical performance in frail elderly people: A randomized, double-blind, placebo-controlled trial. *J. Am. Med. Dir. Assoc. 13*, 720–726. https://doi.org/10.1016/j.jamda.2012.07.005.

Tolboom, N., van der Flier, W.M., Yaqub, M., Boellaard, R., Verwey, N.A., Blankenstein, M.A., Windhorst, A.D., Scheltens, P., Lammertsma, A.A., and van Berckel, B.N.M. (2009). Relationship of cerebrospinal fluid markers to 11C-PiB and 18F-FDDNP binding. *J. Nucl. Med. 50*, 1464–1470. https://doi.org/10.2967/jnumed.109.064360.

Trappe, S., Hayes, E., Galpin, A., Kaminsky, L., Jemiolo, B., Fink, W., Trappe, T., Jansson, A., Gustafsson, T., and Tesch, P. (2013). New records in aerobic power among octogenarian lifelong endurance athletes. *J. Appl. Physiol. 114*, 3–10. https://doi.org/10.1152/japplphysiol.01107.2012.

Trumble, B.C., and Finch, C.E. (2019). The exposome in human evolution: From dust to diesel. *Q. Rev. Biol. 94*, 333–394. https://doi.org/10.1086/706768.
Tsimikas, S., Fazio, S., Ferdinand, K.C., Ginsberg, H.N., Koschinsky, M.L., Santica, M., Moriarity, P.M., Rader, D.J., Remaley, A.T., Reyes-Soffer, G., et al. (2018). NHLBI Working Group recommendations to reduce lipoprotein(a)-mediated risk of cardiovascular disease and aortic stenosis. *J. Am. Coll. Cardiol. 71*(2), 177–192.
Tuchman, A. (2009). Diabetes and the public's health. *Lancet 374*, 1140–1141. https://doi.org/10.1016/S0140-6736(09)61730-X.
United States Census Bureau. (2022). National population by characteristics: 2020–2021 tables>median age and age by sex>annual estimates of the resident population by single year of age and Sex for the United States: April 1, 2020 to July 1, 2021 (NC-EST2021-SYASEX).
Uretsky, S., Rozanski, A., Singh, P., Supariwala, A., Atluri, P., Bangalore, S., Pappas, T.W., Fisher, E.A., and Peters, M.R. (2011). The presence, characterization and prognosis of coronary plaques among patients with zero coronary calcium scores. *Int. J. Cardiovasc. Imaging 27*, 805–812. https://doi.org/10.1007/s10554-010-9730-0.
Urfer, S.R., Kaeberlein, T.L., Mailheau, S., Bergman, P.J., Creevy, K.E., Promislow, D.E.L., and Kaeberlein, M. (2017). A randomized controlled trial to establish effects of short-term rapamycin treatment in 24 middle-aged companion dogs. *GeroScience 39*, 117–127. https://doi.org/10.1007/s11357-017-9972-z.
Urry, E., and Landolt, H.-P. (2015). Adenosine, caffeine, and performance: From cognitive neuroscience of sleep to sleep pharmacogenetics. In *Sleep, neuronal plasticity and brain function*, ed. P. Meerlo, R.M. Benca, and T. Abel, 331–366. Berlin: Springer.
Van Ancum, J.M., Pijnappels, M., Jonkman, N.H., Scheerman, K., Verlaan, S., Meskers, C.G.M., and Maier, A.B. (2018). Muscle mass and muscle strength are associated with pre- and post-hospitalization falls in older male inpatients: A longitudinal cohort study. *BMC Geriatr. 18*, 116. https://doi.org/10.1186/s12877-018-0812-5.
Van Cauter, E., Caufriez, A., Kerkhofs, M., Van Onderbergen, A., Thorner, M.O., and Copinschi, G. (1992). Sleep, awakenings, and insulin-like growth factor-I modulate the growth hormone (GH) secretory response to GH-releasing hormone. *J. Clin. Endocrinol. Metab. 74*, 1451–1459. https://doi.org/10.1210/jcem.74.6.1592893.
van Charante, E., Richard, E., Eurelings, L.S., van Dalen, J-W., Ligthart, S.A., van Bussel, E.F., Hoevenaar-Blom, M.P., Vermeulen, M., van Gool, W. A. (2016). Effectiveness of a 6-year multidomain vascular care intervention to prevent dementia (preDIVA): A cluster-randomised controlled trial. *Lancet 388*, 797–805. https://doi.org/10.1016/S0140-6736(16)30950-3.
Vander Heiden, M.G., Cantley, L.C., and Thompson, C.B. (2009). Understanding the Warburg effect: The metabolic requirements of cell proliferation. *Science 324*, 1029–1033. https://doi.org/10.1126/science.1160809.
van der Helm, E., and Walker, M.P. (2009). Overnight therapy? The role of sleep in emotional brain processing. *Psychol. Bull. 135*, 731–748. https://doi.org/10.1037/a0016570.
Van Dongen, H.P.A., Baynard, M.D., Maislin, G., and Dinges, D.F. (2004). Systematic

interindividual differences in neurobehavioral impairment from sleep loss: Evidence of trait-like differential vulnerability. *Sleep 27*, 423–433.
Van Dongen, H.P.A., Maislin, G., Mullington, J.M., and Dinges, D.F. (2003). The cumulative cost of additional wakefulness: Dose-response effects on neurobehavioral functions and sleep physiology from chronic sleep restriction and total sleep deprivation. *Sleep 26*, 117–126. https://doi.org/10.1093/sleep/26.2.117.
Varady, K.A., and Gabel, K. (2019). Safety and efficacy of alternate day fasting. *Nat. Rev. Endocrinol. 15*, 686–687. https://doi.org/10.1038/s41574-019-0270-y.
Vendelbo, M.H., Møller, A.B., Christensen, B., Nellemann, B., Clasen, B.F.F., Nair, K.S., Jørgensen, J.O.L., Jessen, N., and Møller, N. (2014). Fasting increases human skeletal muscle net phenylalanine release and this is associated with decreased mTOR signaling. *PLOS ONE 9*, e102031. https://doi.org/10.1371/journal.pone.0102031.
Veronese, N., Koyanagi, A., Cereda, E., Maggi, S., Barbagallo, M., Dominguez, L.J., and Smith, L. (2022). Sarcopenia reduces quality of life in the long-term: Longitudinal analyses from the English longitudinal study of ageing. *Eur. Geriatr. Med. 13*, 633–639. https://doi.org/10.1007/s41999-022-00627-3.
Voight, B.F., Peloso, G.M., Orho-Melander, M., Frikke-Schmidt, R., Barbalic, M., Jensen, M.K., Hindy, G., Hólm, H., Ding, E.L., Johnson, T., et al. (2012). Plasma HDL cholesterol and risk of myocardial infarction: A Mendelian randomisation study. *Lancet 380*, 572–580. https://doi.org/10.1016/S0140-6736(12)60312-2.
Voulgari, C., Tentolouris, N., Dilaveris, P., Tousoulis, D., Katsilambros, N., and Stefanadis, C. (2011). Increased heart failure risk in normal-weight people with metabolic syndrome compared with metabolically healthy obese individuals. *J. Am. Coll. Cardiol. 58*, 1343–1350. https://doi.org/10.1016/j.jacc.2011.04.047.
Wade, N. (2009). Dieting monkeys offer hope for living longer. *New York Times*, July 9. https://www.nytimes.com/2009/07/10/science/10aging.html.
Wahlund, L.-O., and Blennow, K. (2003). Cerebrospinal fluid biomarkers for disease stage and intensity in cognitively impaired patients. *Neurosci. Lett. 339*, 99–102. https://doi.org/10.1016/s0304-3940(02)01483-0.
Waks, A.G., and Winer, E.P. (2019). Breast cancer treatment: A review. *JAMA, 321*(3), 288–300. https://doi.org/10.1001/jama.2018.19323.
Walker, M.P. (2009). The role of slow wave sleep in memory processing. *J. Clin. Sleep Med. 5*, S20–S26.
———. (2017). *Why we sleep: Unlocking the power of sleep and dreams.* New York: Scribner.
Wallace, D.F. (2009). *This is water: Some thoughts, delivered on a significant occasion, about living a compassionate life.* New York: Little, Brown.
Wang, C., and Holtzman, D.M. (2020). Bidirectional relationship between sleep and Alzheimer's disease: Role of amyloid, tau, and other factors. *Neuropsychopharmacology 45*, 104–120. https://doi.org/10.1038/s41386-019-0478-5.
Wang, N., Fulcher, J., Abeysuriya, N., Park, L., Kumar, S., Di Tanna, G.L., Wilcox, I., Keech, A., Rodgers, A., and Lal, S. (2020). Intensive LDL cholesterol-lowering treatment beyond current recommendations for the prevention of major vascular events: A systematic

review and meta-analysis of randomised trials including 327 037 participants. *Lancet Diabetes Endocrinol. 8*, 36–49. https://doi.org/10.1016/S2213-8587(19)30388-2.
Wang, Y., and Brinton, R.D. (2016). Triad of risk for late onset Alzheimer's: Mitochondrial haplotype, APOE genotype and chromosomal sex. *Front. Aging Neurosci. 8*, 232. https://doi.org/10.3389/fnagi.2016.00232.
Wang, Y., Jones, B.F., and Wang, D. (2019). Early-career setback and future career impact. *Nat. Commun. 10*, 4331. https://doi.org/10.1038/s41467-019-12189-3.
Warburg, O. (1924). Warburg: The metabolism of cancer cells. Google Scholar.
———. (1956). On the origin of cancer cells. *Science 123*, 309–314. https://doi.org/10.1126/science.123.3191.309.
Watanabe, K., Oba, K., Suzuki, T., Ouchi, M., Suzuki, K., Futami-Suda, S., Sekimizu, K., Yamamoto, N., and Nakano, H. (2011). Oral glucose loading attenuates endothelial function in normal individual. *Eur. J. Clin. Invest. 41*, 465–473. https://doi.org/10.1111/j.1365-2362.2010.02424.x.
Watson, A.M. (2017). Sleep and athletic performance. *Curr. Sports Med. Rep. 16*, 413–418. https://doi.org/10.1249/JSR.0000000000000418.
Watson, J.D. (2009). Opinion | To fight cancer, know the enemy. *New York Times,* August 5. https://www.nytimes.com/2009/08/06/opinion/06watson.html.
Wen, C.P., Wai, J.P.M., Tsai, M.K., Yang, Y.C., Cheng, T.Y.D., Lee, M.-C., Chan, H.T., Tsao, C.K., Tsai, S.P., and Wu, X. (2011). Minimum amount of physical activity for reduced mortality and extended life expectancy: A prospective cohort study. *Lancet 378*, 1244–1253. https://doi.org/10.1016/S0140-6736(11)60749-6.
Westerterp, K.R., Yamada, Y., Sagayama, H., Ainslie, P.N., Andersen, L.F., Anderson, L.J., Arab, L., Baddou, I., Bedu-Addo, K., Blaak, E.E., et al. (2021). Physical activity and fat-free mass during growth and in later life. *Am. J. Clin. Nutr. 114*, 1583–1589. https://doi.org/10.1093/ajcn/nqab260.
WHI (Women's Health Initiative). n.d. About WHI—Dietary Modification Trial. Accessed September 28, 2022. https://sp.whi.org/about/SitePages/Dietary%20Trial.aspx.
WHO (World Health Organization). (2019). Global health estimates: Leading causes of death. https://www.who.int/data/gho/data/themes/mortality-and-global-health-estimates/ghe-leading-causes-of-death.
Willcox, B.J., Donlon, T.A., He, Q., Chen, R., Grove, J.S., Yano, K., Masaki, K.H., Willcox, D.C., Rodriguez, B., and Curb, J.D. (2008). FOXO3A genotype is strongly associated with human longevity. *Proc. Natl. Acad. Sci. 105*, 13987–13992. https://doi.org/10.1073/pnas.0801030105.
Wilson, M.A., and McNaughton, B.L. (1994). Reactivation of hippocampal ensemble memories during sleep. *Science 265*, 676–679. https://doi.org/10.1126/science.8036517.
Winer, J.R., Mander, B.A., Helfrich, R.F., Maass, A., Harrison, T.M., Baker, S.L., Knight, R.T., Jagust, W.J., and Walker, M.P. (2019). Sleep as a potential biomarker of tau and β-amyloid burden in the human brain. *J. Neurosci. 39*, 6315–6324. https://doi.org/10.1523/JNEUROSCI.0503-19.2019.
Wishart, D.S., Tzur, D., Knox, C., Eisner, R., Guo, A.C., Young, N., Cheng, D., Jewell, K.,

Arndt, D., Sawhney, S., et al. (2007). HMDB: The Human Metabolome Database. *Nucleic Acids Res. 35*, D521–526. https://doi.org/10.1093/nar/gkl923.

Wolters, F.J., and Ikram, M.A. (2019). Epidemiology of vascular dementia. *Arterioscler. Thromb. Vasc. Biol. 39,* 1542–1549. https://doi.org/10.1161/ATVBAHA.119.311908.

Wu, G. (2016). Dietary protein intake and human health. *Food Funct. 7*, 1251–1265. https://doi.org/10.1039/c5fo01530h.

Xu, J. (2016). Mortality among centenarians in the United States, 2000–2014. NCHS Data Brief 233. https://www.cdc.gov/nchs/products/databriefs.htm.

Xue, Q.-L. (2011). The frailty syndrome: Definition and natural history. *Clin. Geriatr. Med. 27*, 1–15. https://doi.org/10.1016/j.cger.2010.08.009.

Yamamoto, T., Yagi, S., Kinoshita, H., Sakamoto, Y., Okada, K., Uryuhara, K., Morimoto, T., Kaihara, S., and Hosotani, R. (2015). Long-term survival after resection of pancreatic cancer: A single-center retrospective analysis. *World J. Gastroenterol. 21*, 262–268. https://doi.org/10.3748/wjg.v21.i1.262.

Yamazaki, R., Toda, H., Libourel, P.-A., Hayashi, Y., Vogt, K.E., and Sakurai, T. (2020). Evolutionary origin of distinct NREM and REM sleep. *Front. Psychol. 11*, 567618. https://doi.org/10.3389/fpsyg.2020.567618.

Yan, Y., Wang, X., Chaput, D., Shin, M.K., Koh, Y., Gan, L., Pieper, A.A., Woo, J.A.A., Kang, D.E. (2022). X-linked ubiquitin-specific peptidase 11 increases tauopathy vulnerability in women. *Cell,* 185: 21, 3913-3930.e19. https://doi.org/10.1016/j.cell.2022.09.002.

Yassine, H.N., Braskie, M.N., Mack, W.J., Castor, K.J., Fonteh, A.N., Schneider, L.S., Harrington, M.G., and Chui, H.C. (2017). Association of docosahexaenoic acid supplementation with Alzheimer disease stage in apolipoprotein E ε4 carriers. *JAMA Neurol. 74*, 339–347. https://doi.org/10.1001/jamaneurol.2016.4899.

Yasuno, F., Minami, H., Hattori, H., and Alzheimer's Disease Neuroimaging Initiative (2020). Interaction effect of Alzheimer's disease pathology and education, occupation, and socioeconomic status as a proxy for cognitive reserve on cognitive performance: In vivo positron emission tomography study. *Psychogeriatr. 20*, 585–593. https://doi.org/10.1111/psyg.12552.

Yin, J., Jin, X., Shan, Z., Li, S., Huang, H., Li, P., Peng, X., Peng, Z., Yu, K., Bao, W., Yang, W., Chen, X., Liu, L. (2017). Replationship of sleep duration with all-cause mortality and cardiovascular events. *JAHA 117*. https://www.ahajournals.org/doi/full/10.1161/JAHA.117.005947.

Yoo, S.-S., Gujar, N., Hu, P., Jolesz, F.A., and Walker, M.P. (2007). The human emotional brain without sleep: A prefrontal amygdala disconnect. *Curr. Biol. 17*, R877–878. https://doi.org/10.1016/j.cub.2007.08.007.

Youlden, D.R., Cramb, S.M., and Baade, P.D. (2008). The international epidemiology of lung cancer: Geographical distribution and secular trends. *J. Thorac. Oncol. 3*, 819–831. https://doi.org/10.1097/JTO.0b013e31818020eb.

Youngstedt, S.D., O'Connor, P.J., Crabbe, J.B., and Dishman, R.K. (2000). The influence of acute exercise on sleep following high caffeine intake. *Physiol. Behav. 68*, 563–570. https://doi.org/10.1016/S0031-9384(99)00213-9.

Zelman, S. (1952). The liver in obesity. *Arch. Intern. Med. 90*, 141–156. https://doi.org/10.1001/archinte.1952.00240080007002.

Zethelius, B., and Cederholm, J. (2015). Comparison between indexes of insulin resistance for risk prediction of cardiovascular diseases or development of diabetes. *Diabetes Res. Clin. Pract. 110*, 183–192. https://doi.org/10.1016/j.diabres.2015.09.003.

Zhang, Y., Zhang, Y., Du, S., Wang, Q., Xia, H., and Sun, R. (2020). Exercise interventions for improving physical function, daily living activities and quality of life in community-dwelling frail older adults: A systematic review and meta-analysis of randomized controlled trials. *Geriatr. Nur. 41*, 261–273. https://doi.org/10.1016/j.gerinurse.2019.10.006.

Zheng, Y., Fan, S., Liao, W., Fang, W., Xiao, S., and Liu, J. (2017). Hearing impairment and risk of Alzheimer's disease: A meta-analysis of prospective cohort studies. *Neurol. Sci. 38*, 233–239. https://doi.org/10.1007/s10072-016-2779-3.

Zheng, Y., Lv, T., Wu, J., and Lyu, Y. (2022). Trazodone changed the polysomnographic sleep architecture in insomnia disorder: A systematic review and meta-analysis. *Scientific reports, 12*(1), 14453. https://doi.org/10.1038/s41598-022-18776-7.

Zhou, C., Wu, Q., Wang, Z., Wang, Q., Liang, Y., and Liu, S. (2020). The effect of hormone replacement therapy on cognitive function in female patients with Alzheimer's disease: A meta-analysis. *Am. J. Alzheimers Dis. Other Demen. 35*, 1533317520938585. https://doi.org/10.1177/1533317520938585.

Ziemichód, W., Grabowska, K., Kurowska, A., and Biała, G. (2022). A comprehensive review of daridorexant, a dual-orexin receptor antagonist as new approach for the treatment of insomnia. *Molecules 27*(18), 6041. https://doi.org/10.3390/molecules27186041.

Zuccarelli, L., Galasso, L., Turner, R., Coffey, E.J.B., Bessone, L., and Strapazzon, G. (2019). Human physiology during exposure to the cave environment: A systematic review with implications for aerospace medicine. *Front. Physiol. 10*.

INDEX

Note: Page numbers in *italics* indicate figures and tables.